The neutrophil constitutes the first line of defence in protecting the host from invading bacterial and fungal pathogens. It is a highly potent cytotoxic cell and possesses an armoury of antimicrobial proteins and biochemical pathways that can be used in this protective role.

This book describes the role of the neutrophil in infection and inflammation and provides an up-to-date review of the biochemistry and physiology of this cell, highlighting the mechanisms by which they seek out and destroy pathogenic micro-organisms. The development of these cells during haematopoiesis is described, and the mechanisms that lead to the production of reactive oxidants are reviewed, as are the intracellular signal transduction systems that lead to the cell's activation. The book also discusses recent discoveries concerning the role of cytokines in the regulation of neutrophil function, as well as the importance of the neutrophil as a generator of inflammatory cytokines. Finally there is a description of the biochemical defects that give rise to some of the neutrophil-associated human diseases.

BIOCHEMISTRY AND PHYSIOLOGY
OF THE NEUTROPHIL

BIOCHEMISTRY AND PHYSIOLOGY OF THE NEUTROPHIL

STEVEN W. EDWARDS

Published by the Press Syndicate of the University of Cambridge
The Pitt Building, Trumpington Street, Cambridge CB2 1RP
40 West 20th Street, New York, NY 10011-4211, USA
10 Stamford Road, Oakleigh, Melbourne 3166, Australia

First published 1994

Printed in the United States of America

Library of Congress Cataloguing-in-Publication Data

Edwards, Steven W.
Biochemistry and physiology of the neutrophil / Steven W. Edwards.
p. cm.
Includes bibliographical references and index.
ISBN 0-521-41698-1 (hc)
1. Neutrophils. I. Title.
QR185.8.N47E39 1994
612.1´12 – dc20 93–31732
CIP

A catalogue record for this book is available from the British Library

ISBN 0-521-41698-1 hardback

Contents

Abbreviations

ADCC	antibody-dependent cellular cytotoxicity
ADP	adenosine diphosphate
AM	acetoxymethyl ester
ara-C	cytosine arabinoside
ATP	adenosine triphosphate
BAPTA	*bis*-(*o*-aminophenoxy)-ethane-N,N,N´,N´-tetraacetic acid
B cell	bone-marrow-derived lymphocyte
BH_4	tetrahydrobiopterin
bp	base pairs
BPI	bactericidal/permeability-inducing protein
BSA	bovine serum albumin
C domain	constant domain
C1–C9	complement components
cAMP	cyclic adenosine monophosphate
CAP	cationic antimicrobial protein
CD	cluster of differentiation
cDNA	complementary deoxynucleic acid
CFU	colony-forming unit
CFU-GEMM	granulocyte–erythroid–monocyte–megakaryocyte CFU
CFU_s, CFU_c	CFU developed in the spleen or in culture, respectively
CGD	chronic granulomatous disease
cGMP	cyclic guanosine monophosphate
Cl^-	chloride ion
CMC	critical micellar concentration
CMML	chronic myelomonocytic leukaemia

CR1, CR3	complement receptor types
CSA	colony-stimulating activity
CSF	colony-stimulating factor
da	daltons
DAG	diacylglycerol
DMD	Duchenne muscular dystrophy
DMF	dimethyl formamide
DMPO	5,5-dimethyl-1-pyrroline-1-oxide
DNA	deoxynucleic acid
DPI	diphenylene iodonium
EDRF	endothelium-derived relaxing factor
EGF	epidermal growth factor
EGR-1	early growth phase response gene
EGTA	ethyleneglycol-*bis*-(β-aminoethyl)-N,N,N′,N′-tetraacetic acid
Endo F	Endo-*b*-*N*-acetylglucosaminidase F
EPR, ESR	electron (paramagnetic/spin) resonance spectroscopy
FAB	French–American–British
Fab	fragment antigen-binding
FACS	fluorescence-activated cell sorting
FAD	flavin adenine dinucleotide
Fc	fragment crystallisable
FcγR, FcεR	receptors binding the Fc region of the indicated Ig molecule (γ for IgG, ε for IgE, etc.)
FGF	fibroblast growth factor
fl	femtolitre (10^{-15} litre)
fMet-Leu-Phe	*N*-formylmethionine-leucyl-phenylalanine
FMN	flavin mononucleotide
GAP	GTPase activating protein
G-CSF	granulocyte colony-stimulating factor
GDI	GDP-dissociation inhibition factor
GDP	guanosine 5′-diphosphate
GDS	guanine nucleotide-dissociation stimulator
G-protein	guanosine triphosphate binding protein
GPI	glycosyl-phosphatidylinositol
GM-CSF	granulocyte–macrophage colony-stimulating factor
GNEP	guanine nucleotide exchange protein
GPO	glutathione peroxidase
GR	glutathione reductase

GSH	reduced glutathione
GSSH	oxidised glutathione
GTP	guanosine triphosphate
H	heavy (re chain)
H_1, H_2	histamine receptors
H_2O_2	hydrogen peroxide
5-HETE	5-hydroxy-eicosatetraenoic acid
HLA	human leukocyte antigen
HMM	heavy meromyosin
HMP	hexose monophosphate shunt
HNP	human neutrophil protein
5-HPETE	5-hydroxyperoxy-eicosatetraenoic acid
HPLC	high-performance liquid chromatography
HSP, hsp	heat-shock protein
I_{50}	concentration resulting in 50% inhibition
Ig	immunoglobulin
IL	Interleukin
Ins	inositol
IP_3	inositol trisphosphate
kb	kilobase
K cell	killer cell
K_d	dissociation constant
kDa	kilodaltons
K_m	Michaelis constant
L	light (re chain)
LAM	leukocyte adhesion molecule
LMM	light meromyosin
LMW	low molecular weight
LPS	lipopolysaccharide
LPS-BP	lipopolysaccharide-binding protein
LT	leukotriene
MAC	membrane atttack complex
MAP	microtubule-associated protein
M-CSF	macrophage CSF
MDS	myelodysplastic syndrome(s)
MHC	major histocompatibility complex
Mn-SOD	manganese-dependent superoxide dismutase
mRNA	messenger ribonucleic acid
NADH	nicotinamide adenine dinucleotide (reduced)
NADPH	nicotinamide adenine dinucleotide phosphate (reduced)

NAP-1	neutrophil attractant protein-1 (= Interleukin-8)
NBT	nitroblue tetrazolium
NCF	neutrophil cytosolic fraction
NDGA	nordihydroguaiaretic acid
NK cell	natural killer cell
Nle	norleucine
nM	nanomolar
NMMA	*N*-monomethyl-L-arginine
NO	nitric oxide
·OH	hydroxyl free radical
$^1O_2$	singlet oxygen
O_2^-	superoxide free radical
P_2, P_3, P_4	bis-, tris-, and tetrakisphosphate
PAF	platelet activating factor
PAGE	polyacrylamide gel electrophoresis
PBS	phosphate buffered saline
PC	phosphatidylcholine
PDGF	platelet-derived growth factor
PG	prostaglandin
pg	picogram (10^{-12} gram)
pI	isoelectric point
P_i	inorganic phosphate
PIP	phosphatidylinositol 4-phosphate
PIP_2	phosphatidylinositol 4,5-bisphosphate
PKA	protein kinase A (cAMP-stimulated protein kinase)
PL	phospholipase
pM	picomolar (10^{-12} molar)
PMA	phorbol myristate acetate
PNH	paroxysmal nocturnal haemoglobinuria
PtdIns	phosphatidylinositol
RA	rheumatoid arthritis; also, refractive anaemia
RAEB(-T), RARS	forms of refractive anaemia
redox	reduction–oxidation
RGD	peptide sequence Arg-Gly-Asp
rGM-CSF	recombinant GM-CSF
RNA	ribonucleic acid
RNase	ribonuclease
SC	secretory chain
SCR	single consensus repeat
SDS	sodium dodecyl sulphate
sIg	surface immunoglobulin

SOD	superoxide dismutase
$t_{1/2}$	half-life
T cell	thymus-derived lymphocyte
Tc	cytotoxic T cell
TcR	T-cell receptor
TGA	thymine-guanine-adenine
Th	T helper cell
TLC	thin layer chromatography
TMP/SMX	trimethoprim/sulphamethazole
TNF	tumour necrosis factor
Ts	T suppressor cell
TX	thromboxane
U	unit
V domain	variable domain
VLA	very-late antigen
μl	microlitre (10^{-6} litre)
μm	micrometre (10^{-6} metre)

Preface

At the end of 1980, when I finished writing a book of the cell division cycle with David Lloyd and Robert Poole, I promised myself that I would never write another. During the past few months there have been many times when I wished that I had kept that promise. During this time I have been reassuring my family, colleagues and not least myself, that it was 'almost finished'. Now that it finally is complete, I promise that I will never write another. Well, perhaps not for a few years yet.

In writing this text, I have been primarily aiming at a level where new researchers to the field can obtain an overview of many of the exciting new developments in neutrophil biochemistry and physiology. To achieve this, I have kept the number of references to original work to a fairly low number. In doing so, I do not wish to detract from the hundreds or, more correctly, thousands of original publications that have arisen over the past 10 or 20 years. Instead, I have tried to make this an 'easy read' and so have only quoted key, landmark references or review articles. I hope that most of the important publications and authors are mentioned, and apologise now if I have offended anyone by not directly quoting their articles. To have included all the interesting and relevant publications in this field would have resulted in a text of over twice this size.

Whilst I have attempted to present a balanced view of neutrophil biochemistry and physiology, the book is undoubtedly biased towards my own research interests, where I feel that the most exciting new developments are taking place; that is, towards understanding the processes involved in the respiratory burst, the complexity of the intracellular signal transduction pathways regulating cell activation and the mechanisms responsible for neutrophil priming. In completing these sections I must thank Fiona Watson, Julie Quayle, Becky Stringer, Ann Falconer, Sue Adams, Gordon Lowe, Dave Galvani and Tony Hart for their time spent proofreading and for making useful suggestions. I am particularly grateful for their comments on sections that were less familiar to me; however, any errors that remain are mine. I would also like to thank Becky for Table 2.2 and Figure 7.11, and Gill Ollerhead for the electron micrographs.

When I now reflect on the time that this book has taken, I would like to think that my research group has missed me and suffered because I have not been around as often as I should have. Sadly, this probably has not been the case, and they have coped well without me! My thanks go anyway to them for their understanding. However, I do know that my wife, Te, and son, Jonathan, have missed out because of the many evenings and week-ends that I have been locked away writing. I thank them for their patience and tolerance, and end with the promise that I shall not write another.

June 1993

1
Neutrophils and host defence: The fight against infection

The neutrophil, the subject of this book, plays a key role as part of the immune response to microbial infections. Its major function is the rapid killing of bacteria and fungi before they multiply and spread throughout the body. The neutrophil is only one arm of the immune system, which includes other leukocytes, lymphocytes and molecular components such as complement, antibodies, acute phase proteins and cytokines. These cellular and molecular components of immunity constitute a co-ordinated and sophisticated network that has evolved in order to maximise the survival of the host against the range of pathogens it encounters daily. This chapter describes the role of the neutrophil within this immune network. Other elements of the immune system (the cellular and molecular components) are also briefly described but only with emphasis on how they interact with neutrophil function; thus, the descriptions of these systems focus upon how the cellular and molecular elements assist neutrophil function during infection and how neutrophils themselves may affect and regulate other aspects of the immune response. A more complete description of the immune system may be found in texts such as Davey (1989), Roitt (1990) and Benjamini and Leskowitz (1991).

1.1 The immune system

The immune system protects humans and animals from microbial infections by such infectious agents as bacteria, yeasts and fungi, viruses and protozoa. These differ greatly not only in their size but in their structural and molecular properties, as well as in the ways in which they seek to infect our bodies. Some of these pathogens infect bodily fluids, some penetrate tissues and some even survive and multiply within individual host cells. These intracellular pathogens include viruses, some parasitic protozoa (such as *Plasmodium*, the causative agent of malaria, which infects erythrocytes) and pathogenic bacteria such as *Listeria* and *Legionella*. Infections with these organisms are difficult to counteract because the immune system must recognise the presence of the pathogen within a host cell. On the other hand, many micro-organisms that do not cause disease in humans or animals can survive and multiply within certain parts of our body (e.g. in the gastrointestinal tract). However, if some of these organisms invade other parts of our body, then they are not well tolerated and can cause disease. Perhaps the most striking examples of such organisms are the gut bacteria. In an adult human there may be up to 10^{13} bacteria present in the alimentary canal, but if only small numbers of these enter the bloodstream or urinogenital tract, then they can cause disease.

In the developed countries, infection accounts for less than 2% of all deaths, although the newborn and the elderly are more susceptible to infections than are healthy adults. This is because the immune system is somewhat underdeveloped in newborns, and even more so in premature babies, whose susceptibility to infections is even greater. At the other end of the age spectrum, the immune system becomes defective; this, coupled with a generalised decreased function of our bodily organs and decreased capacity for repair, allows infections to become more common again. Other conditions that lead to a greater susceptibility to infections include poor diet, diseases such as diabetes, disorders of the immune systems (e.g. genetic diseases or haematological disorders) and drug therapy. In patients receiving cytotoxic therapy or radiotherapy for the treatment of cancers, for example, infections are the major causes of morbidity and mortality because the therapy used to kill the tumour also destroys the immune system. Major reasons for the relatively low death rate from infections in the developed countries include the widespread immunisation programs available, good sanitation and adequate diets. There is also an impressive array of antimicrobial agents available, especially antibiotics and antifungal agents, that can be administered to patients if a microbial pathogen evades the immune system.

In the developing countries, however, death from infection is far more widespread, and in some parts of the world 70–80% of all deaths arise from infectious disease. In part, this is geographical: many tropical/subtropical environments allow the proliferation of disease-carrying biting insects (e.g. flies, ticks and beetles) that pass infections (e.g. malaria, sleeping sickness, leishmaniasis) from host to host during blood meals. However, the majority of problems with infections in these developing countries are due to poverty. Infections are common because of poor diet, poor sanitation and contaminated water supplies, lack of effective immunisation programs and unavailability of antibiotics. It is estimated that about 20 million children under the age of 5 years die each year in these countries from diarrhoeal disease. Improved sanitation, clean water and simple therapy would save many of these lives.

The major protection against infection is provided by the skin, which is largely impermeable to water, gases and micro-organisms. It thus serves to protect internal structures and tissues whilst being sufficiently elastic to allow movement. Infectious organisms that penetrate this physical barrier do so via cuts, scratches, burns or bites. Skin, although an effective and efficient barrier, cannot cover the entire body, however: the eyes, the alimentary canal, respiratory tract and urogenital tract are covered not by skin but by a lining of cells (a transparent layer, in the eyes) that permits absorption and secretion of liquids and gases. These parts of the body, prime routes for penetration by infectious organisms, possess specialised structures and features to protect against infection but still allow normal function. The tracts in question secrete mucus, a hydrated gel acting to waterproof and lubricate whilst still allowing the transport of gases, liquids and nutrients. They also are lined with epithelial cells that may be continuously shed and replaced, so as to replace lining cells that may have become damaged or infected; this phenomenon occurs in the gut, but may also occur at lower rates in the respiratory and urogenital tracts. Some epithelial cells are ciliated, and in the lung a major mechanism for the physical removal of micro-organisms and dust particles is the *mucociliary escalator,* in which epithelial cell cilia constantly 'brush' mucus and foreign particles upwards and out of the lungs so that they are cleared via coughing. Mucus and other bodily secretions, such as tears, saliva, nasal secretions and breast milk, also contain constituents that are antimicrobial, such as lysozyme, acid pH, bile and immunoglobulins.

Once a microbe has penetrated these physical barriers, the immune system is then responsible for destroying it before it can multiply and disperse to other parts of the body. The system faced with this task in animals and humans comprises both cellular and molecular components. The cells of the immune system include the *phagocytes* (neutrophils and macrophages) and the *lymphocytes* (primarily T and B cells); these function in close collaboration with the molecular components, which include complement, acute-phase proteins, antibodies and cytokines. Whilst complement (sometimes working alone or sometimes working with antibodies) can kill pathogens, the molecular elements of the immune system usually function to regulate the activity of the cellular components. Thus, the molecular and cellular components act co-ordinately to ensure maximal efficiency of the immune system in recognising and eliminating pathogenic organisms.

1.2 Cellular components of the immune system

1.2.1 Polymorphonuclear leukocytes

The cell types classified as *polymorphonuclear leukocytes* include neutrophils, basophils and eosinophils. These cells are identifiable in blood films because of the unusual morphology of their nucleus, which is irregularly shaped, often multilobed and hence polymorphic. As these are white blood cells, the general description 'polymorphonuclear leukocytes' describes their colour and nuclear morphology (Fig. 1.1a,b). The cytoplasm of these cells has a granular appearance because of the abundance of 'granules' that are, in fact, membrane-bound organelles. Because the distinct properties of these cells and their function in immunity are largely dictated by the constituents of these granules, polymorphic cells are often termed *granulocytes.*

These 'polymorphs' are divided into three subgroups by virtue of the staining properties of their cytoplasmic contents when treated with dye mixtures. Thus, *eosinophils* stain with acid dyes such as eosin and appear red in stained blood films, *basophils* stain with basic dyes and appear blue whilst *neutrophils* stain with both types of dye and their cytoplasm appears purple. The differences between the cytoplasmic contents in these cells types is much more fundamental than is suggested by these simple staining properties. The granules of basophils, neutrophils and eosinophils contain distinct molecular constituents that confer upon the cells their specialised functions during infection and inflammation.

1.2.1.1 Neutrophils

The full name of these cells is *neutrophilic polymorphonuclear leukocytes,* but the terms *neutrophil* and, less-commonly now, *polymorph* are generally used to describe this cell (Fig.1.1a). In fact, most preparations of 'neutrophils' contain about 95–97% neutrophils, the remainder being largely eosinophils, because the commonly-used separation techniques do not efficiently separate these cell types. Neutrophils are the most abundant white cell in the blood, accounting for 40–65% of white blood cells, and are found at concentrations usually in the range $3–5 \times 10^6$ cells/ml blood. This number can increase dramatically (up to tenfold) in cases of infection. They have a relatively short half-life in the circulation (estimated at about 8–20 h), but this may be extended to up to several days if the cells leave the circulation and enter tissues – although it is difficult to measure the lifespan of a tissue neutrophil. Because of the large numbers of neutrophils in the circulation and their relatively short lifespan, vast numbers of neutrophils enter and

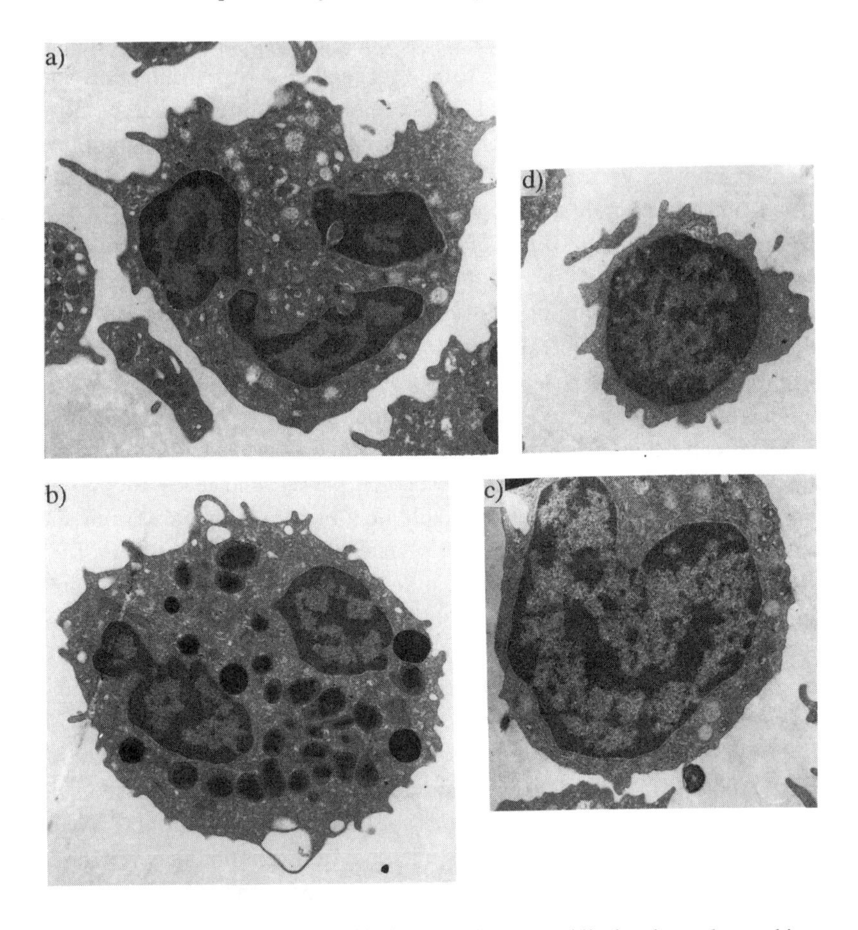

Figure 1.1. Electron micrographs of leukocytes: (a) neutrophil, showing polymorphic nucleus and numerous cytoplasmic granules; (b) eosinophil, showing distinctive granules with 'crystalline' core; (c) monocyte, with horseshoe-shaped nucleus; (d) small lymphocyte. Magnification ×7000.

leave the circulation daily. For example, in an adult human with five litres of blood in the circulation, the total neutrophil pool (not accounting for cells adhering to capillary walls or in tissues) is around 2×10^{10} cells. If these are replaced two or three times a day due to turnover, then the bone marrow, which synthesises these cells, must be able to generate about 5×10^{10} cells every day. In an individual who is infected, this number may increase by an order of magnitude. The huge number of neutrophils that must be produced daily indicates how important these cells are in host protection against infections, and provides clues as to their functional properties. Neutrophils are the first line of defence of the body against bacterial and fungal infections: they are the first cells to be recruited to a site of infection and must respond quickly and potently. These cells are thus highly motile (moving into tissues in response to chemical signals or chemoattractants) and contain an impressive armoury of cytotoxic mechanisms that are capable of killing a range of microbial pathogens.

Neutrophils are about 10 μm in diameter and, whilst in the circulation, are spherical with few, if any, cytoplasmic extrusions. Blood neutrophils are thus said to be in a resting or *non-activated* state. However, once they are *activated,* either by chemical stimuli or by attachment to surfaces, their morphology changes. They may become polarised so that they assume a front end and a rear end, and this polarisation is required before directional movement can occur. They then flatten to assume a classic amoeboid shape with extended pseudopodia and thus become *primed* and ready for action. Priming can be induced by many pathological or physiological stimuli and involves preparing the neutrophil for a state of readiness prior to its full-scale activation. This two-stage activation process probably guards against the non-specific activation of neutrophils: because some of the components that neutrophils use to kill bacteria can also indiscriminately attact host tissues, the possibility of non-specific activation must be minimised. Priming involves movement of some of the cytoplasmic granules, which fuse with the plasma membrane. During this process there is an increase in the number of receptors and some other important proteins on the cell surface, and with more surface receptors – the 'eyes' of the cell – the neutrophil is capable of detecting subtle changes in its microenvironment that may be caused directly or indirectly by the presence of the invading pathogen.

Neutrophils kill their target pathogens by the process of *phagocytosis* (Fig. 1.2). Efficient phagocytic killing is, however, regulated at several steps (Fig. 1.3). Firstly, the neutrophil must be able to recognise the pathogen as 'foreign'. Sometimes the surface properties of the pathogen are so unusual (compared to the surfaces of host tissues) that this recognition is achieved without the involvement of any other factors. More usually, this recognition is aided by the coating of the pathogen with *opsonins,* such as antibodies, complement fragments, acute-phase proteins and fibronectin. This *opsonisation* process is important because neutrophils possess receptors for portions of the opsonin molecules (e.g. immunoglobulin and complement receptors) so that any particle coated with opsonins is labelled as a target for neutrophil phagocytosis (Fig. 1.4).

After the binding of the pathogen to the neutrophil surface via these receptors, the neutrophil must respond by activating its bactericidal arsenal. This activation is achieved through the occupancy of receptors, which trig-

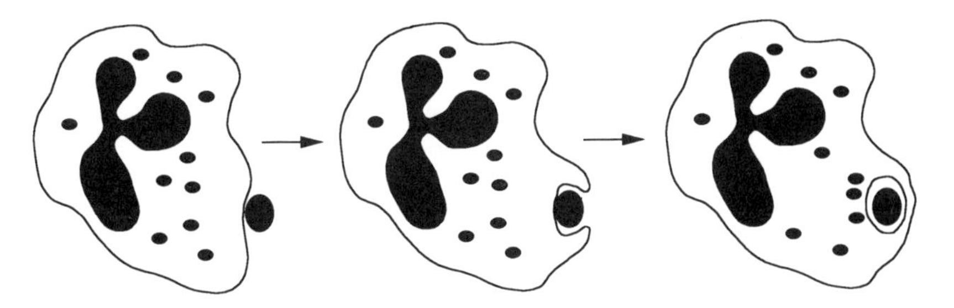

Figure 1.2. Phagocytosis by neutrophils. A bacterium is recognised by receptors of the plasma membrane of a neutrophil. This interaction triggers the formation of pseudopodia around the bacterium so that the bacterium eventually becomes fully enclosed within a phagocytic vesicle. The membrane of this vesicle is derived from the plasma membrane. Cytoplasmic granules then fuse with this vesicle to form a phagolysosome. During this process the granule membranes incorporate into the membrane of the phagolysosome, whilst the contents of the granules are discharged into the vesicle.

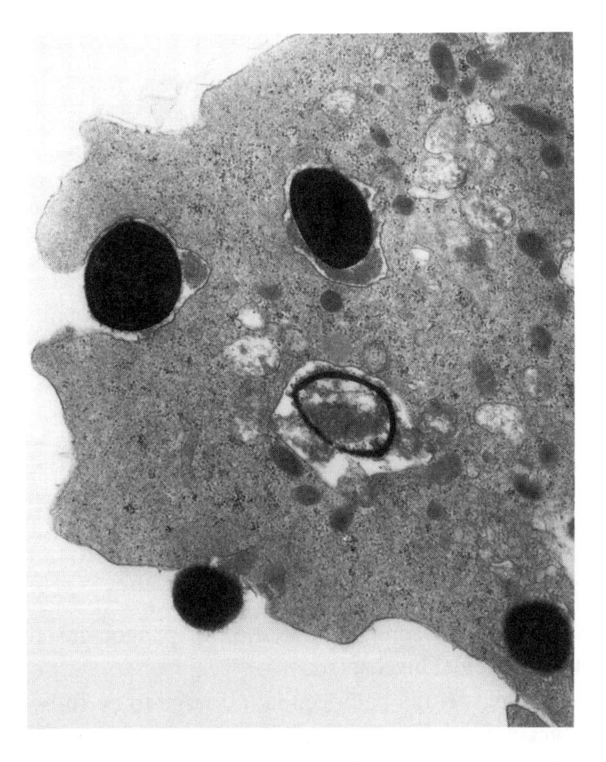

Figure 1.3. Phagocytosis of *Staphylococcus aureus* by a human neutrophil. Neutrophils were incubated with opsonised *S. aureus* and fixed after 15 min incubation. The bacterium in the centre has been lysed, and only the cell wall remains. *Source:* Experiment of Bernard Davies and John Humphreys, reproduced with permission from *Colour Atlas of Paediatric Infectious Diseases,* by Hart and Broadhead (Mosby Year Book Europe).

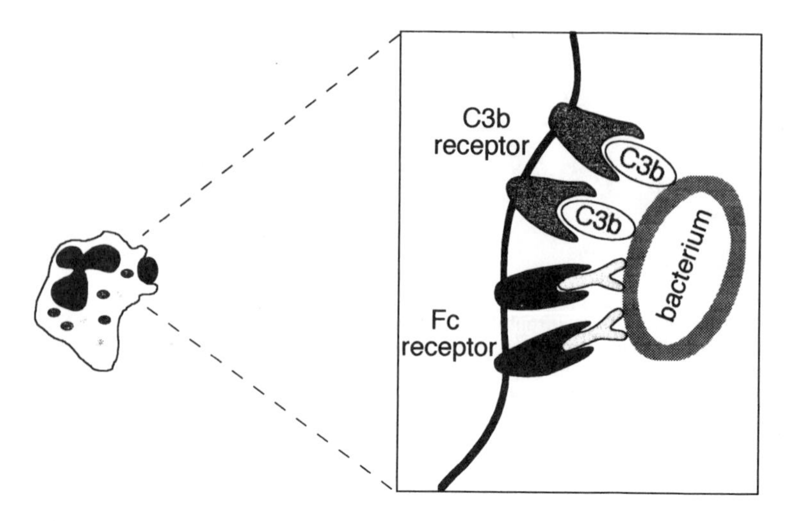

Figure 1.4. Recognition of bacteria by neutrophils. Invading bacteria are opsonised by serum proteins, such as complement fragments (e.g. C3b) and immunoglobulins. The plasma membranes of neutrophils possess receptors for these opsonins (e.g. Fc receptors and complement receptors). Thus, occupancy of these opsonin receptors triggers phagocytosis and activates events such as the respiratory burst and degranulation. Note that the receptors and opsonins are not drawn to scale.

gers the formation of second messenger molecules that directly or indirectly activate specific enzyme systems (e.g. via phosphorylation reactions). These signalling processes and *signal transduction systems* (which 'transduce' a chemical signal from the outside to the inside of the cell) in the neutrophil are complex for several reasons. One such reason for this complexity is that whilst neutrophils function to destroy microbial pathogens, this process is controlled at several stages (Fig. 1.5), including:

i. attachment of the cell to the capillary walls (*margination*) prior to leaving the circulation;
ii. squeezing through gaps between adjacent endothelial cells (*diapedesis*);
iii. migration into tissues (*chemotaxis*);
iv. recognition of the pathogen as 'foreign', initiation of phagocytosis and activation of the bactericidal mechanisms;
v. release of cytotoxic products if the pathogen is too large to be fully enclosed within a phagocytic vesicle (*frustrated phagocytosis*);
vi. release of pro-inflammatory molecules (e.g. chemoattractants) or other immune stimulants (cytokines) if more cells of the immune system (including more neutrophils) must be recruited to the site of infection.

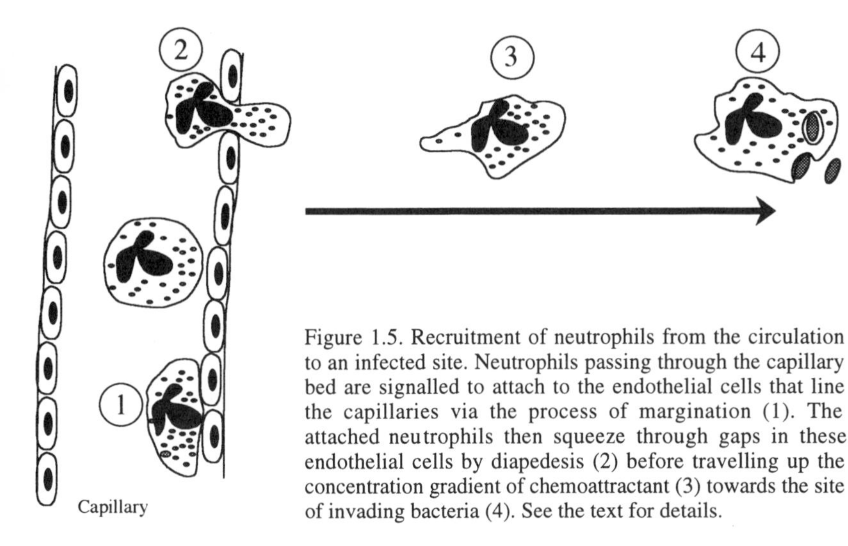

Figure 1.5. Recruitment of neutrophils from the circulation to an infected site. Neutrophils passing through the capillary bed are signalled to attach to the endothelial cells that line the capillaries via the process of margination (1). The attached neutrophils then squeeze through gaps in these endothelial cells by diapedesis (2) before travelling up the concentration gradient of chemoattractant (3) towards the site of invading bacteria (4). See the text for details.

Thus, multiple signalling systems are required in order to specifically activate these various processes. Furthermore, there is an element of overlap or redundancy in these signalling systems: if neutrophil function were controlled by a single signalling system, then a defect in that system would completely abolish all aspects of cell activation. In view of the importance of neutrophils in host protection against infection, such a defect would have devastating consequences, and the host would inevitably have an increased risk of death from infections. Possession of multiple pathways thus guards against a fatal defect that may arise due to mutation.

Another reason for complexity of cell-activation mechanisms resides in the end response of neutrophils – that is, the delivery of cytotoxic products. Whilst these products are highly lethal towards pathogens, they can also attack and destroy host tissues, and this can have deleterious effects on tissue function. Complex intracellular signalling mechanisms to activate these cytotoxic pathways also guards against non-specific activation, which could lead to host tissue damage.

Once the pathogen is enclosed within a phagocytic vesicle, these cytotoxic processes must be activated and delivered to the pathogen (see Fig. 1.2). Here, the plasma membrane and cytoplasmic granules play important roles. The plasma membrane contains an unusual enzyme that is capable of generating a series of reactive oxygen metabolites with broad antimicrobial properties. The enzyme responsible for this is an NADPH oxidase, which although it has a relatively simple task to perform (to transfer a single electron to O_2) has an extraordinarily complex structure and a complex mechanism of activation. The oxidase comprises individual components that are present on the plasma membrane, on the membranes of some granules and in the cytoplasm. An active enzyme complex is assembled by the bringing together of these constituent parts during phagocytosis. Reactive oxidants are thus generated on the membrane (derived from the plasma membrane) that lines the phagocytic vesicle. Hence, the production of these oxidants is efficiently targeted, and they are delivered to the pathogen in high concentrations (Fig. 1.6). This complex structure and complicated activation mechanism also guards against the non-specific activation of the oxidase: the products of the oxidase can also cause considerable damage to host tissues.

The cytoplasmic granules also play an important function in pathogen killing because they contain a range of proteins with cytotoxic properties. These include a range of proteases, hydrolytic enzymes, a peroxidase (myeloperoxidase) and also a number of highly-specialised proteins that affect the permeability of microbial targets. Although the list of antimicrobial proteins present in these granules continues to grow as more constituents are discovered, two facts are apparent:

i. These antimicrobial proteins are 'packaged' within the granules, sometimes in a latent form, and hence are only active when they are released from these stores (Fig. 1.6): this prevents damage to host tissues or, indeed, damage to the neutrophil itself.
ii. By having this broad range of biochemically-distinct antimicrobials, the neutrophil has a degree of overkill, enabling it to attack a variety of microbial targets that may have differing susceptibilities to any one process. This in itself serves two functions: (a) should an organism develop resistance to one of the many different types of pathogens killed, that type may still be effectively killed via an alternative mechanism; (b) should a genetic defect arise that renders a particular antimicrobial system ineffective, then an alternative system will be available and hence will function to protect the host. (Defects in neutrophil function, described in Chapter 8, are usually associated with impaired defence against certain types of microbial pathogens.)

In summary, the importance of the neutrophil in protection against infection is highlighted by the presence of large numbers of these cells in the circulation and their production in vast numbers by the bone marrow. Neutrophil function is regulated via its ability to detect pathogens or signals (generated from host tissue, immune cells or the pathogens themselves) that may be generated during infections, and to leave the circulation and migrate into the infected tissues. Once at the infected site, it recognises the 'foreign'

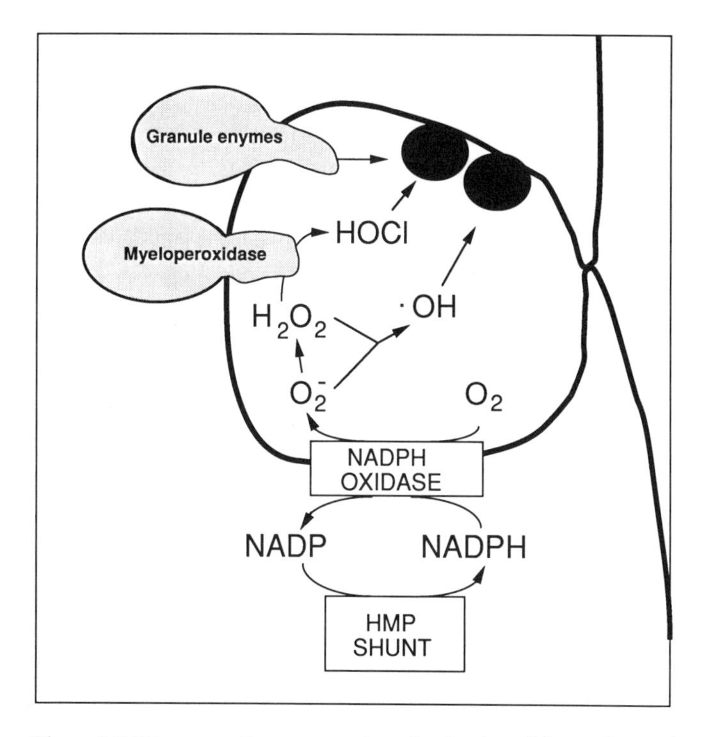

Figure 1.6. Diagrammatic representation of activation of the respiratory burst and degranulation during phagocytosis. Bacteria (represented as dark circles) are phagocytosed and enclosed within a phagocytic vesicle. The hexose monophosphate shunt is activated, and this generates NADPH, which is the substrate for the oxidase. The oxidase itself is then activated via the translocation and assembly of individual component parts, and accepts electrons from NADPH to reduce O_2 to O_2^- (superoxide). This O_2^- then dismutates into H_2O_2, and it is *possible* that ·OH formation occurs (perhaps localised formation on the microbial surface). The cytoplasmic granules then migrate to the phagocytic vesicle and discharge their contents. Hence, myeloperoxidase enters the vesicle and can utilise the H_2O_2 to generate HOCl and related compounds. The other granule constituents (e.g. defensins, proteases) can then attack the phagocytosed bacteria.

invader via its plasma membrane receptors, occupancy of which results in the triggering of phagocytosis and activation of the multiple antimicrobial pathways that lead to pathogen destruction. As we shall see, the neutrophil is highly specialised for this important role. However, we shall also see that the neutrophil can play a more active and direct role in the immune response via its ability to generate soluble mediators (eicosanoids and cytokines), which can result in the recruitment of more immune cells (including more neutrophils) and can also regulate the function of these immune cells.

1.2.1.2 Eosinophils

Eosinophils comprise about 1–7% of the total white cells in human blood and are about the same size as neutrophils (see Fig. 1.1b). They are polymorphonuclear, possessing a multilobed nucleus and densely-staining granules in their cytoplasm. These granules are larger then those found in the cytoplasm of neutrophils and have a distinctive crystalline appearance. A peroxidase (eosinophil peroxidase), genetically and biochemically distinct from the myeloperoxidase of neutrophils, is present in these granules, together with a number of lysosomal enzymes. The crystalloid core of the granules comprises a major basic protein (relative molecular mass = 10 kDa) that is cytotoxic to parasites and to certain mammalian cells, and that also induces the release of histamine from basophils and mast cells. Major basic protein binds to the surface of parasites and damages their cell membranes. Other components of the granules include cationic proteins and a neurotoxin. Eosinophils can also generate reactive oxygen metabolites via an NADPH oxidase identical to that present in neutrophils. The major function of eosinophils appears to be the destruction of parasites such as protozoa, and they probably secrete their cytotoxic products because their targets are often too large to be fully phagocytosed. Numbers of eosinophils in the circulation can become elevated (*eosinophilia*), particularly during parasitic infections, and eosinophils can be found distributed throughout the body, including in regions of the gastrointestinal tract such as the colon.

1.2.1.3 Basophils

Basophils, found only in the blood, are relatively rare, usually comprising <1% of the total white cells in the circulation. Because of the difficulty in obtaining large numbers of these cells, they are not well defined biochemically. They are are related to mast cells, which are primarily found within tissues: both cell types perform similar functions in allergic reactions and immediate hypersensitivity reactions. Their granules contain histamine, serotonin and other components that can mediate allergic/inflammatory responses. *Histamine* (derived from the amino acid histidine via the action of the enzyme histidine carboxylase) is present in human basophils in large quantities (1–3 pg/cell). Its biological function arises from binding to histamine receptors (H_1 or H_2) on target cells. Occupancy of H_1 receptors results in contraction of bronchial and gastrointestinal smooth muscle whereas occupancy of H_2 receptors probably mediates gastric acid secretion by parietal cells and also regulates the functions of some lymphocytes.

One of the distinctive features of basophils is that they possess plasma membrane IgE receptors (FcεR), which bind the fragment crystallisable (Fc) region of IgE with high affinity. This is in contrast to the FcεR on lymphocytes and macrophages, which binds IgE with only low affinity.

Binding of IgE to FcεR on basophils decreases the rate of turnover of the receptor. Thus, once IgE is generated by an immune response, it binds this receptor and the receptor–IgE complex is quite stable. When the antigen to which the IgE was generated is re-encountered, it binds to the IgE attached to the surface of the basophil, and thus rapidly activates the cell. This activation results in the *degranulation* (discharge) of histamine, serotonin and other granule constituents that give rise to the classical allergic/hypersensitivity reactions. After IgE–antigen binding, the receptors are *capped* (i.e. cluster at one portion of the cell surface) and become internalised. Neither the receptors nor IgE reappear on the cell surface; instead they are degraded.

1.2.2 Mononuclear phagocytes

Monocytes, or mononuclear phagocytes, constitute about 7% of the total white cells in blood. They are the largest circulating leukocyte, possessing a distinctive horseshoe-shaped nucleus (see Fig. 1.1c). In the bloodstream, monocytes have a half-life of only a few days, but if they are recruited into tissues, then their half-life increases to weeks or even months. During this recruitment, monocytes differentiate into macrophages, a process involving a selective change in gene expression due to transcription and translation. This differentiation not only greatly increases the lifespan of tissue macrophages but also results in the acquisition of their specialised functions. For example, when monocytes are recruited into tissues they develop specialised features required for their function within these tissues. Thus, liver macrophages (*Kupffer cells*), lung (alveolar) macrophages, peritoneal macrophages and bone macrophages (*osteoclasts*) each have specialised functions. The range of macrophage functions is extremely diverse and includes: antigen presentation (to T cells); wound healing; tumour-cell destruction; bone remodelling; phagocytosis of microbial pathogens, cell debris or immune complexes; production of tissue- and immune-cell-regulators such as cytokines and growth factors. Apart from their presence within tissues, macrophages are also found lining the membranes of the gut, the urogenital tract, the lungs, lymph nodes and the spleen, as well as in synovial membranes. Hence, they are involved in host protection and host repair.

Apart from their ability to secrete growth factors and cytokines, certain types of macrophages are avidly phagocytic. These possess a variety of cytotoxic processes, such as a range of proteases, and also have the ability to generate reactive oxygen metabolites (via the NADPH oxidase) and reactive nitrogen metabolites, such as nitric oxide (NO). This NO is generated via a NO synthase from arginine and can, at low concentrations, serve as a tissue- or cell-regulatory molecule; high concentrations of this molecule are microbicidal and cytotoxic. The ability of macrophages to secrete immunoregulatory compounds and cytotoxic products depends upon the differential gene activation that has occurred during their recruitment into tissues.

1.2.3 Lymphocytes

There are three major classes of lymphocytes (see Fig. 1.1d):

i. *B lymphocytes,* which secrete antibodies;
ii. *T lymphocytes,* which are subdivided into three types of cells with cytotoxic and regulatory functions;
iii. *large granular lymphocytes,* which have potent cytotoxic activity.

1.2.3.1 B lymphocytes

Both B and T lymphocytes exist as clones, each clone possessing a unique receptor (an immunoglobulin-like molecule) that recognises only one epitope on the surface of an antigenic structure. These receptors are called *surface immunoglobulins* (sIg) and are present at up to 10^5 copies per cell. Because of the large numbers of epitopes ($>10^8$) that are potentially present on the surfaces of the range of pathogens that the immune system must recognise and dispose of, there must be $>10^8$ different receptors and hence $>10^8$ different lymphocyte clones. However, after cell development, only a few individual cells of a particular clone exist (as *naïve* or *virgin lymphocytes*) because large numbers of each of these individual clones cannot be accommodated in the circulation. Thus, when a naïve or virgin lymphocyte first encounters the antigen to which it possesses a receptor, there are too few cells available to mount an effective challenge. When a lymphocyte receptor first binds its antigen, receptor occupancy leads to the rapid growth and division of the virgin lymphocyte via a process assisted by the secretion of cytokines by other types of (helper) lymphocytes. This process of *clonal expansion* leads to the development of two types of lymphocytes: some of the cells differentiate into *effector cells* (e.g. antibody-secreting cells or cytotoxic lymphocytes) whereas others develop into *memory cells* (Fig. 1.7).

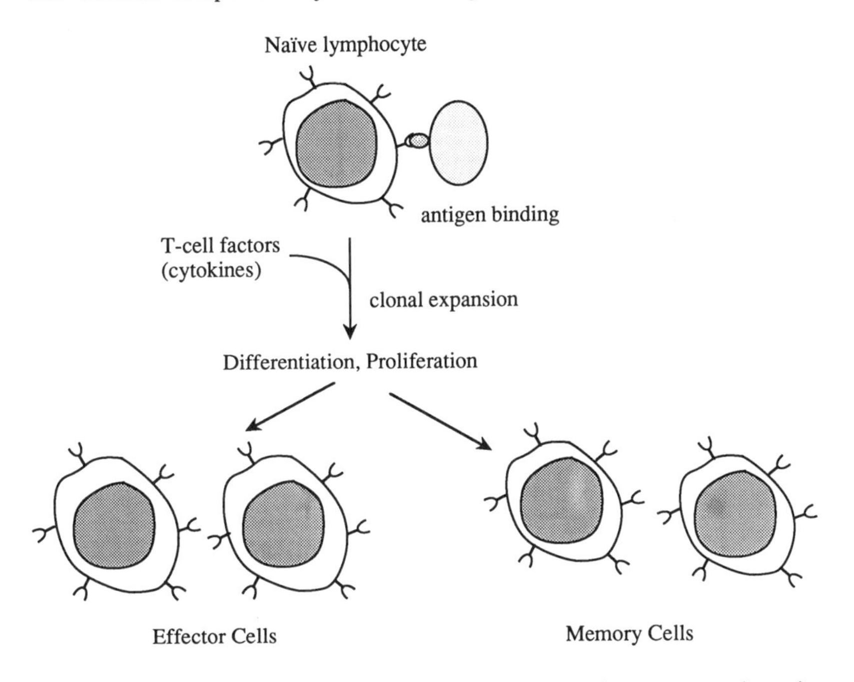

Figure 1.7. Lymphocyte activation. When naïve lymphocytes first encounter the antigen that is recognised by their receptor, they are stimulated to differentiate and proliferate. This clonal expansion is aided by the production of cytokines. Two cell types develop from this process: the *effector* cells (i.e. either antibody-secreting plasma cells or cytotoxic T cells) and *memory* cells. Both cell types possess virtually the same receptor that was expressed on the naïve lymphocyte.

The receptor present on the effector cells and memory cells is virtually the same as that present on the virgin lymphocyte, so the effector and memory cells recognise and bind the same epitope. After the antigen is eliminated, the effector cells die, but the memory cells remain. If the same antigen is encountered again, then the immune response is activated more quickly and more potently because of the large numbers of memory cells of a clone compared to the small numbers of virgin cells that originally existed. Thus, the antigen is usually cleared before an infection develops. This immunological memory is the basis of immunisation.

When B lymphocytes are stimulated by antigen they develop into plasma cells. The receptor of plasma cells can be secreted from the cell because some of these molecules lack the hydrophobic domain that normally anchors the receptor to the plasma membrane. These secreted molecules are *immunoglobulins* or *antibodies* and have a common structure (Fig. 1.8). They are found in a variety of bodily fluids and in many tissues.

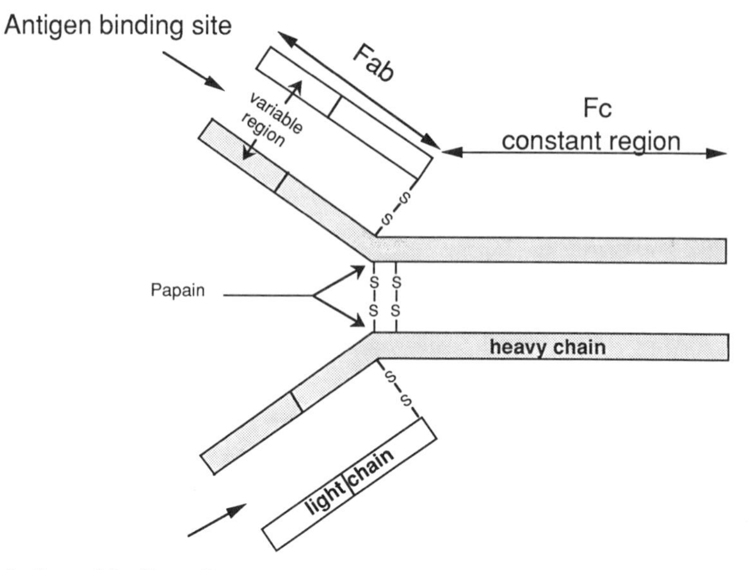

Figure 1.8. Diagrammatic representation of an antibody molecule. An antibody molecule comprises two heavy chains (dark shading) and two light chains (unshaded). These chains are held together by disulphide (-S—S-) bonds. The site of papain cleavage (yielding Fab and Fc portions of the molecule) is shown. Each antibody molecule has two potential antigen-binding sites.

1.2.3.2 T lymphocytes

T cells exist as three separate subsets of cells:

(i) *Cytotoxic T cells (Tc).* These cells lyse host cells infected with intracellular parasites such as viruses. Viral glycoproteins (arising from particle disintegration or from de novo biosynthesis) become expressed on the surface of the infected cells in association with major histocompatability complex (MHC) class I molecules, and this MHC–antigen complex is recognised by the T-cell receptor (TcR). Two accessory molecules on the surface of the T cells assist in this process: CD8, which aids the binding of the receptor to the target, and CD3, which is involved in the generation of signalling molecules that activate the T cells. Naïve or virgin T cells that encounter their antigen for the first time are then activated to proliferate and differentiate via a mechanism that requires Interleukin-2 (IL-2), secreted by T helper cells (see next paragraph). Activation of the T cell results in lysis of the target via the secretion of pore-forming molecules (e.g. perforin) by

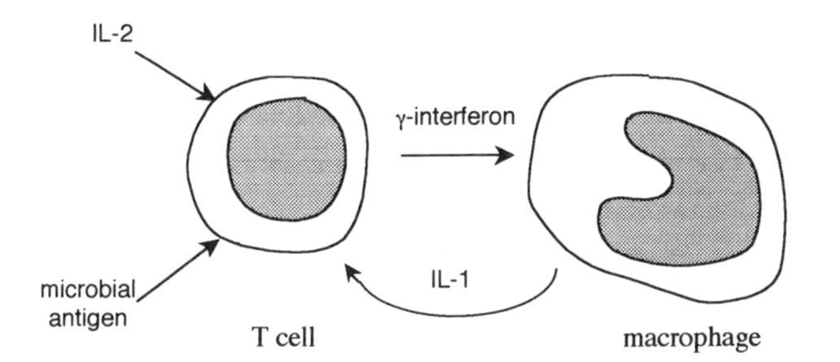

Figure 1.9. Interactions between macrophages and T helper cells. Stimulation of T helper cells results in the secretion of γ-interferon, which then stimulates macrophages. The γ-interferon-stimulated macrophages then secrete Interleukin-1, which stimulates the T helper cells to secrete more γ-interferon. Thus, a cytokine loop develops.

the T cell. Because of their surface properties, cytotoxic T cells are defined as $CD8^+$.

(ii) *T helper cells (Th).* The major function of these cells appears to be the secretion of cytokines upon activation of their receptor. Antigen-presenting cells, such as macrophages, phagocytose viral and other pathogenic particles, which are then degraded within phagolysosomes. Some of these partly-degraded molecules are expressed on the surface of macrophages in association with MHC class II molecules. This antigen–MHC complex is recognised by the TcR on the T helper cells, and two other molecules on the T helper cells assist in this process: CD4 aids receptor binding whilst CD3 leads to the generation of intracellular signalling molecules that activate the T cell. IL-1 generated by the macrophage is required for activation of the T cell, which in turn secretes γ-interferon, which activates the macrophage (Fig. 1.9). The activated T cell thus generates a series of cytokines, including IL-2, which activates cytotoxic T cells. T helper cells are designated $CD4^+$.

(iii) *T suppressor cells (Ts).* These cells regulate the function of all activated T cells, most activated B cells and some activated macrophages. They perform this role by the secretion of (suppressor) cytokines and thus help to down-regulate the immune response when the antigen has been cleared.

1.2.3.3 Large granular lymphocytes

(i) *Natural killer (NK) cells.* These are present in the bloodstream in very low numbers ($<1\%$ of the total white-cell population). They have membrane features that suggest they belong the lymphocyte family of cells, but their cytoplasm also contains large numbers of granules. Although their biochemical characterisation has been impeded by their low abundance in blood, they have been shown to be capable of antibody-independent killing of host cells, especially those infected with viruses. Their killing response is somewhat non-specific: a range of host cells infected with a variety of viruses may be killed. The cytotoxic mechanisms are thought to be initiated by receptors on the surface of NK cells that bind viral glycoproteins expressed on the surface of infected cells. A space forms between the NK cell and the target cell, into which a variety of proteins are secreted by the NK cells. One such protein is *perforin,* a cylindrical molecule that is inserted into the target cell membrane. This forms a pore in the target cell, thereby affecting ion transport, which subsequently leads to the lysis of the target by osmotic shock.

(ii) *Killer (K) cells.* These large granular lymphocytes utilise antibody-dependent cellular cytotoxicity (ADCC) in order to kill their targets. Hence, the efficient function of K cells requires antibodies generated by B cells. The killing mechanisms employed by these cells are largely unknown as they are even rarer in the blood than NK cells. Receptors on the plasma membrane of the K cells recognise the Fc regions of antibodies bound to the target; hence any host cell coated with antibodies is a potential target for this killing mechanism.

1.3 Molecular components of the immune system

1.3.1 Properties of antibodies

Antibodies or immunoglobulins can be cleaved with papain (a protease) to yield three fragments (see Fig. 1.8): two *fragment antigen-binding* (Fab) sections, which retain the ability to bind antigen, and a *fragment crystallisable* (Fc) section. In the NH_2-terminal portions of the Fab fragments – the regions that bind the antigen – the amino-acid sequences of immunoglobulins are different in antibodies secreted by different B-cell clones. The Fc regions, on the other hand, are much more homologous in structure and are almost identical in immunoglobulins secreted by different B-cell clones. The Fc regions do not bind antigen but are recognised by specific receptors on the surface of phagocytic cells (Fc receptors). Thus, if an antibody binds its antigen (and each immunoglobulin molecule can bind two molecules of antigen), then the Fc regions cluster and are exposed. Phagocytic cells (e.g. neutrophils and macrophages) recognise these exposed Fc regions of antibodies by virtue of their Fc receptors, and thus particles that are opsonised

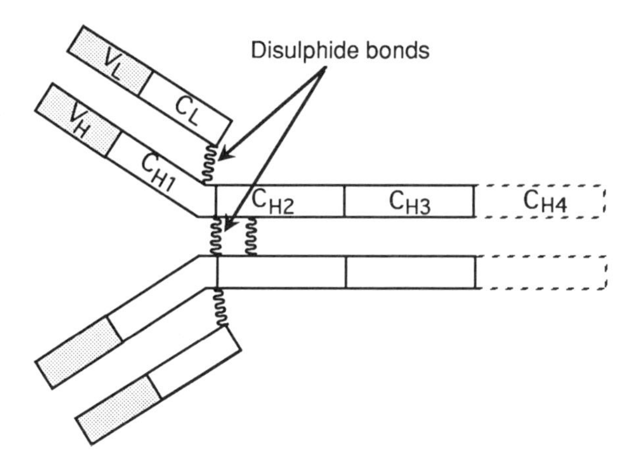

Figure 1.10. Generalised structure of the variable and constant domains within antibodies. The variable regions (dark shading) of either the light or heavy chains are indicated as V_L or V_H, respectively. The light chains also possess one constant region (C_L), whereas the heavy chains possess either three or four constant regions (C_{H1}–C_{H4}), depending upon the class of immunoglobulin (see text for details).

in this way (or labelled as 'foreign') are readily phagocytosed (see Fig. 1.4). Apart from their ability to opsonise targets, immunoglobulins can also activate complement (which will lyse antibody-coated targets), can neutralise and precipitate antigens and can cause degranulation of mast cells.

Treatment of antibodies with reagents that reduce disulphide bonds (e.g. mercaptoethanol) splits the molecule into two light (L) chains (relative molecular mass = 25 kDa) and two heavy (H) chains (of 50–70 kDa). Thus, in the intact molecule, these light and heavy chains must be held together by disulphide bonds between adjacent cysteine residues in the chains. A closer examination of the structure of these chains reveals that each L chain has two domains whilst each H chain has four or five domains, depending upon the immunoglobulin class to which the antibody molecule belongs (Fig. 1.10). The first (NH_2-terminal) domain of the heavy and light chains is highly variable (V domain) in antibodies secreted from different clones; the second and subsequent domains are much more constant in different antibodies (C domains). Thus, the specificity with which different antibodies bind to their antigens resides in the structure of the V domains. In each V domain there are three hypervariable loops of up to 15 amino acids. Each antigen-binding site thus comprises six hypervariable loops (three from the L chain and three from the H chain); the size, shape, charge and hydrophobicity of the amino acids within these loops dictates the three-dimensional structure of the antigen-binding site. This structure is thus complementary to the epitope on the antigen, providing the unique 'fit'.

1.3.1.1 Immunoglobulin classes

Whilst all immunoglobulins conform to the generalised pattern described above (i.e. each molecule comprises two L chains and two H chains), there are five different classes of antibodies that represent variations on this common theme. The L chains of these molecules may be κ or λ (classified according to their structure); the H chains are classified as α, δ, ε, γ or μ in IgA, IgD, IgG, IgE and IgM, respectively. Furthermore, some of these antibodies possess ancillary protein chains that form polymeric antibody structures.

(i) *IgG.* This is the predominant immunoglobulin in blood, lymph, cerebrospinal fluid and peritoneal fluid. The two H chains are termed γ chains, each having a relative molecular mass of 50 kDa (Fig. 1.11), whilst the L chains may be κ or λ and are each 25 kDa. The heavy chains comprise three constant regions. The molecules of IgG may thus exist as $\kappa_2\gamma_2$ or $\lambda_2\gamma_2$. Each IgG molecule (relative molecular mass = 150 kDa) has a valency of 2 (i.e. each molecule can bind two molecules of antigen). Its relatively small size and simple structure enables it to enter tissues and even cross the placental barrier, thus conferring passive immunity from the mother to the foetus. It is the most abundant immunoglobulin in serum, comprising about 70–80% of the total, and has an average half-life of 23 days.

IgG is a very efficient opsonin (coating bacteria and labelling them as 'foreign' prior to phagocytosis), and its Fc region is recognised by specific IgG receptors (FcγR) on the surfaces of neutrophils, monocytes and some other immune cells. Subclasses of IgG (e.g. IgG_1, IgG_2, IgG_3 and IgG_4) exist in human serum that have about 90% homology with each other in the constant regions but only about 60% homology with immunoglobulins of different classes (e.g. IgA, IgM). The occurrence of these different subclasses in human serum is IgG_1 (70%) > IgG_2 (20%) > IgG_3 (7%) > IgG_4 (3%). In contrast to the other IgG subclasses, IgG_3 has a low half-life in serum (7 days), but it is much more efficient in the fixation of complement than are the other subclasses. IgG_2 can bind to polysaccharides and is a T-cell-independent antibody generated in response to infections by bacteria that possess an extracellular polysaccharide capsule. In addition to serving as an opsonin, IgG functions in antibody-dependent cellular cytotoxicity (ADCC), complement fixation, the neutralisation of toxins and bacterial immobilisation.

(ii) *IgM.* Usually the first type of antibody to be generated during an immune response, IgM is also known as *macroglobulin* (Fig. 1.12). It is a large

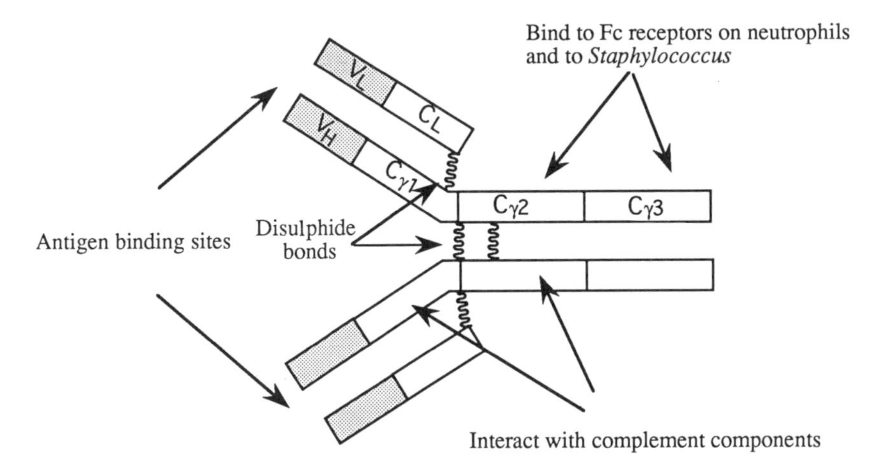

Figure 1.11. Schematic representation of an IgG molecule. The heavy chains comprise three constant regions ($C_{\gamma 1}$–$C_{\gamma 3}$); the molecular sites involved in binding to antigen, complement or to Fc receptors are indicated.

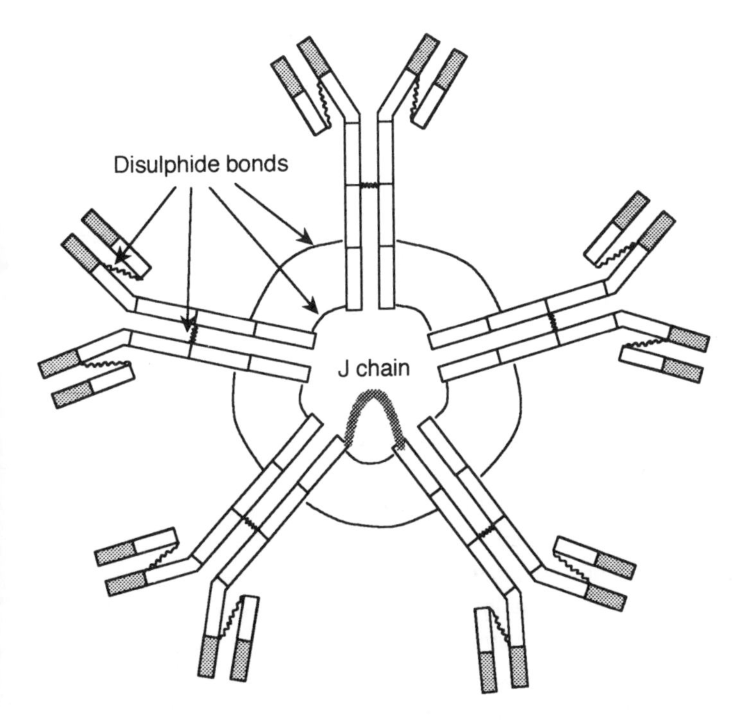

Figure 1.12. Schematic representation of an IgM molecule. Each IgM molecule comprises five IgG molecules joined by disulphide bonds and a J chain. Although the molecule has a predicted valency of 10 (i.e. a single IgM molecule can bind 10 molecules of antigen), this number is not reached in practice.

molecule (relative molecular mass = 900 kDa) firstly because it comprises five molecules of the basic IgG structure held together by a J chain of 15 kDa, and secondly because each immunoglobulin molecule contains an extra C domain on the heavy chain. It thus has a predicted valency of 10, but in practice a valency of 5 is more common because of geometric considerations (i.e. each IgM molecules can bind five molecules of antigen). Its large size restricts its entry into tissues, and because its Fc regions form a centralised core, they are inaccessible to Fc receptors. Thus, IgM is not a good opsonin for phagocytic cells; however, it is a very efficient antibody for the fixation of complement.

(iii) *IgA*. This comprises two immunoglobulin molecules joined by a J chain and a secretory piece (Fig. 1.13). IgA is the major antibody found in bodily secretions such as saliva, mucus, sweat, gastric fluid, tears and colostrum. This dimeric molecule, with a relative molecular mass of 165 kDa and a predicted valency of 4, is the major form of IgA in secretions; in serum the molecule is largely monomeric. Dimeric IgA is secreted by plasma cells associated with epithelial cells of, for example, intestinal villi, tear glands, lactating breasts and bronchial mucosa. Its major function is thus to protect the host from pathogens invading the mucus membranes, which serve to connect the body with the outside world. The function of serum IgA is unknown.

The 15-kDa J chain is synthesised by the same B-cell clone that produces the IgA molecule. The IgA molecules are transported across the epithelial cells and enter the lumen, this transport being mediated via another protein called the *secretory chain* (SC). The IgA molecules that are finally secreted are dimers of relative molecular mass of 400 kDa. The heavily glycosylated SC (80 kDa), synthesised and secreted by the epithelial cells, binds to the IgA molecules via non-covalent bonds. The IgA molecule thus has a valency of 4 (i.e. a single molecule has four antigen-binding sites), with all four sites recognising the same antigen.

IgA has a short half-life in serum (6 days) and comprises about 12–20% of the total serum immunoglobulins. However, because of its presence in bodily fluids, it is the most abundant immunoglobulin present in the body. It comprises three constant domains, and neutrophils, monocytes and some other immune cells possess receptors for IgA (FcαR). Neither of the two IgA subclasses, IgA_1 and IgA_2, can fix complement via the classical pathway. Instead, these antibodies neutralise antigens at mucosal surfaces, in the absence of complement fixation (which would be pro-inflammatory), and the neutralised antigens are cleared.

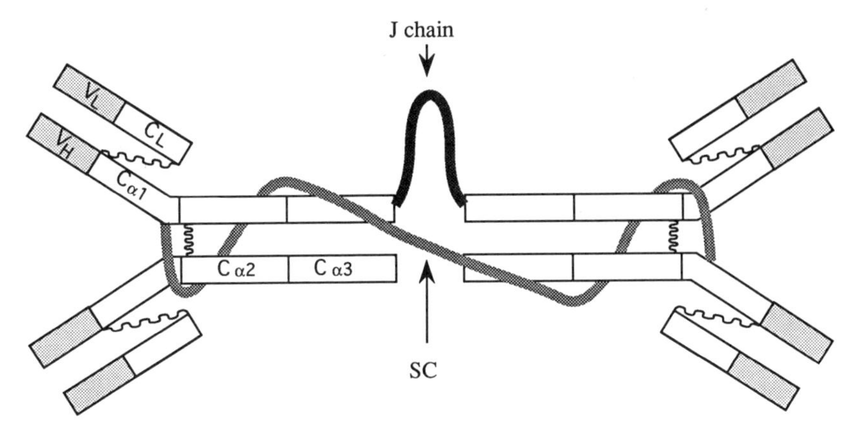

Figure 1.13. Schematic representation of an IgA molecule. Each IgA molecule comprises immunoglobulin molecules joined to each other via a J chain. The heavy chains possess three constant regions ($C_{\alpha 1}$–$C_{\alpha 3}$). The secretory chain (SC) is secreted by epithelial cells and binds to the IgA dimer via disulphide bonds (indicated by wriggly lines).

(iv) *IgD.* This has a relative molecular mass of 180 kDa and is present in serum only at very low levels because it not secreted in large quantities by B cells. Secreted IgD also has a short half-life in serum and is primarily found as a surface component of B cells.

(v) *IgE.* Present in serum only at low levels, this has an extra constant domain on each H chain that enables the molecule to bind with high affinity to the receptor on basophils and mast cells (Fc εR); it is thus larger (200 kDa) than IgG. Once bound to this receptor the antibody may remain attached for many weeks or months. When the antigen to which it was originally raised is re-encountered, it combines with and cross-links the IgE-receptor molecules on the cell surface. This process induces the degranulation of molecules responsible for allergic/hypersensitivity reactions (e.g. histamine and serotonin).

1.3.2 Complement

The complement system comprises twenty plasma proteins present in the blood and in most bodily fluids. They are normally present in an inactive form but become activated via two separate pathways: the *classical pathway,* which requires antibody, and the *alternative pathway,* which does not. Once the initial components of complement are activated, a cascade reaction is initiated whereby the products of each step in the sequence activate the next component.

1.3.2.1 The classical pathway

During complement activation via the classical pathway, nine major complement components (designated C1–C9) become activated in a sequential process, the product of each activation step being an enzyme that catalyses a subsequent step in the cascade. The purpose of the cascade is twofold: firstly, a sequential activation process decreases the possibility of non-specific activation; secondly, the initial response is amplified so that large numbers of complement molecules become activated in response to small amounts of initial signal. The order of events is as follows.

i. The first step in the activation cascade involves the activation of the first component of complement, C1 (Fig. 1.14) – in fact, a complex of three proteins (C1q, C1r and C1s) held together by Ca^{2+}. C1 is activated by antibodies bound to cell surfaces (e.g. surfaces of pathogens or in some cases to surfaces of host cells). The C1q (a polymer of six identical subunits) binds either to the Fc region of at least two adjacent IgG molecules that are closely bound to a surface, or to a single IgM molecule; hence IgM is a very efficient activator of complement. Because C1q can only bind to IgG molecules that are adjacent, monomeric IgG (e.g. IgG unbound in bodily fluids or in the circulation) does not fix complement. IgG_4, IgA, IgD and IgE do not have C1q receptors and hence do not fix complement.
ii. Activated C1q releases C1r, which in turn activates C1s; the latter has proteolytic activity and cleaves C4 (a glycoprotein, relative molecular mass = 180 kDa) to release a fragment (C4a) and generate activated C4 (C4b). Some C4 binds to the membrane adjacent to the antibody, whilst some binds to C1s; this activated C1s–C4 complex cleaves C2.
iii. The cleaved C2 interacts with activated C4 to form $C\overline{42}$ (the overbar indicates an active molecule), a C3 convertase. This acts upon C3 (a β-globulin) to produce C3b and a smaller fragment, C3a (a soluble molecule whose properties are discussed in §3.9.3.1). This is one part of the complement activation pathway that leads to amplification: a single $C\overline{42}$ complex can activate hundreds of C3 molecules.
iv. C3b can bind to the membranes adjacent to the site of its production so that the target may become coated with C3b molecules. Some C3b also combines with $C\overline{42}$ to form $C\overline{423}$, which is a C5 convertase. This activates C5 to form C5a and C5b.

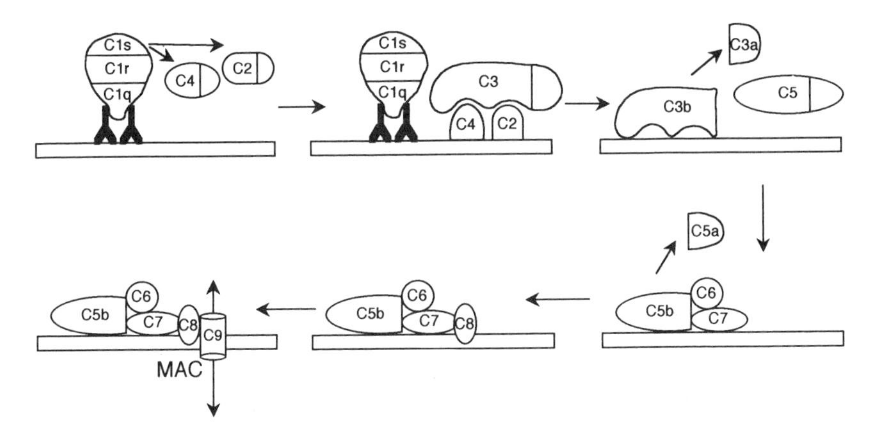

Figure 1.14. Complement activation via the classical pathway. The sequential activation of complement following antibody deposition onto a surface is shown. C9 forms a pore in the membrane, eventually leading to cell death by osmotic lysis. See text for details.

v. C5a is a small molecule whose properties are discussed in section 3.2. C5b binds stoichiometrically to C6 and C7 to form $C\overline{567}$ on the cell membrane.
vi. This $C\overline{567}$ complex directs C8 and C9 to form a membrane attack complex (MAC) that creates a transmembrane channel whose size depends upon the number of C9 molecules inserted into the membrane. This channel results in the osmotic lysis of the target.

1.3.2.2 The alternative pathway

Activation of complement via the alternative pathway requires no antibody–antigen complexes. It may be activated by endotoxin (LPS) on the surface of many Gram-negative organisms, by zymosan of the cell walls of many yeasts or by aggregated IgA.

In serum, there are always trace amounts of C3b that, once attached to a recognition site on a pathogen, may combine with serum factor B to form C3bB (Fig. 1.15). This complex is activated by factor D, which cleaves factor B whilst it is attached to C3b, forming C3bBb. This is a C3 convertase that converts C3 into more C3b and C3a (which is soluble); the C3b can then combine with more factor B and then factor D to amplify the amount of C3bBb produced. However, the C3bBb complex can dissociate, and its stability is regulated by the serum protein properdin (factor P). The released C3b may be inactivated by factor H and factor I to regulate the levels of C3b produced. C3b laid down on the surfaces of cells can lead to activation

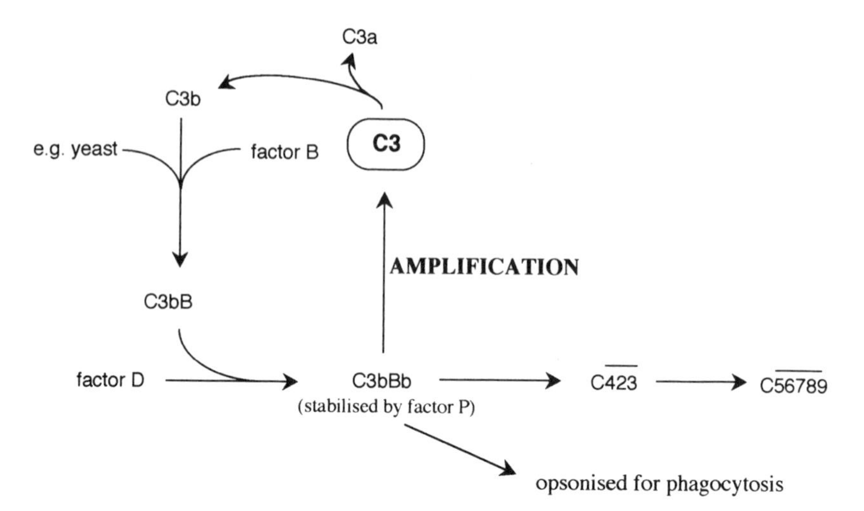

Figure 1.15. Complement activation via the alternative pathway. Always present in serum are trace amounts of C3b, which may attach to recognition sites on, for example, yeast cell walls. C3b may combine with serum factor B to form C3bB, and this complex is acted upon by factor D to form C3bBb. This latter complex is a C3 convertase and may act upon C3 to form more C3b to amplify the process. The C3bBb-coated particles may activate other complement components (C4–C9) or be recognised by complement receptors on neutrophils.

of complement components C5–C9, resulting in target cell death as described above. Alternatively, the target surface becomes coated with C3b, an opsonin, and hence is targeted as 'foreign' for subsequent destruction by phagocytes.

1.3.2.3 Interactions of complement components with neutrophils

Neutrophils possess receptors for several components of complement (described in detail in §3.9.3). When C3 is activated (via either the classical or alternative pathway), soluble C3a is generated whilst C3b is deposited on the target cell surface. Because neutrophils possess receptors for C3b (CR1, or complement receptor 1), C3b-coated particles are recognised as 'foreign' by neutrophils and are phagocytosed. However, if C3b is deposited on target cells following activation of the classical pathway, then it is rapidly cleaved by factors H and I to generate C3bi (see §3.9.3). C3bi is recognised by a second type of complement receptor on neutrophils, CR3 (Mac-1), and again C3bi-coated particles become opsonised targets for phagocytosis.

Two of the soluble factors generated during complement activation, C3a and C5a, also play important roles in inflammation. Both are termed *ana-*

phylotoxins, and they induce the symptoms of inflammation (e.g. release of histamine, serotonin, hydrolytic enzymes, platelet activating factor, arachidonic acid metabolites and reactive oxygen metabolites) from cells that possess receptors for these molecules. C4a structurally resembles C3a, but has such low affinity for its receptor (probably the C3a receptor) that it likely plays a negligible role in inflammation. Neutrophils also possess receptors for C3a and C5a: C3a can cause secretion; C5a induces a powerful chemotactic response (at concentrations as low as 10^{-10} M) and, at higher concentrations, can stimulate reactive oxidant production and degranulation.

1.3.3 Acute-phase proteins

These are generated by the liver during infections and other forms of inflammatory challenge, as part of the *acute-phase response.* This response to infection is characterised by fever, sleep, adrenotrophic hormone release, decreased plasma iron and zinc levels, elevated neutrophils in the bloodstream and enhanced cytokine production. These changes, part of the body's response to combat infection, occur within hours. The elevated temperatures may inhibit the replication of some bacteria and viruses and may also enhance the function of some immune cells.

The acute-phase response is characterised by the enhanced (two- to three-fold) synthesis of a number of proteins, such as antiproteinases, complement components, fibrinogen and ceruloplasmin, which are normally synthesised at fairly low rates. The rate of synthesis of some proteins is enhanced 100–1000-fold – for example, C-reactive protein, α-macroglobulin and acid-1-glycoprotein. Thus, serum levels of these proteins serve as indicators of disease.

The precise function of many acute-phase proteins is not known. C-reactive protein binds lipids, whilst α-macroglobulin and ceruloplasmin can scavenge some reactive oxygen metabolites. However, many acute-phase proteins are glycoproteins and can bind to bacterial surfaces; hence, they may serve as non-specific opsonins for phagocytosis, and their synthesis is stimulated by IL-1 and IL-6.

1.3.4 Cytokines

It is now recognised that the function of the immune system is carefully regulated by *cytokines,* small (8.5–40 kDa) proteins secreted by immune cells and by tissue cells such as endothelial cells and fibroblasts. T helper cells and macrophages are major sources of cytokines during inflammatory activation, but neutrophils may also generate a range of cytokines that can affect either their own function (i.e. *autocrine activation*) or the function of other immune cells (i.e. *paracrine effects*).

A striking observation of the effect of cytokines on immune and tissue cells is that the generation of one cytokine will almost inevitably initiate a cytokine cascade, which can lead to amplification of the immune response. Cytokines can often act in combinations to activate cells, by binding to specific cytokine receptors on the plasma membranes of these target cells. This receptor occupancy can lead to some rapid (≤1 h) effects, resulting from the generation of intracellular signalling molecules that lead to enzyme activation. Alternatively, cytokines can induce longer-term (≥24 h) effects, which result from activated gene expression. Often exposure of a cell to a cytokine will induce the expression of the same cytokine (sometimes together with the corresponding cytokine receptor), and this will lead to amplification of the initial signal. Sometimes exposure of a cell to one cytokine will induce the synthesis of several other cytokines, resulting in diversification of the original signal. Some of the many and varied functions of cytokines include activation of B and T cells, enhancement of monocyte and neutrophil functions (such as receptor expression, NADPH oxidase activity and phagocytic killing), regulation of haematopoiesis (the process regulating the production of blood cells in the marrow; see §2.1) and tumour-cell destruction. Some cytokines down-regulate a function induced by another cytokine (e.g. IL-4 can inhibit IL-1 production); hence, they have both positive and negative effects on immune-cell responses. They are thus a new class of locally-produced hormones that regulate cell development and function.

All of the major cytokines have now been cloned, the genes sequenced, recombinant proteins expressed and monoclonal antibodies generated; these can be used either to neutralise their functions or else to measure their levels in biological samples. Thus, the biological effects of cytokines have been extensively studied in vitro, and their increasing use in clinical studies is leading to the characterisation of their in vivo effects. Of interest is the finding that many cytokine inhibitors, agonists and receptor agonists exist. Indeed, some pathogens encode genes that are cytokine agonists, which may provide a mechanism by which these organisms evade the immune system.

The first cytokines to be discovered were the *interferons,* detected in the supernatants of T cells stimulated with either antigen or mitogen. Indeed, these were the first cytokines to be used in clinical trials and the first to be registered as drugs. Three types have been identified:

i. interferon-α (at least 23 different subtypes of which are known to be generated by leukocytes),
ii. interferon-β (generated primarily by fibroblasts) and
iii. interferon-γ or immune interferon (generated by T cells).

All interferons inhibit the multiplication of certain viruses, some affect tumour growth and all regulate immune-cell function. Of these interferons, two are important in neutrophil physiology: interferon-α, generated by neutrophils exposed to G-CSF (§7.3.4), and interferon-γ, which has profound effects on neutrophil function, such as enhancement of biosynthesis, increasing the activity of the respiratory burst, up-regulation of some plasma membrane receptors and enhancing the phagocytic killing of these cells (see §§7.2, 7.3.5). Furthermore, interferon-γ is being used successfully for the treatment of chronic granulomatous disease (§8.2), a condition intimately associated with impaired neutrophil and monocyte respiratory burst activity.

Other cytokines known to regulate immune function are the following:

Interleukins (IL-1–IL-13 have now been characterised);

colony-stimulating factors (CSFs), such as granulocyte colony-stimulating factor (G-CSF), granulocyte–macrophage CSF (GM-CSF) and macrophage CSF (M-CSF);

tumour necrosis factor (TNFα) and

several *growth factors,* such as erythropoeitin and transforming growth factor (TGFβ).

Some of these are involved in haematopoiesis (e.g. IL-1, -3, -5, -6; GM-, M-, G-CSF); their role is described in Chapter 2. Others (e.g. IL-1, -6, -8; TNFα; GM-, M-, G-CSF) are implicated in inflammation either directly (e.g. pure IL-1 can cause some symptoms of inflammation) or indirectly, via their ability to activate immune cells that participate in the inflammatory response (e.g. lymphocytes, neutrophils and macrophages); some of these effects are described in Chapters 2 and 3. Such cytokines as IL-4, interferon-α and IL-10 may be involved in immunosuppression; others, such as IL-1, IL-6, TNFα and TGFβ, are involved in tissue remodelling.

Thus, the role of cytokines in inflammation, immune-cell function and tissue repair is varied and complex. Later sections of this book explore the cytokines generated by neutrophils during inflammatory challenge (§7.3.4), the regulation of neutrophil function by cytokines (§7.2.1) and human diseases associated with neutrophil dysfunction in which cytokines may play important roles (§§8.2.5, 8.8).

1.4 Neutrophil function during inflammation

The varied roles that neutrophils play during the inflammatory response are summarised in Figure 1.16. Many of these individual aspects of neutrophil function are discussed in detail throughout the rest of this book. In summary, neutrophil function is regulated at several steps:

i. Chemotactic factors (Chapter 3), which may be derived from bacteria, from damaged host cells, from complement activation or due to cytokine production by tissues or immune cells, are generated at the site of infection.
ii. These chemoattractants are 'sensed' via plasma membrane receptors on the surface of neutrophils as they pass through the capillary bed. The chemoattractants may also affect the function of endothelial cells lining the capillaries, and these cells may themselves secrete factors that signal the circulating neutrophils. Alternatively, some factors secreted during inflammation (e.g. prostaglandins) may affect vascular permeability, which then aids the passage of serum proteins into the inflamed site.
iii. The neutrophils then roll onto and attach to the endothelial cells (via processes again regulated by the expression of receptors on the surfaces of neutrophils and endothelial cells) during the process of *margination.*
iv. The neutrophils then squeeze through gaps between adjacent endothelial cells by *diapedesis,* a process requiring gross morphological changes in the neutrophil and regulated by subtle changes in the arrangement of the cytoskeletal network (Chapter 4).
v. Neutrophils then migrate up the chemoattractant gradient by the process of chemotaxis until they reach the invading micro-organisms.
vi. By the time the neutrophils reach the microbe, two events may have occurred. Firstly, the microbes may have become opsonised by coating with serum proteins such as complement fragments, immunoglobulins or acute-phase proteins (Chapter 3). Secondly, the neutrophils will have become primed in that they will have increased surface expression of some plasma membrane receptors required for opsonophagocytosis (Chapter 7) and for NADPH oxidase activity (Chapter 5), and will have activated protein biosynthesis.
vii. The primed neutrophils will then, via occupancy of these receptors, activate subtle signal transduction systems (Chapter 6) that control phagocytosis, degranulation and activation of the NADPH oxidase. They will thus phagocytose and kill the microbes.

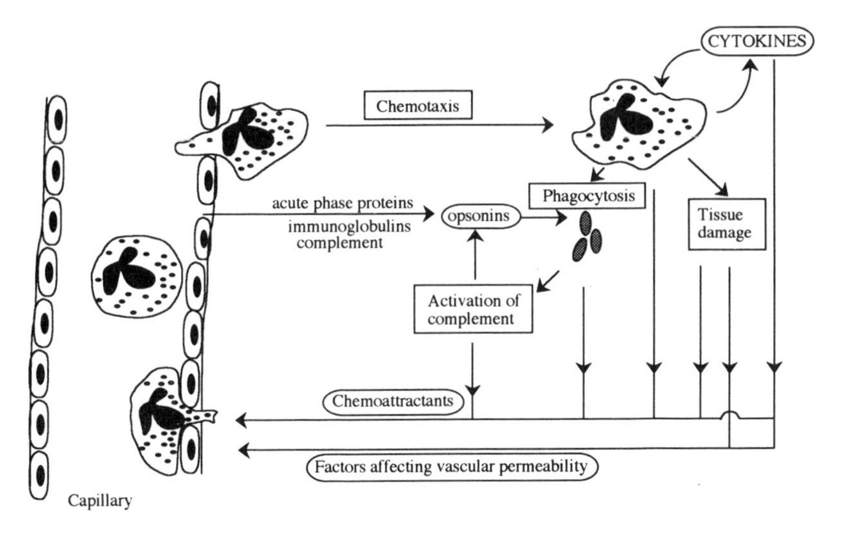

Figure 1.16. Activation of neutrophil function during acute inflammation. Bacterial invasion of tissues leads to the formation of chemoattractants, which may be derived following complement activation or from damaged tissues, or generated by the bacteria, neutrophils or other immune cells. These chemoattractants are 'sensed' by receptors on neutrophils circulating in the capillaries, and the neutrophils then attach to the endothelial cells. The vascular permeability increases, which assists in the passage of cells and serum proteins into the tissue, and the neutrophils leave the circulation and move towards the invading bacteria by chemotaxis. Serum proteins (e.g. complement fragments, immunoglobulins and acute-phase proteins) then opsonise the bacteria, which are thus recognised as 'foreign' by the neutrophils and are subsequently phagocytosed. The activated neutrophils may then generate chemoattractants (either directly, or indirectly following neutrophil-dependent host-tissue damage) or cytokines, which can augment the function of neutrophils or other immune cells. Thus, neutrophil infiltration and activation can result in an amplification of the inflammatory response.

viii. Some neutrophil products may then augment the inflammatory response: for example, chemoattractants, such as leukotriene B_4 (LTB_4), will signal the recruitment of more neutrophils to the infected site, whereas secondary cytokines generated by the activated neutrophils will cause the activation and recruitment of more immune cells, including neutrophils. Some of these neutrophil-derived cytokines may also affect the rate of production of neutrophils by the bone marrow (Chapter 2). Other neutrophil-derived products, such as released degradative enzymes and reactive oxidants, may cause further tissue damage, which may exacerbate the generation of pro-inflammatory signals.

ix. If the infection is cleared, then neutrophil function is down-regulated and the cells may die and eventually be cleared by phagocytosis via tissue macrophages. If the infection is not cleared, then either more neutrophils will be recruited or these cells will be replaced by macrophages and lymphocytes, as a phase of chronic inflammation then develops.

1.5 Bibliography

Benjamini, E., & Leskowitz, S. (1991). *Immunology. A Short Course*, 2nd Ed. Wiley-Liss, New York.

Davey, B. (1989). *Immunology. A Foundation Text.* Open University Press, Milton Keynes, U.K.

Gallin, J. I., Goldstein, I. M., & Snyderman, R. (1988). *Inflammation: Basic Principles and Clinical Correlates.* Raven Press, New York.

Galvani, D. W., & Cawley, J. C. (1992). *Cytokine Therapy.* Cambridge University Press, U.K.

Hallett, M. B., ed. (1989). *The Neutrophil: Cellular Biochemistry and Physiology*. CRC Press, Boca Raton, Fla.

Mimms, C. A., Playfair, J. H. L., Roitt, I. M., Wakelin, D., & Williams, R. (1993). *Medical Microbiology*. Mosby Europe Ltd., U.K.

Roitt, I. M. (1990). *Essential Immunology*, 7th Ed. Blackwell Scientific Publ., Oxford, U.K.

2
The development and structure of mature neutrophils

2.1 Haematopoiesis

The process responsible for the production of blood cells is known as *haematopoiesis,* which occurs in the bone marrow. In an adult human, the bone marrow weighs about 2.6 kg, which accounts for 4.5% of the total body weight. Although the bone marrow is dispersed throughout the body, it is nevertheless a larger organ than the liver, which weighs about 1.5 kg. About 55–60% of all cells produced by the bone marrow are neutrophils. The marrow is thus highly proliferative, with mitoses observed in 1–2.5% of all nucleated cells. The cellularity of the marrow varies considerably with age: for example, in the young about 75% of the marrow comprises cells, whereas in adults this figure is decreased to about 50% and in the elderly only 25% of the marrow is cellular.

2.1.1 The bone marrow

The bone marrow is comprised of the cells that divide and develop into mature blood cells and also of stromal cells consisting of fibroblasts, macrophages and adipocytes. There are three major cellular types in the marrow (Fig. 2.1); these give rise, by the processes of division and differentiation, to the eight major blood cells.

i. *Pluripotent* or *multipotent stem cells* have the ability to divide and differentiate into blood cells of all lineages, but are also capable of self-renewal. Thus, a pluripotent stem cell can differentiate to form two cells that are more mature (i.e. myeloid- or lymphocyte-like), or else it can divide to form two identical, uncommitted stem cells. Indeed, experiments with irradiated mice have shown that as few as 30 stem cells can replace all of the cells of the bone marrow.

ii. *Progenitor cells* are committed to a single, specific lineage. These have the ability to proliferate (and hence increase in number) but they can also differentiate into mature cells.
iii. *Mature (or nearly-mature) cells* have the characteristic morphological features associated with the circulating, mature cells.

Stem cells are detected by their ability to form colonies when incubated either in vitro in the presence of suitable growth-promoting substances (see §2.1.2) or in the spleens of irradiated mice. In these experimental systems the progenitor cells are termed *colony-forming units* (CFU), and defined as CFU_S and CFU_C if they develop in the spleen or in culture, respectively. Production of cells of any one particular lineage therefore requires (a) the differentiation of stem cells into committed progenitor cells, (b) amplification of these progenitor cells via cell division and (c) differentiation of progenitor cells into mature cells.

2.1.2 Regulation of haematopoiesis

Because the bone marrow is spread throughout the body, there must be highly-sophisticated mechanisms for co-ordinating its function within these spatially-separated locations. Furthermore, the numbers of circulating, mature cells, particularly neutrophils, may need to be rapidly increased during infections, and this increase must be controlled by sophisticated regulatory processes. For example, normal human blood contains about 4000–6000 neutrophils/μl, but during infections numbers can rise to 10 000–20 000/μl. At neutrophil counts of <1000/μl there is a serious risk of infection. Experiments in the 1960s by Donald Metcalf and his colleagues (reviewed in Metcalf 1985) showed that the survival, proliferation and differentiation of immature haematopoietic cells was dependent upon humoral factors with *colony-stimulating activity* (CSA) – that is, factors that could induce the formation of colonies during culture of immature cells in vitro. Early sources of CSA were conditioned media from blood leukocytes, splenic leukocytes, placenta, some tumour cells and serum from endotoxin-treated animals. Improvements in protein purification techniques and, in particular, the advent of gene cloning techniques leading to the production of large quantities of recombinant proteins for structural and functional studies, have led to our present understanding that haematopoiesis is regulated by the activities of four major *colony-stimulating factors* (CSFs):

i. Interleukin-3 (IL-3 or multi-CSF),
ii. granulocyte–macrophage CSF (GM-CSF or CSF-α),
iii. granulocyte CSF (G-CSF or CSF-β) and
iv. macrophage CSF (M-CSF or CSF-1).

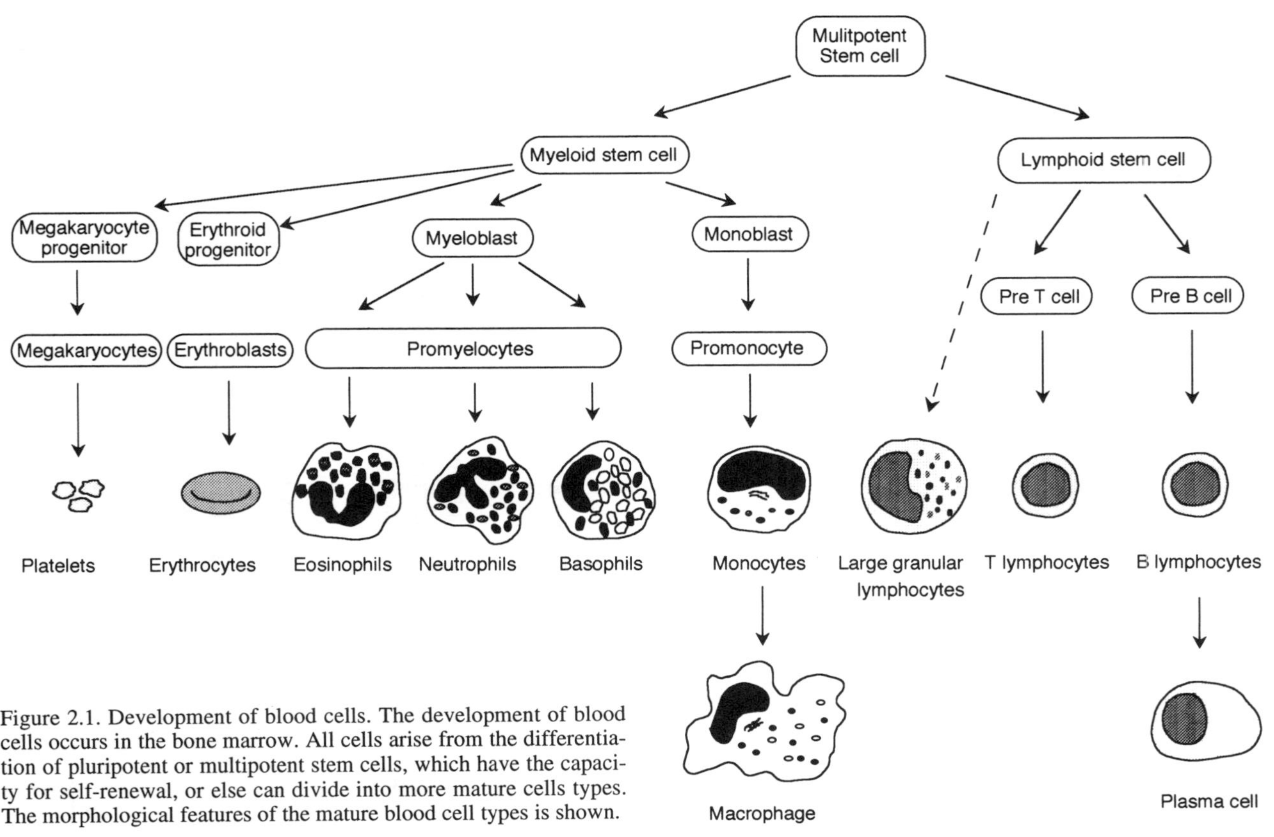

Figure 2.1. Development of blood cells. The development of blood cells occurs in the bone marrow. All cells arise from the differentiation of pluripotent or multipotent stem cells, which have the capacity for self-renewal, or else can divide into more mature cells types. The morphological features of the mature blood cell types is shown.

Unlike IL-3, which is the least specific of the CSFs and affects the replication of pluripotent stem cells and precursor cells of all lineages, M-CSF, GM-CSF and G-CSF are more specific in that they affect only monocyte and granulocyte development (Table 2.1). These CSFs regulate haematopoietic cell development in in vitro models, and there is now considerable evidence to suggest that they also perform similar functions in vivo. The genes for these CSFs have been identified and cloned, and the amino acid sequences have been deduced or directly determined. Recombinant proteins have been expressed and used in experimental or clinical studies, and neutralising antibodies (i.e. antibodies that block their activity) have been produced. They have been shown to be synthesised by a variety of cells including fibroblasts, endothelial cells, marrow stromal cells and lymphocytes. They appear to function synergistically with each other or sometimes with other cytokines (such as some Interleukins) to regulate the proliferation and differentiation of either stem cells or progenitor cells during haematopoiesis.

Whilst these different CSFs exhibit little sequence homology to each other, they do share some common features:

i. The purified components are all glycosylated proteins, although the biological importance of the carbohydrate moiety is uncertain because recombinant proteins may be non-glycosylated and yet still possess full activity.
ii. The tertiary structures of the mature proteins are maintained by disulphide bridges, and mutations in recombinant proteins show that the positions of the cysteine residues are essential for biological activity.
iii. Whilst the proteins have little sequence homology, the CSF genes share some common structural features.
iv. GM-CSF and IL-3 are adjacent on chromosome 5 and their transcription may be functionally linked.
v. The receptors for three of the CSFs have some sequence homology in that the extracellular domains possess matching cysteine residues and a common TRP-SER-X-TRP-SER motif (where X is any amino acid).

2.1.3 Regulation of haematopoiesis by CSFs

It is believed that all blood cells arise from the division and differentiation of multipotent stem cells. These stem cells have considerable capacity for self-renewal by cell division but may also differentiate into progenitor cells of lymphocytic or myeloid lineages (see Fig. 2.1). Experiments with colony-

Table 2.1. *Properties of the major colony-stimulating factors*

Component	Other name	Native protein (glycoprotein) (kDa)	Recombinant protein (non-glycosylated) (kDa)	Colonies stimulated	Chromosomal location	Produced by
G-CSF	CSF-β	19.6	18.8	neutrophils, BFU-E	17q11.2-21	endothelial cells, macrophages, fibroblasts
GM-CSF	CSF-α	22	14.3	neutrophils, monocytes, neutrophil/monocyte, eosinophils, BFU-E	5q23-31	T lymphocytes, fibroblasts, epithelial cells, endothelial cells, macrophages
M-CSF	CSF-1	70–90 (dimer)	21 (subunit)	monocytes	1p13-21	macrophages, fibroblasts, endothelial cells
IL-3	multi-CSF	20	14.6	stem cells, BFU-E, neutrophil/monocytes, megakaryocytes	5q23-31	T-lymphocytes, keratinocytes

Note: BFU-E = erythroid precursor or 'burst-forming unit'; see text for details.

forming units indicate that a primitive progenitor of the myeloid lineage is a cell type termed CFU-GEMM (granulocyte–erythroid–monocyte–megakaryocyte colony-forming unit), a progenitor cell with the potential to differentiate further into a number of mature cell types (Fig. 2.2). These may differentiate into granulocyte–macrophage progenitor cells (CFU-GM), which may themselves develop via granulocyte, monocyte or eosinophil lineages into mature cells. These developmental processes are controlled by the four major CSFs, which act synergistically with each other or with other cytokines. It should also be noted that as cells develop the characteristics of mature cells, their proliferative ability decreases; hence mature cells are often incapable of cell division (Fig. 2.3).

2.2 Properties of the CSFs

2.2.1 IL-3

Interleukin-3 (IL-3 or multi-CSF) is produced by activated T4 helper lymphocytes and also by keratinocytes located within dermal tissues. It is the least restricted of the CSFs in terms of its specificity in directing blood-cell development: it can stimulate the proliferation of granulocytes, macrophages, erythrocytes, megakaryocytes, mast cells and multipotent stem cells. IL-3 is probably important in regulating the proliferation and differentiation of multipotent stem cells and committed progenitor cells, rather than in acting upon primitive stem cells or the more mature cell types. Its ability to enhance the survival, proliferation and differentiation of the more mature type of multipotential cell is probably augmented by the activities of IL-1 and IL-6. It synergises with G-CSF, GM-CSF, M-CSF and IL-5 in the production of granulocytes, monocytes and eosinophils (see Fig. 2.2). IL-3 infusion into some primates does not lead to enhanced numbers of mature cells in the circulation (such as neutrophils), but rather results in increases in the number of multipotent stem cells in the marrow. On the other hand, it has been reported that IL-3 infusion into humans may increase cell numbers in the circulation.

IL-3 also has effects on mature cells. Those reported include the enhancement of the cytotoxicity of macrophages, stimulation of the proliferation of tissue-derived mast cells and (in the presence of endotoxin) the stimulation of the tumouricidal activity of monocytes, possibly via enhanced production of tumour necrosis factor (TNF). However, there are no reported effects of IL-3 on mature neutrophil function.

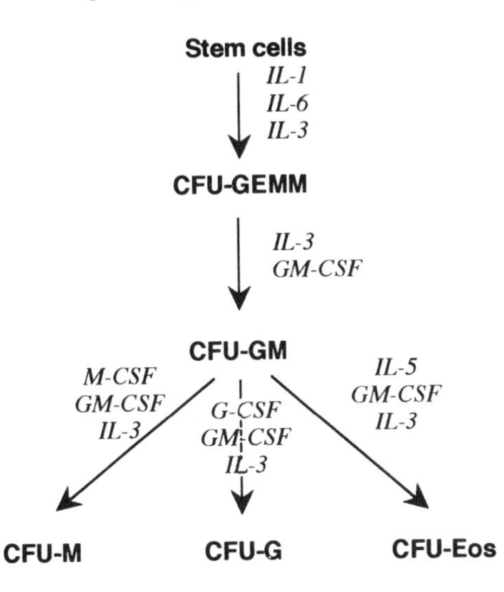

Figure 2.2. Role of cytokines in blood-cell development. *Abbreviations:* CFU, colony-forming unit; CFU-GEMM, granulocyte–erythroid–monocyte–megakaryocyte CFU; CFU-GM, granulocyte–macrophage CFU; CFU-M, macrophage CFU; CFU-G, granulocyte CFU; CFU-Eos, eosinophil CFU; IL, Interleukin. See text for details.

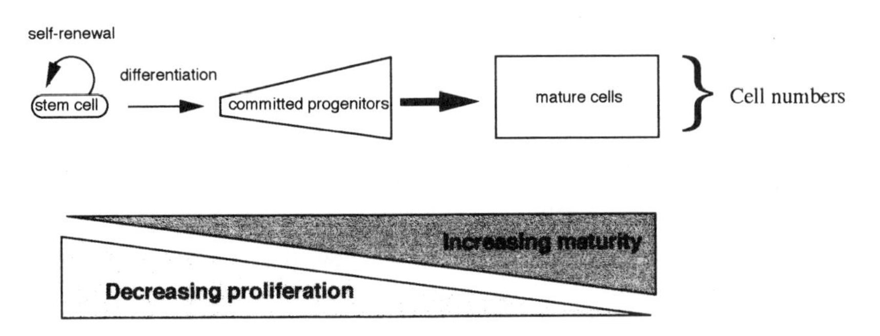

Figure 2.3. Maturation of bone marrow cells. Stem cells have the capacity for self-renewal, but also differentiate into more mature cell types. As the cells become more mature, their ability to proliferate declines and their number in the marrow increases.

2.2.2 M-CSF

Macrophage CSF, which stimulates the proliferation and differentiation of monocytes, is produced by macrophages, endothelial cells and fibroblasts. It is a homodimer glycoprotein of relative molecular mass 40–80 kDa. Its production by macrophages is stimulated by the addition of IL-3, GM-CSF and

γ-interferon. At least three different molecular forms of M-CSF are generated, and molecular cloning and sequencing experiments reveal that these forms are generated via alternative mRNA processing. These molecules are membrane associated, and as such they are able to act on adjacent cells via cell–cell contact. Alternatively, proteolytic cleavage can release soluble M-CSF, which may act at some distance from the cell that produced it. M-CSF stimulates the proliferation of alveolar, peritoneal, bone-marrow-derived and splenic macrophages, and also the proliferation and development of mature macrophages from GM-CFU. Its action is greatly enhanced when it synergises with other cytokines such as IL-1, perhaps because this latter cytokine up-regulates the expression of M-CSF receptors.

M-CSF also has other functions apart from its role in haematopoiesis. For example, it is produced by the uterine epithelium during pregnancy with maximal levels occurring just before parturition, and M-CSF receptors are present on placental trophoblasts. M-CSF affects the function of mature macrophages, stimulating their ability to secrete TNF, GM-CSF, prostaglandins, interferons and plasminogen activator. Phagocytosis and tumour-cell killing by macrophages are also enhanced by M-CSF, as is their anti-microbial activity towards *Listeria monocytogenes* and *Candida albicans*. M-CSF is chemotactic for macrophages, and treated cells are larger, exhibit membrane ruffling and have an increased number of cytoplasmic vacuoles.

2.2.2.1 M-CSF receptors

These are encoded by the c-*fms* proto-oncogene, which is a member of a superfamily of genes containing the PDGF receptor and the ligand for the c-*kit* proto-oncogene. M-CSF receptors have a high affinity for the ligand and a relative molecular mass of 165 kDa. They comprise an extracellular ligand-binding domain, a single membrane-spanning segment and an intracellular domain containing tyrosine kinase activity. The sequence of events following M-CSF binding to its receptor is probably (i) aggregation of receptors, (ii) stimulation of tyrosine kinase activity, (iii) intermolecular phosphorylation and (iv) internalisation and subsequent degradation of receptors. The substrate(s) for tyrosine kinase activity are unknown but may include components in the plasma membrane and cytoplasm, such as the c-*raf* proto-oncogene and phosphatidylinositol-3-kinase.

2.2.3 G-CSF

Granulocyte CSF was first identified (as CSF-β) in placenta-conditioned medium and was shown to induce the differentiation of murine myelomonocytic leukaemia cells (WEHI-3B cells) into more mature, postmitotic cells. The use of antisera, which neutralised its activity, showed that G-CSF activity was distinct from that of GM-CSF, which had previously been partially purified during the late 1970s. This new CSF was named G-CSF by virtue of its ability to stimulate the formation of granulocytic colonies in vitro. In 1983 Nicola and colleagues purified the molecule and showed it to be a 24–25 kDa hydrophobic glycoprotein containing neuraminic acid and at least one internal disulphide bond that was essential for function. Unlike IL-3 or GM-CSF, murine G-CSF is functional on human cells and vice versa. It is produced by a variety of cells including nonhaematological cells such as squamous carcinoma cells and hepatoma cell lines.

Souza and colleagues (1986) probed a cDNA library from the tumour cell line 5637 using degenerate oligonucleotide probes synthesised from knowledge of the partial NH_2-terminal sequence of the purified protein. Using this approach they isolated a clone encoding a predicted protein of 174 amino acids. At about the same time, Nagata and colleagues (1986) used a similar approach to screen a cDNA library from CHU-2 cells and isolated a clone for G-CSF that encoded a 177-amino-acid protein. The shorter (174-amino-acid) clone had three amino acids (Val-Ser-Glu) deleted between Leu and Val at the NH_2 terminus, and the recombinant protein generated from this clone had much greater activity than that encoded by the 177-amino-acid clone.

G-CSF is encoded by a single human gene located on chromosome 17q11-22. It comprises five exons and occupies a total of 2.3 kb (kilobases) of the genome. At the 5′ terminus of intron 2 there are two donor splice sequences present in tandem, 9 bp (base pairs) apart. Alternative splicing at these sites can thus give rise to two different mRNAs for G-CSF. Sequence studies have shown that the human and murine genes are highly conserved, with 69% homology in the coding and non-coding regions of the DNA and 73% homology in predicted amino acid sequences. Furthermore, four of the five cysteine residues in the murine and human proteins are located at conserved sites. Four of these cysteines form two disulphide bridges that are essential for biological activity, whilst the fifth (Cys 17) can be substituted without altering activity.

G-CSF may be related to IL-6 because both the number and size of introns and exons in these two genes are similar. Furthermore, amino acid residues 20–85 in G-CSF and 28–91 in IL-6 have 26% homology, and the positions of four cysteine residues in these regions are precisely conserved. Whilst the IL-6 gene is located on chromosome 7p15, it may be that the two genes arose from duplication and have since diverged.

The effects of G-CSF on neutrophil production during haematopoiesis are augmented via its interaction with other cytokines such as GM-CSF. Furthermore, the co-operative action of G-CSF with IL-4 augments the proliferation of neutrophil colonies (possibly by IL-4 increasing the number of G-CSF receptors), and the combination of G-CSF with IL-3 or IL-6 is particularly effective on the stimulation of primitive progenitor cells. Recombinant G-CSF has been shown to stimulate both granulopoiesis and the release of mature neutrophils into the bloodstream of laboratory animals and humans with neutropenias that have arisen either from disease or therapy (e.g. after cytotoxic or radiotherapy for the treatment of malignant disease). G-CSF is generally well tolerated by patients and can decrease the period of neutropenia in a dose-dependent manner, such that normal numbers of mature neutrophils are achieved more rapidly. When used in combination with standard-dose cyclophosphamide/doxorubicin/etoposide in patients with small carcinoma of the lung, G-CSF treatment was found to decrease (i) the period of neutropenia, (ii) the incidence of fever, (iii) the requirement for antibiotics and (iv) the incidence of hospitalisation.

G-CSF increases the number of progenitor cells in the bloodstream tenfold. It has been used in the treatment of patients with myelodysplastic syndromes (MDS; §8.8) where it can increase neutrophil counts and sometimes improve neutrophil function in these patients. Because some leukaemic cells are able to proliferate rather than differentiate in response to G-CSF, this CSF may potentially induce a leukaemic transformation in these patients; however, its combined use with cytotoxic agents such as cytosine arabinoside appears to decrease this possibility. No doubt clinical trials already underway will establish the optimal treatment regimen for G-CSF, so that the beneficial effects of this cytokine for the treatment and management of haematological disorders can be realised.

2.2.3.1 Regulation of G-CSF expression

G-CSF expression is controlled at both the transcriptional and posttranscriptional levels. A sequence of 300 nucleotides upstream of the initiation codon is conserved in both the murine and human genes, and this appears to contain three regulatory sites. G-CSF (and some other cytokine genes) may be constitutively transcribed by cells such as blood monocytes, fibroblasts and endothelial cells, but the mRNA may be short-lived ($t_{1/2} < 15$ min). The mRNA contains poly-AUUUA sequences in the untranslated region, and this motif is usually associated with mRNA instability. Indeed, such regions have also been identified in mRNA for GM-CSF, IL-1, IL-6, interferons, TNF, some growth factors, c-*jun*, c-*fos*, c-*myc* and c-*myb*. Upon the addition of endotoxin to monocytes, the stability of G-CSF mRNA increases and G-CSF production accumulates and becomes detectable. IL-1, TNF, phorbol esters and cycloheximide are also known to increase G-CSF mRNA stability. The mechanisms by which these agents perform this function are unknown, but they may interfere with an RNase activity that recognises the poly-AUUUA regions of the mRNA. Because of the effects of cycloheximide (which is an inhibitor of protein biosynthesis), it may be that this putative RNase is short-lived.

G-CSF is produced by monocytes, macrophages, vascular endothelium, fibroblasts, mesothelial cells and neutrophils in response to agents such as endotoxin, TNF, IL-1, phorbol esters, GM-CSF, IL-3, IL-4 and γ-interferon. Production of the G-CSF protein (unlike its mRNA) is not constitutive, but is inducible in normal cells. However, in malignant cell lines, such as bladder carcinoma, hepatoma, squamous cell carcinoma, melanoma and sarcoma, secretion of G-CSF may be constitutive. In normal humans G-CSF levels in the circulation are usually <30 pg/ml, but in conditions of stress (e.g. infection, cytotoxic drug therapy, transplantation) levels can rise to about 2000 pg/ml. In patients with cyclic neutropenia, there appears to be an inverse relationship between the neutrophil count and the period of neutropenia associated with enhanced G-CSF levels. This observation suggests that there may be a 'sensor' that detects neutrophil numbers and then regulates G-CSF production. Alternatively, it has been proposed that G-CSF can only regulate neutrophil production during periods of stress, and can play only a minor role in the constitutive production of these cells.

2.2.3.2 G-CSF receptors

Binding of G-CSF to cells of the neutrophil lineage, such as myeloblasts through to mature cells, occurs via specific, high-affinity receptors. Receptors are not detected on cells of the erythroid or megakaryocytic lineages, although some monocytic cells bind G-CSF with low affinity. In human neutrophils there are about 700–1500 receptors per cell that bind G-CSF with a K_d of 900 pM, and receptor expression is down-regulated by exposure to GM-CSF, endotoxin, fMet-Leu-Phe, TNF, PMA and G-CSF itself, whilst IL-2 and IL-6 have no effect on expression. The human G-CSF receptor is a single-chain polypeptide of 812 or 759 amino acids (relative molecular mass = 100–150 kDa). The two molecular forms appear to arise from alternative processing. There is no endogenous kinase activity associated with this receptor.

G-CSF receptors are also found on human myeloid leukaemia cells, leukaemic cell lines, human placenta, trophoblastic cells, vascular endothelial cells and on small lung carcinoma cells, but the function of these is unknown. The K_d of receptor binding on acute or chronic myeloid leukaemia cells (300 pM) is similar to that on mature neutrophils, but only some of these immature cells respond to G-CSF. Because some leukaemic cells proliferate in response to G-CSF, its usefulness in the treatment of certain types of haematological disorders is restricted.

The receptor comprises a ligand-binding extracellular domain, a single transmembrane domain and a cytoplasmic domain. The extracellular domain, like that of the receptors for GM-CSF, erythropoietin, IL-3, IL-4, IL-6 and IL-7, contains some common structural motifs, such as four cysteine residues, W-S-X-W-S near the transmembrane domain and several repeats of a sequence related to the fibronectin type III domain. G-CSF also contains a region homologous to the prolactin receptor, and shows some similarity to the NCAM family of adhesion receptors. The transmembrane and cytoplasmic domains have 50% homology with the IL-4 receptor, indicating that they may share common signal transduction pathways. There is no evidence for endogenous kinase activity, but the signal transduction pathways responsible for cell activation are, as yet, unknown. G-CSF binding to its receptor has no effect on intracellular Ca^{2+}, the transmembrane potential or intracellular pH, but G-CSF exposure accelerates the rate of membrane depolarisation after fMet-Leu-Phe stimulation without affecting the Ca^{2+} rise. The G-CSF receptor may be linked to cell activation via G-proteins, which activate tyrosine kinase activity and elevate cAMP levels.

In addition to stimulating the production of mature neutrophils during haematopoiesis, G-CSF also affects the function of mature neutrophils. These effects are discussed in §7.2.1.

2.2.4 GM-CSF

This CSF was originally defined by its ability to stimulate the production of GM-CFU from bone marrow progenitor cells, and it was first purified from mouse-lung-conditioned medium in 1977. This molecule was shown to be a glycoprotein of relative molecular mass 23–29 kDa. Partial amino acid sequencing of this protein (after treatment with neuraminidase) was performed prior to the synthesis of oligonucleotides that were then used to screen a cDNA library to isolate GM-CSF clones. Human GM-CSF was first purified from the cell line Mo, and was found to be identical to the previously-described 'neutrophil migration inhibitory' activity of T cells. The human protein was also shown to be a 22-kDa glycoprotein, and this molecule was subsequently cloned. The cDNA clones from human and murine sources are 60% homologous, encoding polypeptides of 144 and 141 amino acids, respectively. Both molecules contain a 17-amino-acid leader sequence, and the human (recombinant) rGM-CSF was first expressed in monkey COS cells. Whilst the murine and human molecules possess 54% homology at the amino acid level, unlike the species cross-reactivity observed with G-CSF, there is no crossover in species-specificity of human and murine GM-CSF. The position of four cysteine residues is conserved in molecules from different species. The disulphide bridge between cysteine residues 51 and 93 is essential for activity, but the bridge formed between cysteines 85 and 118 may be removed without loss of biological activity. Glycosylation occurs at N- and O-linked sites, but non-glycosylated recombinant proteins show enhanced activity over the natural, glycosylated form.

The GM-CSF gene occupies 2.5 kb of both the human and murine genomes and comprises four exons and three introns. In the mouse, the gene is located on chromosome 11 and is adjacent to the gene for IL-3. The human GM-CSF gene is located on the long arm of chromosome 5q21-q32, 9 kb downstream of the IL-3 gene. Interestingly, the genes for IL-4, IL-5, M-CSF, c-*fms* and EGR-1 (early growth phase response gene) are also located on the long arm of chromosome 5. Deletions in this region of the genome are detected in patients with such haematological disorders as the myelodysplastic syndromes, acute leukaemias and refractory anaemias.

GM-CSF predominantly affects cells of the myeloid lineage, and it stimulates the growth of granulocytic and macrophage colonies in vitro. It has effects on progenitor cells such as multipotent blast cells, which give rise to colonies capable of development into cells of the myeloid lineages. At high concentrations of GM-CSF, macrophage colonies are preferentially produced, whereas at lower concentrations granulocyte colonies predominate. Very early erythroid and eosinophilic progenitor cells and possibly CFU-GEMM respond to GM-CSF, which also synergises with G-CSF in the stimulation of primitive multipotent cells.

Recombinant GM-CSF (produced in *Escherichia coli,* yeast or COS cells) has been tested for its ability to affect haematopoiesis in primates and humans. Because of its relatively short half-life in the circulation, daily administration (usually via intravenous infusion) is required. Administration results in a transient neutropenia, monocytopenia and eosinopenia within 30 min of administration, presumably because of the ability of GM-CSF to stimulate the expression of adhesins and hence increase the numbers of leukocytes adhered to the capillary endothelium in the marginated pool. Additionally, these leukocytes may accumulate in the lungs after GM-CSF administration, which may contribute to the decrease in the observed numbers of circulating cells. Circulating cell numbers return to normal by 2 h, and this is followed by a rapid increase in neutrophil count, which may result from the release of cells from the marginated pool or else the release of mature cells from the bone marrow. Increases in neutrophil numbers as great as 50–100-fold have been reported (especially in neutropenic patients), and increases in the numbers of circulating monocytes, eosinophils and sometimes immature cells (such as myeloblasts, promyelocytes and myelocytes, especially at high GM-CSF doses) may also be observed.

In addition to its effects on haematopoietic cells, GM-CSF can also affect the function of mature cells. GM-CSF treatment increases the survival, cytotoxicity and eicosanoid formation by eosinophils, and can increase the tumouricidal activity, cytokine expression, surface antigen expression and oxidative metabolism of macrophages. It is chemotactic for endothelial cells, can induce the proliferation of some tumour cells, stimulates histamine release from basophils and affects the viability and function of Langerhans cells. Its effects on mature neutrophils are described in §§7.2.1, 7.3.4.

2.2.4.1 GM-CSF expression

GM-CSF is undetectable in the serum of normal humans, and no normal cells have been shown to express this protein constitutively. Some transformed cells may constitutively express GM-CSF, and it is actively synthesised and secreted by antigen- and lectin-stimulated T cells and by endothelial cells and fibroblasts exposed to TNF, IL-1 or endotoxin. Other sources of GM-CSF include stimulated B lymphocytes, macrophages, mast cells and osteoblasts, whilst TNF and IL-1 can stimulate its production by acute myeloid leukaemia cells. Some solid tumours and synovial cells from rheumatoid joints may also express GM-CSF and this may be important in disease pathology.

Regulation of expression may occur at both the transcriptional and post-transcriptional levels. The mRNA for GM-CSF contains (in common with those of some other cytokines) conserved regulatory sequences in the 3´ untranslated region, which may affect its rate of translation. The gene is constitutively transcribed in monocytes, endothelial cells and fibroblasts, but the mRNA is unstable and so does not accumulate to levels sufficient to allow translation into significant amounts of protein. Activation of these cells results in the increased expression of GM-CSF protein, which arises from both an enhanced rate of transcription (as detected in nuclear runoff experiments) and also an increased stability of the mRNA, perhaps by mechanisms analogous to those described above during activation of G-CSF expression (§2.2.3.1).

2.2.4.2 *GM-CSF receptors*

The murine receptor for GM-CSF was first characterised by the binding of ^{125}I-labelled GM-CSF to myelomonocytic cells. Scatchard analysis revealed the presence of both low-affinity (K_d = 1 nM) and high-affinity (K_d = 20 pM) binding sites. Human neutrophils express only high-affinity receptors (K_d = 50 pM) with about 800–1000 receptors per cell; interestingly, mature neutrophils possess greater numbers of receptors than do immature cells. Receptors are present on neutrophils, monocytes and eosinophils, and some cells (such as HL-60, U937, TF1 and acute non-lymphocytic leukaemia cells) also possess low-affinity receptors. GM-CSF is active on mature neutrophils at concentrations of 1–100 pM, with a half-maximal response observed at 10 pM.

Ligand binding induces receptor internalisation and neutrophil stimulants such as fMet-Leu-Phe, and PMA can down-regulate the number of GM-CSF receptors. The receptor has an apparent molecular mass of 84 kDa in human cells and 130 kDa in murine cells. cDNA clones for the receptor have been obtained after screening a placenta library (placental membranes possess GM-CSF receptors), and these clones predict an 85-kDa protein of 400 amino acids. Sequence homology of GM-CSF receptors with the receptors for G-CSF, growth hormone, IL-3, IL-4, IL-6, IL-7, prolactin and IL-2 receptor β-chain suggests that they belong to a gene superfamily. None of these receptors possesses kinase domains, but the position of four cysteine residues is conserved, as is the Trp-Ser-X-Trp-Ser motif.

GM-CSF and IL-3 have been shown to compete for receptors in some types of cells (e.g. eosinophils and KG-1 cells), indicating some structural homology between GM-CSF and IL-3 receptors, perhaps because they share certain subunits or adapter proteins. GM-CSF occupancy results in phosphorylation of certain proteins, and because the receptor possesses no inherent kinase activity, receptor occupancy must be linked to kinase activity via the generation of second messenger molecules. Pretreatment of cells with pertussis toxin abolishes the effects of GM-CSF, indicating the involvement of G-proteins in signal transduction. Priming of neutrophil functions with GM-CSF involves the activation of phospholipases A_2 and D.

2.3 Interactions of CSFs and cytokines during haematopoiesis

Whilst the individual CSFs described above can all affect haematopoiesis by acting alone on different types of stem or progenitor cells, they are much more potent when they act together on a particular cell type. Such synergistic actions are possible because many cell types possess receptors for more than one CSF, and the interaction of one type of CSF with a cell may subsequently lead to the up- (or down-)regulation of the expression of another type of CSF receptor. Experimental evidence suggests that such interactions of CSFs may occur in vivo during haematopoiesis because, whilst G-CSF alone has a fairly low ability to stimulate granulocyte proliferation in vitro, it is the most potent of all CSFs tested in increasing the numbers of blood neutrophils when administered in vivo. This suggests that G-CSF may stimulate the production of other CSFs or regulatory cytokines, which can then synergise with G-CSF.

Other cytokines also appear to be involved in the regulation of haematopoiesis, but it is difficult to assess if these directly affect the activities of stem and progenitor cells. Alternatively, they may mediate their effects via the stimulation of stromal cells to secrete a secondary factor that is then actually responsible for the observed effects. Other cytokines implicated in stimulation of haematopoiesis include the following:

IL-1, which has no proliferative effect when acting alone but modulates the responses of haematopoietic cells to other CSFs (possibly by enhancing CSF-receptor expression), enhances the differentiation and survival of early progenitor cells observed in response to IL-3 and stimulates the production of G- and GM-CSF by T lymphocytes, endothelial cells, macrophages and fibroblasts;

IL-4, which has no effect alone but enhances the proliferative response to G- and GM-CSF, whilst inhibiting the responsiveness to IL-3;

IL-5, which selectively stimulates the proliferation of eosinophil precursors and the biological activities of mature eosinophils and is produced only by activated T cells (thus eosinophil development and function is entirely T-cell dependent);

IL-6, which has no proliferative effect when acting alone but accelerates the formation of multilineage granulocyte-macrophage colonies induced by IL-3.

CSF levels in normal human serum are usually low or undetectable. During infections, the immune system responds by activation of leukocytes, and often this is accompanied by an increase in numbers of specific leukocytes in the circulation in order to combat the infective agent. Thus, it is not surprising that many pathogens activate CSF production, which then results in increases in numbers of progenitor cells in the bone marrow and spleen, and subsequent increases in numbers of mature leukocytes in the circulation.

Pathogens known to stimulate CSF production include *Salmonella typhimurium, Mycobacterium lepraemurium, Brucella abortus* and *Schistosoma mansonii.* Additionally, non-viable bacteria or bacterial products, such as *Nocardia rubra* cell-wall fragments, muramyl peptides and bacterial endotoxins, can also induce CSF production.

Undoubtedly, complex interactions between immune cells regulate not only the activities of immune cells, but also their rate of formation during infections. During infections, neutrophils are usually the first cells to be recruited to the infective site, and their cytotoxic processes become activated to attack the pathogenic organisms. The neutrophil itself is an active producer of many cytokines (§7.3.4) that may have autocrine effects but will also affect the functions of other cells within its microenvironment. IL-1 production by neutrophils, and perhaps by resident and infiltrating macrophages, will stimulate the activity and clonal expansion of sensitised T lymphocytes, which may have been presented with antigens by macrophages. The activated T lymphocytes (and activated macrophages and neutrophils) may then secrete CSFs, which may then stimulate the release of mature cells from the bone marrow and accelerate haematopoiesis. Neutrophils in the circulation will adhere to endothelial cells (in a process promoted by cytokines) and migrate into the infected tissues. Local concentrations of CSFs and other cytokines will then up-regulate their responsiveness and augment their ability to destroy the pathogens.

When the infection is cleared, the rate of haematopoiesis declines and returns to basal levels. This may be achieved merely by a decline in CSF production, but probably more complex regulatory controls exist. Prostaglandins are known to inhibit colony formation by progenitor cells in vitro and can also suppress the formation of M-CSF. Also, α-, β- and γ-interferons and TNF can exert effects on colony formation, whilst ferritin, transferrin and lactoferrin can decrease CSF production. Leukocytes themselves may be able regulate to their own rate of formation in that G-CSF exposed neutrophils can secrete α-interferon, which can down-regulate haematopoiesis (§7.3.4.5). Some doubt has been expressed as to whether CSFs regulate basal or stress-induced (e.g. during infections) haematopoiesis. Evidence supporting their role in regulating blood-cell development during infections is strong, and experiments using CSF-neutralising antibodies also favour their involvement in basal haematopoiesis. For example, anti-G-CSF antibodies can induce neutropenia in dogs, and mice with osteopetrosis have a genetic defect in chromosome 3 that prevents transcription of M-CSF mRNA. These mice have defective macrophage production, and their macrophages cannot develop into the bone-remodelling cells, the osteoclasts.

2.4 Development of neutrophils

Much of the activity of the bone marrow is dedicated towards the production of neutrophils with about 0.9×10^9 neutrophils produced per kilogram body weight per day (in an average person). In the circulation there are about 4×10^6 mature neutrophils per millilitre of blood, but large numbers of mature cells are also present in the bone marrow. Neutrophils leave the circulation and enter tissues, but it is difficult to estimate the pool size of these tissue cells. Furthermore, many cells adhere to the walls of blood vessels (the so-called *marginating pool*), and so the number of cells free in the circulation can vary quite rapidly (and reversibly) by changes in the numbers of marginating cells. Numbers of neutrophils within the circulation can increase up to tenfold very rapidly (e.g. during infections), and this increase may result initially from the release of the mature cells stored in the marrow. If high numbers of circulating neutrophils need to be sustained, then the production of neutrophils in the marrow can also be increased. If mature cells are not recruited into tissues, they are thought to exist in the circulation for about 12–24 h before being cleared. If they do migrate into tissues, then it is believed that they survive for several days, but their ultimate fate is unknown.

The mature neutrophil has a diameter of 10–12 μm and a volume of about 350 fl (see Fig. 1.1a). The cytoplasm accounts for about 65% of the total volume of the cell, with the nucleus comprising about 20% of the volume. The cytoplasm is packed with a group of organelles, termed the *granules,* which give the cell one of its names (granulocyte), and these granules account for about 15% of the cell volume. There are a number of types of cytoplasmic granules, termed *azurophilic* (or primary), *specific* (or secondary) and *gelatinase-containing* (or tertiary). There may also be other types of subcellular organelles, termed *secretory vesicles* (§2.5). The composition and function of these granules varies, and they are synthesised at different stages in the development of the neutrophil. Because the cytoplasm contains very few distinguishable mitochondria, these cells do not utilise mitochondrial respiration or oxidative phosphorylation to fulfil their energetic requirements; rather, they derive their energy from glycolysis, which is oxygen independent. Thus, neutrophils can function efficiently at the low O_2 tensions that may be present within inflammed tissues. Golgi apparatus is identifiable, and the cell contains only small amounts of endoplasmic reticulum.

The development of mature neutrophils in the bone marrow occurs via the differentiation of multipotential stem cells into progenitor cells that are committed to neutrophilic lineages. As this process proceeds the ability of

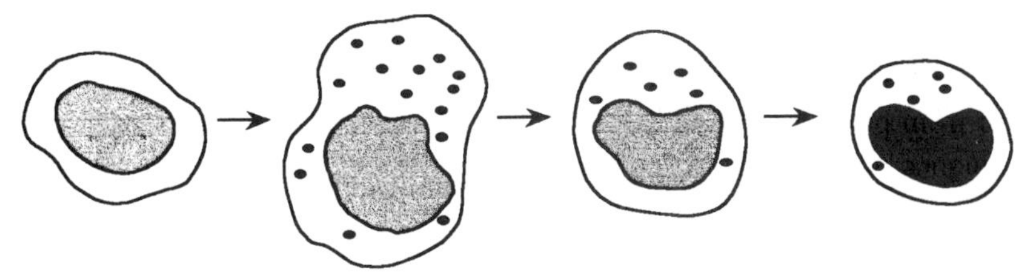

Figure 2.4. Neutrophil development. Morphological features of neutrophils and neutrophil precursors.

cells to self-renew decreases, and the decision to self-renew or to differentiate may be stochastic (i.e. governed by the laws of probability). Differentiation is characterised by the appearance of specific markers and by the disappearance of others (e.g. particular enzymes or plasma membrane markers). This differential gene expression equips the developing neutrophil with its functional armoury, but may also result in the expression of CSF-receptors or other receptors, which will allow for the expansion of the neutrophil precursor pool in response to granulocytic CSFs.

The stages of neutrophil development have been extensively and elegantly described by Dorothy Bainton (1980). The earliest identifiable cells of the neutrophil lineage are the myeloblasts, and the developmental stages are shown in Figure 2.4.

Identification of these cell types is based on the morphological analysis of cells, staining for peroxidase and analysis by light and electron microscopy. Of every 100 nucleated cells in the marrow, 2 are myeloblasts, 5 are promyelocytes, 12 are myelocytes, 22 are metamyelocytes and band cells and 20 are mature neutrophils. These values are somewhat variable and may differ between individuals. Thus, about 60% of all marrow cells are of the neutrophil lineage.

2.4.1 *Myeloblasts*

These cells are relatively undifferentiated and have a large nucleus, distinguishable nucleolus but few, if any, cytoplasmic granules. Myeloblasts arise from a precursor pool of stem cells, and both the rough endoplasmic reticulum and the Golgi apparatus stain for peroxidase, indicating that this enzyme is beginning to be synthesised. This cell type is capable of proliferation.

2.4.2 *Promyelocytes*

The promyelocytic stage of development is characterised by the acquisition of large numbers of peroxidase-containing granules. These vary in shape and size but are mostly spherical (~500-nm diameter, density 1.23 g/ml), but they may be ellipsoid, crystalline or connected to filaments. They are termed *azurophilic* because of an early observation that they stain red-purple with azure dyes. They are sometimes termed *primary granules* because they are the first to appear during neutrophil development. Peroxidase (which is in fact the enzyme myeloperoxidase) is also present in the rough endoplasmic reticulum, the Golgi and secretory vesicle, indicating that this enzyme is synthesised in large quantities in these cells. Promyelocytes are capable of proliferation, and the cultured cell line HL-60 was originally isolated from a patient with promyelocytic leukaemia. This cell line is thus commonly used to study the molecular processes regulating neutrophil maturation, because it can be induced to differentiate into mature, neutrophil-like cells in culture.

2.4.3 *Myelocytes*

These cells are characterised by the fact that they synthesise and accumulate large numbers of peroxidase-negative granules, termed the *specific* (or *secondary*) *granules*. Specific granules are either spherical (200-nm diameter) or rod-shaped (130 × 1000 nm). Mitoses can be observed in these cells, and probably three divisions can occur, with the granules being equally distributed between the daughter cells during division. Because synthesis of azurophilic granules ceases during this stage, the number of these per cell decreases during progressive divisions.

2.4.4 Metamyelocytes, band cells and segmented cells

These more mature cell types are incapable of division and are identified by their nuclear morphology, granule content and accumulation of glycogen particles. In a mature neutrophil there are 200–300 granules, with the specific granules being about twice as abundant as azurophilic granules.

2.5 Morphology of the mature neutrophil

Microscopic examination of the mature neutrophils reveals two striking features: a single multilobed nucleus and a dense, granular appearance of the cytoplasm (see Fig. 1.1a). The nucleus typically comprises two to four segments, and within this organelle the chromatin is coarsely clumped. Until recently, this abnormal chromatin structure was taken as evidence that the nucleus was transcriptionally inactive; however, it is now appreciated that the mature neutrophil does perform active transcription (§7.3), although rates of biosynthesis are somewhat lower than those observed in cells such as monocytes. There is no detectable nucleolus, so there can be only limited synthesis of ribosomal RNA in these cells.

In the cytoplasm, there are only small amounts of Golgi apparatus and endoplasmic reticulum; this, again, originally led to the erroneous conclusion that mature neutrophils were biosynthetically inert. It is also difficult to detect structures in the cytoplasm that can be identified as mitochondria, and any that are present probably play little or no role in neutrophil physiology: any mitochondrial structures detected probably represent vestigial organelles from earlier developmental stages of the neutrophil, where their role was much more important. The mature neutrophil thus derives the energy it requires for cell function via glycolysis, which is an oxygen-independent process.

The most abundant organelles within the cytoplasm are the granules, which are membrane-bound organelles containing an array of antimicrobial proteins. As discussed above (§2.4), three major types have been identified to date: azurophilic, specific and gelatinase-containing granules. Additionally, newly-described structures called *secretory vesicles* have been identified.

The contents of these vesicles include albumin, but their membranes are probably more important for neutrophil function. Indeed, apart from containing degradative enzymes that may be extracellularly secreted from the neutrophil or else discharged into phagocytic vesicles, the membranes of many types of these granules and vesicles contain important molecules. These include certain receptors (such as CR1, CR3, the fMet-Leu-Phe receptor; Table 2.2) and the cytochrome b of the NADPH oxidase (§5.3.2.1). Thus, when the granules and vesicles fuse with the plasma membrane during cell activation, the number of receptors and cytochrome b molecules on the cell surface (or on the membrane of the phagocytic vesicle) increases very rapidly (within minutes). Some of these granules and vesicles are thus intracellular stores of receptors that can be supplied to the cell surface upon appropriate neutrophil activation or priming with cytokines (see §7.2.1).

2.6 The antimicrobial granule enzymes

2.6.1 Role of granule proteins in microbial killing

In addition to the O_2-dependent processes utilised by neutrophils for microbial killing (Chapter 5), a wide range of antimicrobial proteins exist whose activity is largely independent of the requirement for molecular O_2. These so-called *oxygen-independent proteins* were in fact the first antimicrobial processes of neutrophils to be discovered, when it was found that crude subcellular fractions of lysed neutrophils could kill a range of Gram-positive and -negative bacteria. The first of these antimicrobial proteins to be identified, termed *phagocytin,* was isolated in a protein fraction prepared from acid-extracted neutrophil homogenates. It was then gradually appreciated that the cytoplasmic granules of neutrophils played a role in microbial killing because they possessed a variety of lysosomal enzymes, such as acid hydrolases, as well as phagocytin. Furthermore, these granules fused with the newly-formed phagocytic vesicles, discharging their contents; hence the ingested microbe was exposed to these destructive proteins (Fig. 2.5; see also Fig. 1.6).

Interest in these granule-associated antimicrobial proteins diminished somewhat when it became clear that the toxic oxygen metabolites generated during the respiratory burst played an important role in microbial killing during infections. The discovery that the biochemical defect in patients with CGD resided in an abnormality of the respiratory burst enzyme seemed to confirm the importance of efficient O_2-dependent microbicidal mechanisms in combating infections. This led to a great deal of interest aimed at dis-

Table 2.2. *Major components of neutrophil granules*

	Peroxidase-positive	Peroxidase-negative		
Component	Azurophil	Specific	Gelatinase-containing	Secretory vesicle
Antimicrobial				
Myeloperoxidase	+			
Defensins	+			
BPI	+			
Cathepsin G	+			
Elastase	+			
Proteinase 3	+			
Azurocidin	+			
Lysozyme	+	+		
Hydrolases				
β–glucuronidase	+		low	
α-mannosidase	+			
α-fucosidase	+			
β-glucosamidase	+			
Phopholipase A_2	+			
Phopholipase C	+			
Phopholipase D	+			
Receptors				
fMLP		+	+	+
CR3		+	+	+
gp150,95		+	+	+
CD45			+	
CR1				+
Laminin		+	+	
Others				
Cytochrome b		+	+	+
β_2-microglobulin			+	
Gelatinase		+	+	
Collagenase		+		
Diamine oxidase			+	
H^+-ATPase			+	
Vitamin B_{12}-binding protein		+		
Tetranectin			+	
DAG lipase			+	
Alkaline phosphatase	+		latent	
Albumin				+
Acetyl CoA:lysoPAF acetyl transferase			+	

covering the mechanisms leading to the generation of reactive oxidants, often at the expense of research into the nature and properties of the antimicrobial granule proteins. However, it was noted that neutrophils of patients even with severe forms of CGD (in terms of their respiratory burst defect) could still kill many types of pathogens almost as effectively as neutrophils from normal healthy controls. Furthermore, experiments in which neutrophils were incubated under anaerobic conditions, such that they were supplied with insufficient levels of O_2 to generate reactive substantial levels of reactive oxidants during a respiratory burst, could also kill some (but not all) pathogens as effectively as those incubated under aerobic conditions. *Salmonella typhimurium* and *Escherichia coli,* for example, are killed as effectively during incubation under anaerobic conditions, whereas *Staphylococcus aureus* is killed only under aerobic conditions. Thus, there is much debate as to which antimicrobial system is the most important: the O_2-dependent or -independent system.

Patients with CGD have an increased susceptibility to infections by organisms such as *S. aureus*, *Aspergillus*, *S. typhimurium, Serratia marcescens* and sometimes *Candida.* This observation, taken alone, would indicate that these organisms are killed by O_2-dependent processes involving the generation of reactive oxygen species. Indeed, *S. aureus* can be killed by H_2O_2 alone (or, better, by H_2O_2 + myeloperoxidase) in vitro; however, it is probably not realistic to consider the O_2-dependent and -independent antimicrobial systems working in isolation within phagolysosomes. For example, reactive oxygen metabolites generated within phagolysosomes may activate other antimicrobial proteins, and so an inability to generate oxidants may indirectly impair microbial killing. Furthermore, the NADPH oxidase is a proton pump, and in normal neutrophils the phagolysosomes become slightly alkaline for 15–45 min after activation. Thereafter they become acidic (pH $\approx$ 6.0), and the alkalisation is thought to be linked to respiratory burst activity. CGD neutrophils do not become alkaline, rather becoming initially acid, and this perturbation of phagolysosomal pH may impair microbicidal activity by not producing the optimal conditions for the activity of the antimicrobial granule proteins. Thus, it is likely that during phagocytosis there is a close interplay between the O_2-dependent and -independent antimicrobial systems, such that an optimal killing potential against a wide range of microbial targets (which themselves have different degrees of resistance to individual killing processes) is achieved.

The antimicrobial neutrophil proteins are located within intracellular granules. Approximately 30% of all neutrophil granules contain myeloperoxidase, and these azurophilic granules are the first granule types to be

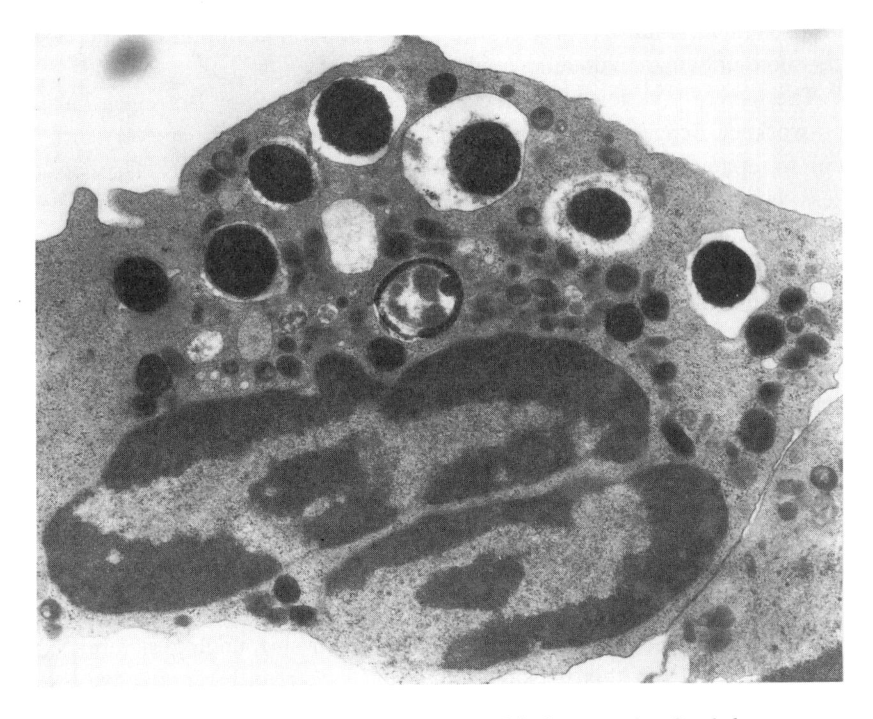

Figure 2.5. Electron micrograph of a human neutrophil phagocytosing *Staphylococcus aureus.*

synthesised during neutrophil maturation (at the promyelocyte stage). There are about 1500 azurophilic granules per mature neutrophil. In addition to myeloperoxidase, they contain the defensins, cathepsin G, azurocidin, BPI (bactericidal/permeability-inducing protein), hydrolases, elastase and collagenase (see Table 2.2). These granules are extremely heterogeneous in terms of their size, shape and protein content.

The peroxidase-negative granules produced later in neutrophil development are termed *secondary* or *specific granules* (because they stain only with specific dyes). They are twice as abundant in the cytoplasm of the mature neutrophil as the azurophilic granules. Specific granules contain much of the lysozyme, lactoferrin, vitamin-B_{12}-binding protein, adhesin receptors, fMet-Leu-Phe receptors, and cytochrome b_{-245}.

The *gelatinase-containing (tertiary) granules* are often difficult to separate from specific granules by density-gradient centrifugation. The membranes of these granules contain CR3, CD11c, CD45, an H^+-ATPase, the fMet-Leu-Phe receptor and the cytochrome b; other enzymes present include diacylglycerol lipase, acetyl CoA–lysoPAF acetyl transferase, gelatinase, β-glucuronidase (at lower levels than found in azurophilic granules), β_2-microglobulin and diamine oxidase.

The secretory vesicles have recently been discovered by Borregaard and co-workers (Sengeløv, Nielson & Borregaard, 1992). These are very difficult to separate from the plasma membrane on density gradients. They possess 'latent' alkaline phosphatase activity (i.e. subcellular fractions must be incubated with detergents such as Triton to release activity) and albumin, whilst the membranes contain CR1, CR3 and the fMet-Leu-Phe receptor. They are endocytic vesicles but can be rapidly translocated to the plasma membrane.

Thus, the specific and gelatinase-containing granules and the secretory vesicles all contain stores of preformed membrane proteins that are translocatable to the plasma membrane durng activation to enhance neutrophil responsiveness. The ease of mobilisation of these subcellular stores (in terms of the types and concentrations of stimuli required for the stimulation of their translocation) is secretory vesicles > gelatinase-containing granules > specific granules.

The contents of the granules are biologically inert within the neutrophil cytoplasm because of the impermeability of their surrounding membrane. Their release from this packaged state can occur during phagocytosis, whereby they are secreted into the phagocytic vesicle (Fig. 2.6), or else by secretion from the cells, so that they may act extracellularly. Movement and discharge of granules, termed *degranulation,* appears to be regulated by separate control mechanisms such that degranulation of specific versus azurophilic granules can occur independently. For example, if *S. typhimurium* is opsonised with IgG and C3, both the defensins and lactoferrin (markers for degranulation of azurophilic and specific granules, respectively) are detected within the phagolysosomes. However, if these bacteria are opsonised only with C3, then the appearance of lactoferrin within phagolysosomes is impaired – that is, degranulation of specific granules does not occur. Neutrophils may also degranulate and release their granule contents in response to soluble agonists, and again the discharge of different granule types is regulated independently. For example, addition of low concentrations of fMet-Leu-Phe or PMA, or incubation with cytokines, results in degranulation of specific granules. Thus, the contents of these granules are discharged whilst the membranes of these granules fuse with the cytoplasmic granules, thereby increasing the number of receptors and cytochrome b_{-245} molecules expressed on the cell surface. Low levels of secretion of azurophilic granules is observed upon stimulation of neutrophils with higher concentrations (e.g. 1 μM) of fMet-Leu-Phe, but this secretion is enhanced if the cells are first primed with GM-CSF or else co-incubated with cytochalasin B, which in-

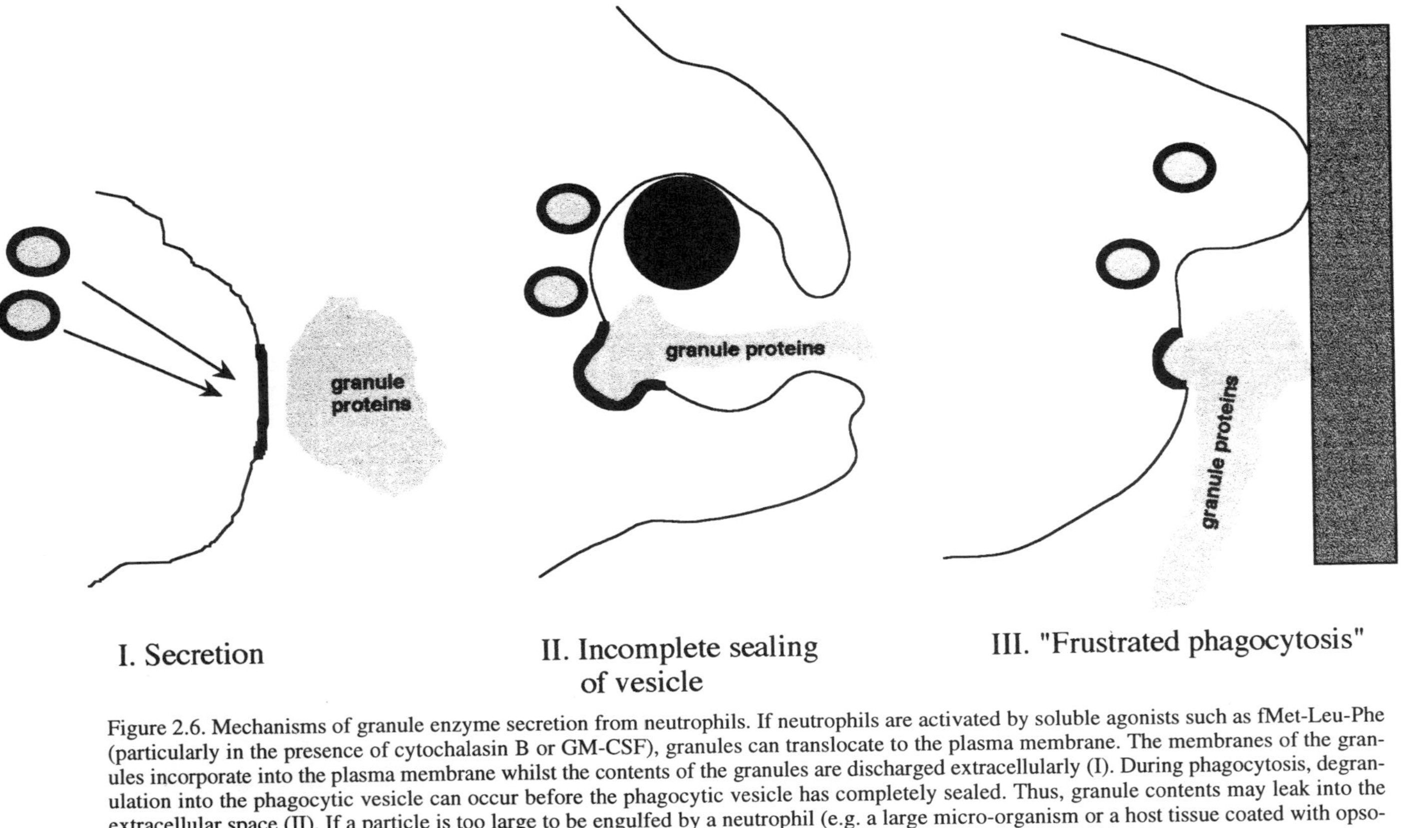

Figure 2.6. Mechanisms of granule enzyme secretion from neutrophils. If neutrophils are activated by soluble agonists such as fMet-Leu-Phe (particularly in the presence of cytochalasin B or GM-CSF), granules can translocate to the plasma membrane. The membranes of the granules incorporate into the plasma membrane whilst the contents of the granules are discharged extracellularly (I). During phagocytosis, degranulation into the phagocytic vesicle can occur before the phagocytic vesicle has completely sealed. Thus, granule contents may leak into the extracellular space (II). If a particle is too large to be engulfed by a neutrophil (e.g. a large micro-organism or a host tissue coated with opsonins), then granule enzymes may be released during 'frustrated phagocytosis' (III).

terferes with the microfilaments (§4.1.3.1). Thus, cytochalasins – usually cytochalasin B, although cytochalasin D may be more useful because it does not affect glucose transport – are used to enhance the extracellular release of granule contents in response to stimulation. Furthermore, granule components may be released extracellularly if vesicles do not adequately seal during phagocytosis, or if the target organism is too large to be fully enclosed (i.e. *frustrated phagocytosis;* Fig. 2.6).

Neutrophil granules contain a multitude of components, only some of which have been isolated and characterised. Some of these have direct antimicrobial activities – that is, it has been shown that the purified components are capable of microbial killing in vitro. Other components, however, are not directly antimicrobial, and it may be that these (such as the acid hydrolases that degrade carbohydrates, proteins and nucleic acids) function in the digestion of killed organisms, rather than in their destruction. Other granule proteins do not possess catalytic functions and yet are extremely antimicrobial (e.g. defensins, BPI), whereas still others may be *bacteriostatic* (i.e. prevent microbial growth) or else play as-yet-undefined roles in neutrophil function (e.g. lactoferrin, vitamin B_{12}-binding protein).

2.6.2 *Myeloperoxidase*

2.6.2.1 *Properties*

Myeloperoxidase was first discovered in 1941 by Agner, who isolated the enzyme from canine pus. It is present in neutrophils from a variety of species (excepting chickens) at levels reported in the range 2–5% of the total cellular protein or 2–4 μg per 10^6 cells. The enzyme is green due to its distinctive absorption spectrum, and gives purified neutrophil suspensions and pus a distinctive green colour. Indeed, the first name given to this protein was *verdoperoxidase.* The enzyme is synthesised early in the development of the neutrophils, at the promyelocyte stage, and packaged into azurophilic granules. Hence, this enzyme is present in the promyelocytic cell line HL-60, which has been used as the cellular source for the study of molecular controls regulating its transcription, translation and processing. Because of its unique location within these granules and its ease of assay, myeloperoxidase is the marker enzyme commonly used to assess the subcellular location of azurophilic granules and to trace the function of these organelles during degranulation.

Myeloperoxidase is a glycosylated, arginine-rich, extremely cationic protein with a relative molecular mass of 150 kDa, comprising two pairs of polypeptide chains. The heavy (α) subunit of 59 kDa and light (β) subunit of 13.5 kDa are arranged as a protomer, and the two protomers linked by a single disulphide bond between the two heavy subunits along their long axes. The two protomers may be separated by reduction and alkylation into hemimyeloperoxidase, and each protomer (i.e. α–β pair) still retains enzymatic activity. The haem and carbohydrate moieties both bind to the heavy subunit; thus, the enzyme is said to have an $\alpha_2\beta_2$ structure. Each molecule contains two iron-binding prosthetic groups that are linked to the heavy subunit. Originally, these iron atoms were thought to be present in haem groups, but magnetic circular dichroism, resonance Raman spectroscopy, and electroparamagnetic resonance spectroscopy (EPR) have revealed that these Fe atoms are in fact present in chlorin groups.

In some preparations of myeloperoxidase, a 39-kDa band is observed after SDS-PAGE under non-reducing conditions. This band is termed the α'-subunit, which has also been reported in a patient with chronic myelogenous leukaemia. Thus, the enzyme may exist in $\alpha_2\beta_2$, $\alpha\alpha'\beta_2$ or $\alpha'_2\beta_2$ forms. Ion-exchange chromatography has revealed three different isoforms of myeloperoxidase that exhibit differences in size of the α chain and different susceptibility to 3-aminotriazole, but these have similar enzymatic activities and spectroscopic properties. Moreover, the distribution of the enzyme within azurophilic granules is not homogeneous: 'light' granules contain large amounts of this protein whereas 'dense' granules contain somewhat less.

2.6.2.2 *Biosynthesis*

Histochemical studies of bone marrow samples show that peroxidase-containing granules are detectable in promyelocytes. The human promyelocytic leukaemia cell line HL-60 grows easily in culture, and the cells resemble promyelocytes both structurally and functionally. Furthermore, they can be induced to differentiate in vitro upon addition of various agents, such as retinoic acid and phorbol esters, and these differentiated cells resemble more mature forms of neutrophils. HL-60 cells possess almost the same amount of myeloperoxidase (4.4 μg per 10^6 cells) as mature neutrophils, and the enzyme purified from these cells has the same subunit structure. The cells thus actively synthesise the enzyme only until they are induced to differentiate. This cell line has been extensively used to study the molecular events controlling the expression of enzymes such as myeloperoxidase, and also to investigate the molecular controls that lead to a cessation of their expression.

The biosynthesis of myeloperoxidase has been characterised by *pulse-chase experiments*. This is achieved by incubating cells with a radioactive protein precursor, such as [^{35}S]-methionine, for a short time and then adding a vast excess of non-radioactively-labelled precursor. Thus, during the 'pulse' period only newly-synthesised precursor molecules become radiolabelled, and during the 'chase' period the appearance of labelled molecules that have been processed can be followed. This procedure can be made specific for a particular protein if an antibody exists that can be used to immunoprecipitate both precursor and mature forms of the protein. Thus, after immunoprecipitation the only molecules analysed are radioactively-labelled forms of the protein under investigation.

The primary translation product for myeloperoxidase (which is encoded by a single gene) is an 80-kDa protein that is rapidly glycosylated to give a 92-kDa glycoprotein, termed *promyeloperoxidase.* This contains five N-linked, high-mannose, oligosaccharide side chains. Glucose residues on these side chains are rapidly cleaved by glucosidase I to yield an 89-kDa molecule, which is packaged into a prelysosomal compartment. Although some of this 89-kDa protein may be secreted, most is processed further to yield a 75-kDa protein that is then cleaved into the heavy and light subunits. During this process the heavy chain retains four high-mannose oligosaccharide chains whereas the light subunit is deglycosylated.

When HL-60 cells are induced to differentiate into more mature cells by the addition of agents such as retinoic acid or PMA, the biosynthesis of myeloperoxidase ceases. Unlike that from undifferentiated cells, mRNA isolated from differentiated cells is no longer capable of synthesising myeloperoxidase in an in vitro reticulocyte lysate translation system. Furthermore, mRNA levels for myeloperoxidase (determined in Northern transfers using a myeloperoxidase cDNA probe) are decreased within 3 h of addition of differentiating agents and are undetectable by 24 h. This decrease in mRNA levels is likely to result from a termination in transcription of this gene, although it may also be possible that the mRNA is degraded at a much faster rate.

2.6.2.3 The myeloperoxidase gene

Several groups have reported the isolation of cDNA clones for myeloperoxidase from cDNA libraries constructed from mRNA purified from HL-60 cells (Morishita et al. 1987; Weil et al. 1987). If such libraries are induced to express the proteins for which the cDNA molecules encode, then they can be screened for particular genes using antibodies. Indeed, this has been the approach used to isolate cDNA clones for myeloperoxidase from HL-60 cells. Full-length clones have been described and sequenced, and their identity confirmed by a variety of means:

i. Messenger RNA prepared from these clones can be translated in vitro and the translated proteins recognised by specific myeloperoxidase antiserum.
ii. The size of the translated protein (75 kDa) is about the same size as the primary translation product, which was determined from pulse-chase experiments.
iii. The predicted amino acid sequence of the cDNA clones matched the actual amino acid sequence data obtained from studies of the purified protein.

Northern blot analysis has revealed that myeloperoxidase transcripts are only present in cells of the granulocytic lineage, and mRNA species of 3.0–3.3 and 3.5–4.0 kb are detected. No transcripts are detected in mature bloodstream neutrophils nor in HL-60 cells that have been induced to differentiate into more mature cells. The gene is located on chromosome 17q22-23 and comprises 12 exons and 11 introns, and the 5′-flanking region contains sequence homology to elastase and regions in the 5′-promoter region of the c-*myc* proto-oncogene. This is of interest because some patients with acute promyelocytic leukaemia have a translocation in this region of the chromosome: acquired myeloperoxidase deficiency is commonly associated with acute myeloproliferative disease.

2.6.3 Bactericidal/permeability-inducing protein (BPI)

Experiments by Peter Elsbach and colleagues in the 1970s showed that although *E. coli* lost viability very quickly after incubation with neutrophils, these non-viable organisms still retained several important biochemical functions, such as membrane transport and macromolecular biosynthesis. As these functions are associated with the inner plasma membrane of the bacteria, these observations suggested that the lethal 'hit' on *E. coli* by neutrophils occurred on the outer membrane. Because disrupted neutrophils also affected the bacteria in this way, it was concluded that the process was independent of the respiratory burst; hence these workers investigated the granule proteins for the source of this activity (reviewed in Elsbach & Weiss, 1983).

The activity was attributed to the protein known as BPI, a 55–60-kDa protein located in azurophilic granules. This protein is lysine rich, has a pI > 9.6 and accounts for about 1% of the total neutrophil protein. It is synthesised at the promyelocyte stage of neutrophil development; thus, biosynthetic studies may be performed using the promyelocytic cell line HL-60. BPI is active against a large number of Gram-negative organisms, such as

E. coli, S. typhimurium, S. typhi, Neisseria gonorrhoeae, Shigella boydii and *Pseudomonas aeroginosa* at concentrations of 1–5 $\times 10^{-8}$ M, but inactive against Gram-positive bacteria, fungi and mammalian cells at concentrations of up to 100-fold greater. It has only been detected (using immunological assays) in cells of the myeloid lineage.

The activity of BPI has been located to a 25-kDa aminoterminal fragment homologous to the hepatocyte-derived lipopolysaccharide-binding protein (LPS-BP), which is one of the acute-phase proteins. LPS-BP and the 25-kDa fragment are immunologically *cross-reactive* (i.e. antibodies raised against one molecule recognise the other) and both bind to the lipid A region of LPS: unlike BPI, however, LPS-BP does not possess microbicidal activity. These observations predict that BPI activity is associated with the presence of LPS on susceptible organisms and, indeed, Gram-negative organisms with short LPS chains (*rough* strains) are more susceptible than those with long chains (*smooth* strains) to the toxic effects of BPI. Activity is optimal at neutral pH and largely independent of ionic strength, although high concentrations of Ca^{2+} and Mg^{2+} (mM) inhibit its effects (see later in this section). It has been shown (using immunological staining) to bind to *E. coli* within phagolysosomes.

BPI has been purified and characterised from both human and rabbit neutrophils, and the two proteins from these different species are highly homologous: the first 17 amino-terminal amino acids are 80% homologous and the two molecules are immunologically cross-reactive. BPI is tightly associated with the azurophilic granule for two reasons:

i. Whereas high salt concentrations or low pH solubilise myeloperoxidase and elastase from azurophilic granules, these treatments release very little BPI.
ii. Upon stimulation of neutrophils with fMet-Leu-Phe plus cytochalasin B, large quantities of myeloperoxidase and elastase, but very low levels of BPI, are released extracellularly.

The amino-terminal domain contains cationic/amphiphilic regions that possess the antibacterial activity of the molecule, and a slightly acidic carboxy-terminal domain contains three hydrophobic regions, which may anchor BPI to the granule membrane and prevent its secretion. These two domains are separated by a hydrophilic, proline hinge region. The region 209–214 amino acids (Leu-Val-Ala-Pro-Pro-Ala) is an elastase cleavage site that cleaves the active 25-kDa fragment. Thus, elastase activation within the phagolysosome may free this active fragment of BPI, which then binds to the target organisms.

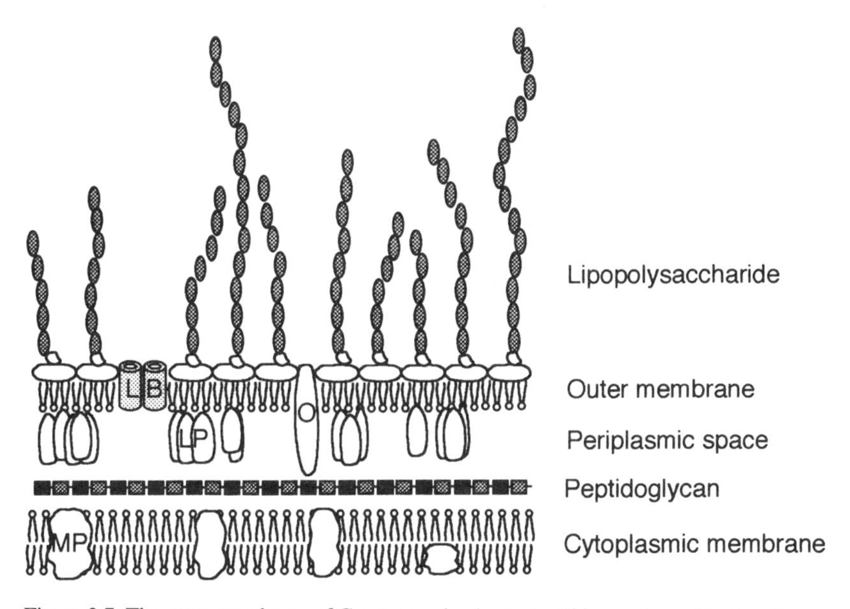

Figure 2.7. The outer membrane of Gram-negative bacteria. *Abbreviations:* LB, LamB protein; LP, lipoprotein; O, OmpA protein; MP, membrane protein. The peptidoglycan backbone consists of alternating residues of *N*-acetylglucosamine and *N*-acetylmuramic acid, which are cross-linked via short peptides.

The outer membrane of Gram-negative organisms provides a barrier against hydrophobic molecules because of the asymmetric arrangement of the hydrophobic bilayer (Fig. 2.7). Most of the phospholipids occupy only the inner leaflet, whereas the outer leaflet comprises the lipid portion of LPS. The external polysaccharide chain of LPS thus produces a hydrophilic barrier against hydrophobic substances, and this barrier is stabilised by cations such as Ca^{2+} and Mg^{2+}, which cross-link the LPS chains (Fig. 2.8a). BPI added to *E. coli* rapidly binds to the outer membrane and increases its permeability to hydrophobic substances. This insertion into the outer membrane of the bacterial envelope occurs via electrostatic and hydrophobic interactions. These interactions result in an irreversible loss of colony-forming ability within 30 sec exposure (at 37 °C) and a reversible increase in the permeability of the outer membrane to hydrophobic molecules. These events are accompanied by a reversible activation of enzymes that degrade phospholipids and peptidoglycans. Respiration is inhibited by over 90% within 20 min of exposure.

The highly basic BPI molecule is envisaged to interact initially with the charged surface sites (i.e. the charged sugar conjugates or phosphate

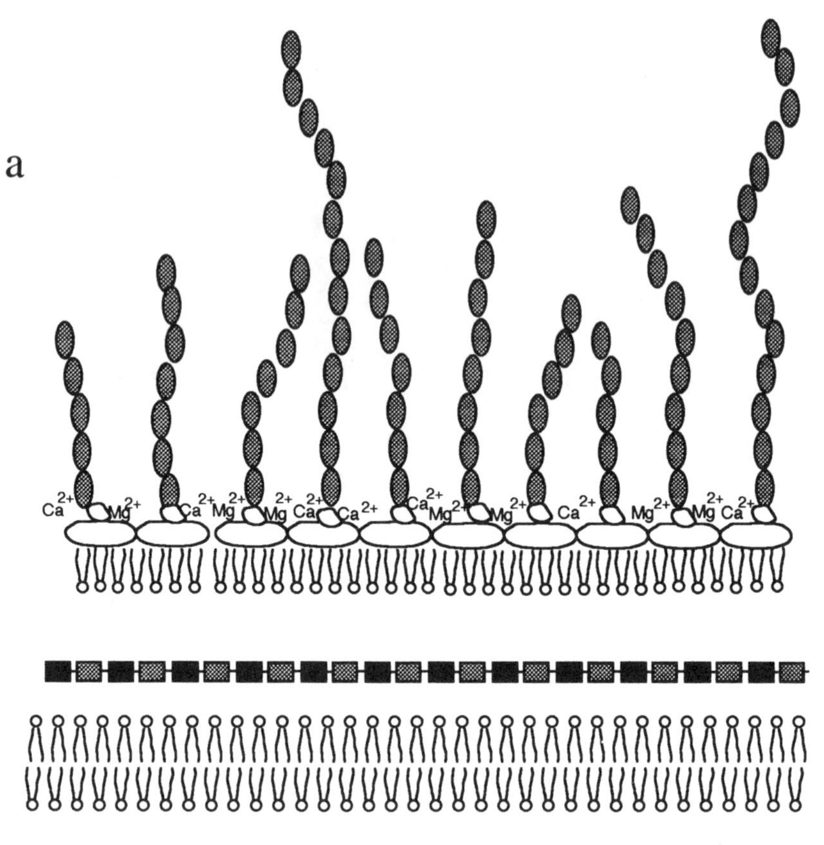

Figure 2.8. Proposed mechanism of action of BPI: (a) The external polysaccharide chain of lipopolysaccharide (LPS) is a hydrophilic barrier stabilised by Ca^{2+} and Mg^{2+}, which cross-link the LPS chains. (b, *facing*) BPI is highly basic and initially interacts by displacing the cations that bind to charged sites on the surface. See text for details.

groups) by displacement of the Mg^{2+} and Ca^{2+} normally located at these sites (Fig. 2.8b). Indeed, binding of BPI can be reversed by addition of high concentrations of Mg^{2+} and Ca^{2+}. This binding and displacement of cations then destabilises the LPS layer. At saturation, the ratio of binding of BPI to LPS is 1 : 5, which is equivalent to 5×10^5 molecules of BPI bound per cell. As the LPS chain length is increased, the efficacy of BPI is decreased, presumably because of the decreased ability to gain access to the bases of these LPS chains. Therefore, smooth strains of bacteria are more resistant to the effects of BPI, requiring some 20-fold higher concentrations to achieve loss of viability compared to rough strains, which are killed at $1–5 \times 10^{-8}$ M.

The sequences of events following BPI binding that lead to loss of viabil-

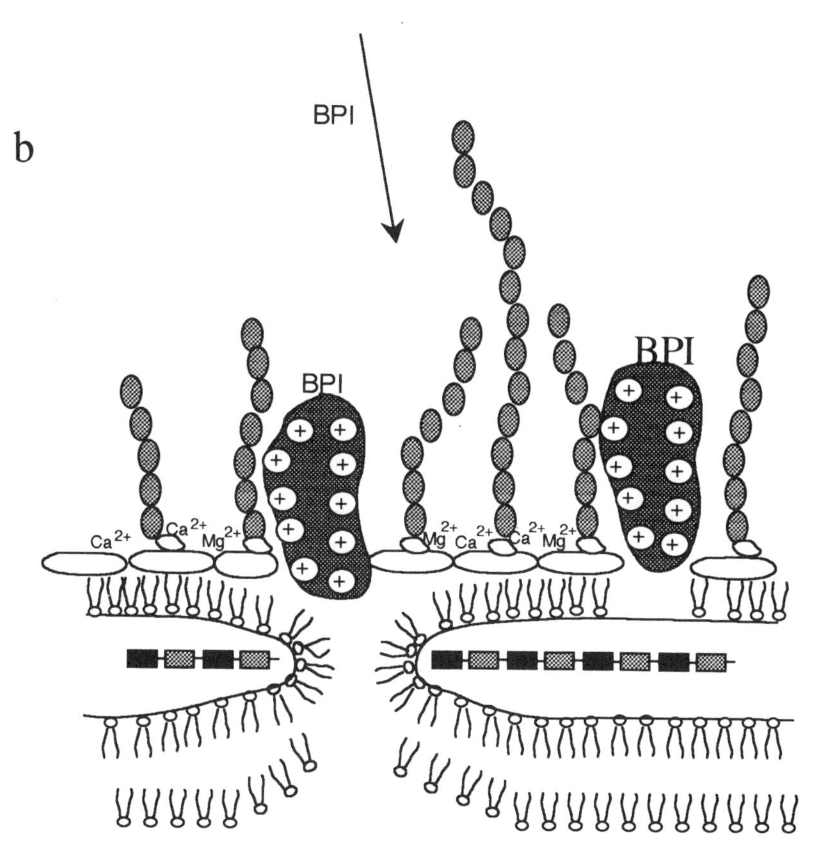

Figure 2.8 (*cont.*)

ity are unknown. Binding of BPI to LPS has a destabilising effect and is accompanied by the hydrolysis of bacterial phospholipids, which follows the activation of phospholipase A_2. Peptidoglycan degradation may also be activated via activation of specific enzymes. Such activities in themselves, however, do not account for loss of viability.

2.6.4 Defensins

Zeya and Spitznagel (1966) described the antimicrobial properties of arginine- and cysteine-rich cationic proteins, which they obtained by acid extraction of neutrophil granule homogenates. Extensive studies by Lehrer and colleagues (Lehrer & Ganz, 1990; Lehrer, Ganz & Selsted, 1991) purified these compounds from rabbit and human neutrophils and characterised their molecular properties. These small proteins are termed *defensins* and are in fact a family of proteins with related structures and functions. Defensins are 29–34-amino-acid molecules (relative molecular mass = 3.5–4 kDa) containing four to ten arginine residues and six conserved cysteine residues that form three intramolecular disulphide bonds. The molecules are cyclic because the cysteine residues near the NH_2- and COOH- termini form a bridge. X-ray crystallographic studies show that defensins have a spatial separation of charged and hydrophobic molecules, which makes the molecules amphiphilic; this may assist in their ability to insert into target microbial membranes.

There are four major human defensins, HNP-1–4, which are present in azurophilic granules. Defensins HNP-1–3 differ in only a single amino acid (Fig. 2.9), and HNP-4 is about 100-fold less abundant than the others. They constitute about 5–7% of the total neutrophil protein and 30–50% of the azurophilic granule protein, and to date have only been detected in neutrophils, being absent from monocytes, lymphocytes and eosinophils. Azurophilic granules are extremely heterogeneous in terms of their density and distribution of enzymes, although the significance of this heterogeneity is unknown. Defensins are present in 'dense' or 'classical' azurophilic granules; the 'light' granules contain most of the myeloperoxidase and elastase but are low in defensins. After phagocytosis defensins can be detected in phagolysosomes, and if neutrophils are activated to release granule enzymes (e.g. by stimulation with fMet-Leu-Phe plus cytochalasin B), defensins can be detected in the extracellular medium. Some cells of the mouse small intestine contain mRNA encoding *cryptdin,* a protein with molecular features analogous to those of typical defensins.

In vitro, purified defensins (50 μg/ml) are active against a wide range of Gram-positive and -negative bacteria, some fungi (e.g. *Cryptococcus neoformans*) and some enveloped viruses (e.g. *Herpes simplex*). Interestingly, human defensins only kill bacteria that are metabolically active (i.e. incubated with the defensins in a nutrient-rich medium), but kill metabolically inactive *C. neoformans.* Rabbit and rat defensins, which have structural similarities to the human defensins but are more cationic, are capable of killing metabolically inert organisms. HNP-1 (and, to a lesser extent, HNP-2) is a chemoattractant for monocytes, and some defensins are cytotoxic for cultured mammalian cells. HNP-4 inhibits cortisol function of cultured adrenal cells, whilst others induce mast-cell degranulation. Some defensins can function as non-specific opsonins for bacteria, and may inhibit the ac-

NHP-1	A C Y C R I P A C I A G E R R Y G T C I Y Q G R L W A F C C
HNP-2	C Y C R I P A C I A G E R R Y G T C I Y Q G R L W A F C C
HNP-3	D C Y C R I P A C I A G E R R Y G T C I Y Q G R L W A F C C
HNP-4	V C S C R L V F C R R T E L R V G N C L I G G V S F T Y C C T R V

Figure 2.9. Amino acid sequences of human defensins. The conserved positions of six cysteine residues are shown in hatched boxes. *Abbreviations:* A, alanine; C, cysteine; D, aspartic acid; E, glutamic acid; F, phenylalanine; G, glycine; H, histidine; I, isoleucine; K, lysine; L, leucine; M, methionine; N, asparagine; P, proline; Q, glutamic acid; R, arginine; S, serine; T, threonine; V, valine; W, tryptophan; Y, tyrosine.

tivity of protein kinase C. These observations imply that, in addition to their role in infections, defensins may have important functions in the regulation of inflammation and, perhaps, in antineoplastic activity.

The mechanism of action of defensins is largely unknown. Incubation with defensins results in the formation of voltage-regulated ion channels that permeabilise the outer and inner membranes of metabolically-active *E. coli*. Because the target bacteria must be metabolically active for defensins to exert their effects, it may be that the transmembrane electromotive force is involved in the mechanism of action.

The cDNAs for two human and three rabbit defensins have been cloned, and analysis of the predicted amino acid sequences indicate a remarkable sequence of events involved in the processing of the primary translation products. Whereas the mature proteins isolated from neutrophil granules are 29–34 amino acids, the initial translation product encoded by the mRNA is 94–95 amino acids. This precursor *prodefensin* possesses a large hydrophobic signal sequence that is proteolytically removed during processing. Furthermore, whereas the mature protein is very cationic, the cleaved fragment of the prodefensin is negatively charged. Because the net charge of the primary translation product is neutral, the prodefensin is not toxic and can be transported through the cytoplasm to the granules during processing without any harmful effects on the neutrophil. Mature neutrophils do not actively synthesise defensins, so studies of defensin biosynthesis and processing can only be performed on cell lines such as HL-60. Radiolabelling of newly-synthesised proteins by incorporation of [^{35}S]-methionine and analysis of defensins and defensin precursors using antibodies have confirmed that the initial defensin precursor comprises 94 amino acids. This is processed into the mature protein via sequential proteolytic cleavage into 75-, 56- and 39-amino-acid products. The 75-amino-acid precursor is detected in the cytoplasm and endoplasmic reticulum whereas the 56-amino-acid and smaller proteins are detected in the granules. The reasons for such a complex processing scheme are unknown.

2.6.5 Proteases

Neutrophil granules possess several proteases that exhibit maximal activity at neutral or slightly alkaline pH, including cathepsin G (a chemotrypsin-like protein), collagenase, elastase, proteinase 3, 'azurophil granule protein 7' and azurocidin. Some of these have been shown to be directly microbicidal whereas others have potent protease activity but do not kill bacteria in in vitro experiments. Interestingly, proteins such as cathepsin G retain microbicidal activity even when their catalytic activity is destroyed by heat treatment, indicating that their ability to kill their targets is independent of protease activity.

2.6.5.1 Cathepsin G

Cathepsin G, a cationic, glycosylated protein of relative molecular mass ~27 kDa, exists in four isoforms (25–29 kDa) that are identical in amino acid sequence but differ in levels of glycosylation. It is a component of azurophilic granules and present in human neutrophils at 1.5–3 $\mu g/10^6$ cells, but at lower levels in monocytes. cDNA has been cloned and sequenced (and the amino acid sequence predicted), and the gene has been localised to chromosome 14q11.2. The gene comprises five exons and four introns, a structure similar to that of the elastase gene.

Concentrations of cathepsin G of 25 $\mu g/ml$ are effective against a range of Gram-positive and -negative bacteria and some fungi, and it is very effective against *Neisseria gonorrheae.* The microbicidal activity is quite dependent upon the ionic strength of the incubation medium, decreasing as the ionic strength or serum concentrations are increased. Its activity is optimal at around neutral pH but is inhibited by 5–10 mM Mg^{2+} and >0.1 M NaCl. Its action on susceptible bacteria is to inhibit respiration and energy-dependent transport systems, followed by the inhibition of protein-, RNA- and DNA-biosynthesis. These may be directly-toxic effects or else may be secondary to perturbation of membrane structure/function. If cathepsin G is heated to 90–100 °C, it loses its proteolytic activity but still retains its microbicidal activity, indicating that its cytotoxic effect is independent of proteolysis. Gram-positive organisms are more sensitive than Gram-negative ones: for example, 10^{-6} M cathepsin G kills >90% of 2×10^6 *S. aureus,* whereas *E.coli* and *P. aeroginosa* require 2.5-fold more protein to achieve this level of toxicity.

2.6.5.2 *Azurocidin*

This is a 29-kDa protein that has NH_2-terminal sequence homology with elastase and cathepsin G. However, it contains glycine and not serine at the predicted catalytic site, and so lacks protease and peptidase activity. Purified azurocidin kills a range of organisms (e.g. *E. coli, S. faecalis,* and *C. albicans*) in vitro. It functions optimally at pH 5.5 and in conditions of low ionic strength.

2.6.5.3 *Elastase*

Elastase is a cationic glycoprotein present in azurophilic granules of human neutrophils at 1–4 $\mu g/10^6$ cells. Its cDNA has been cloned and its amino acid sequence predicted. The protein exists as four isoenzymes that differ in the levels of glycosylation. There are only a few reports indicating that elastase is microbicidal in its own right, but it is clear that it can enhance the activity of other neutrophil antimicrobial proteins. For example, elastase potentiates the lytic effects of lysozyme and cathepsin G in vitro and cleaves the active 25-kDa aminoterminal portion of BPI from the intact holoprotein. This activity detaches BPI from the azurophilic granule so that it is free to bind bacteria within phagolysosomes. Elastase is probably functional in the degradation of cell walls of killed bacteria, and its capacity to enhance the effects of other microbial systems may result from its ability to elicit limited proteolysis of the surface of the phagocytosed organism, thus exposing target sites for attack by other neutrophil antimicrobial systems: for example, limited proteolysis of bacterial outer surfaces may expose the peptidoglycan backbone of the cell walls to lysozyme. On the other hand, the ability of elastase to enhance the microbicidal effects of cathepsin G and the myeloperoxidase–H_2O_2 system are independent of its catalytic activity because heat-inactivated elastase is equally effective in this role.

2.6.6 *Lysozyme*

Lysozyme is a 14.4-kDa cationic protein (pI > 10) with the ability to kill a wide range of Gram-positive bacteria. It is present in both azurophilic and specific granules of neutrophils and is also found in the granules of monocytes and macrophages, in blood plasma, tears, saliva and airway secretions. In human neutrophils it is present at 1.5–3 $\mu g/10^6$ cells. Since its discovery in 1922 by Fleming, it has been widely studied by protein biochemists, and its three-dimensional structure has been precisely defined. It exerts its effects by cleavage of the β1–4 glycoside bond between *N*-acetylglucosamine and *N*-acetylmuramic acid in the peptidoglycan backbone of bacterial cell walls. Because this peptidoglycan structure in Gram-negative bacteria is protected by an outer membrane, these organisms are not very susceptible to the lytic effects of lysozyme.

Lysozyme is extremely active against such bacteria as *Bacillus subtilis, B. megaterium* and *Micrococcus lysodeikticus.* Indeed, the susceptibility of this latter organism to lysozyme forms the basis of a laboratory assay for this enzyme. However, in many species, such as rabbits, cattle and rhesus monkeys, although the neutrophils may be deficient of lysozyme the host has no increased susceptibility to infections. Gram-negative bacteria, which are normally resistant to lysozyme, can become sensitive when first pretreated with antibody plus complement or with reactive oxidants, which may expose the peptidoglycan backbone. *C. albicans* is also sensitive to lysozyme, especially under conditions of low ionic strength. The susceptibility of an organism to lysozyme depends upon a number of factors:

the complexity of the peptidoglycan network – that is, the extent of cross-linking and the length of cross-bridges that must be penetrated by lysozyme in order to gain access to the glycoside bonds;

the degree of *O*-acetylation of peptidoglycans;

physical barriers (e.g. the outer membrane of Gram-negative bacteria) that prevent access of lysozyme to the peptidoglycan.

2.6.7 Lactoferrin

Lactoferrin is a 78-kDa, slightly basic (pI = 8.7) glycoprotein present in the specific granules of neutrophils. It is present at 4–8 μg/10^6 cells and is a member of the transferrin family of iron-binding proteins. In addition to its presence in neutrophils, it is also detectable in serum, tears, semen and human milk. Lactoferrin contains two iron-binding sites and hence may have bacteriostatic properties by depleting the available iron in the bacterial microenvironment. Thus, inhibition of bacterial growth by lactoferrin may be envisaged as resulting from inhibition of the biosynthesis of haemoproteins, such as cytochromes and catalase. Whereas for most organisms lactoferrin results in growth inhibition (which may be reversed by addition of excess iron), some organisms, such as *S. mutans* and *Vibrio cholera,* are in fact killed by lactoferrin. Furthermore, it has been proposed that lactoferrin may regulate the production of the hydroxyl free radical during activation of the respiratory burst (§5.4.2.5). Thus, iron-deficient lactoferrin may decrease ·OH formation by chelating catalytic iron from the microenvironment of the respiratory burst, or else iron-saturated lactoferrin may accelerate its formation by providing catalytic iron. However, these putative roles are controversial; indeed, many researchers have questioned whether any ·OH at all is generated by activated neutrophils. It has also been proposed that lactoferrin may play a role in neutrophil adhesion as well as in the regulation of haematopoiesis.

cDNA for lactoferrin has been cloned, and the gene is present on chromosome 3q1-23, the region of the genome encoding both transferrin and the transferrin receptor. DNA sequence analysis reveals that homologies between lactoferrin and transferrin exist.

2.6.8 CAP57 and CAP37

Two granule proteins termed *cationic antimicrobial proteins* (CAPs) with relative molecular masses of 57 and 37 kDa have been described. These are arginine rich and active against *S. typhimurium* and other Gram-negative organisms at concentrations of 5 μg/ml, whereas *P. mirabilis, P. vulgaris* and *S. marcescens* are resistant. CAP37 is a monocyte chemoattractant. It is appears likely that CAP57 is in fact BPI whereas CAP37 is azurocidin.

2.6.9 Hydrolases and other enzymes

It is unlikely that the hydrolases present within neutrophil granules are directly cytotoxic; rather, they may play important roles in the digestion of killed organisms within phagolysosomes. In addition to the proteases, neutrophil granules contain a wide range of enzymes degrading carbohydrates, lipids and nucleic acids, but the role of these is not defined. Granule neutral lipases and phospholipases are present, but most pathogenic bacteria possess few, if any, neutral lipids. Phospholipases may thus be important in the digestion of phospholipids of killed bacteria, but there is no available evidence that these enzymes enhance the effects of other antimicrobial systems. Neutrophil granules also possess acid nucleases (although at 20-fold lower levels than in monocytes), but these appear to have little effect on the degradation of either chromosomal or plasmid DNA of phagocytosed and killed *E. coli.*

2.7 Bibliography

Agner, K. (1941). Verdoperoxidase: A ferment isolated from leukocytes. *Acta Chem. Scand.* (Suppl. 8) **2**, 1–62.

Almeida, R. P., Melchior, M., Campanelli, D., Nathan, C., & Gabay, J. E. (1991). Complementary DNA sequence of human neutrophil azurocidin, an antibiotic with extensive homology to serine proteases. *Biochem. Biophys. Res. Commun.* **177**, 688–95.

Bainton, D. F. (1980). The cells of inflammation: A general view. In *The Cell Biology of Inflammation,* vol. 2 (Weissman, G., ed.), pp. 1–25, Elsevier/North Holland, New York.

Demetri, G. D., & Griffin, J. D. (1991). Granulocyte colony-stimulating factor and its receptor. *Blood* **78**, 2791–808.

Devarajan, P., Mookhtiar, K., Van Wart, H., & Berliner, N. (1991). Structure and expression of the cDNA encoding human neutrophil collagenase. *Blood* **77**, 2731–8.

Elsbach, P., & Weiss, J. (1983). A reevaluation of the roles of the O_2-dependent and O_2-independent microbicidal systems of phagocytes. *Rev. Infect. Dis.* **5**, 843–53.

Fleming, A. (1922). On a remarkable bacteriolytic ferment found in tissues and secretions. *Proc. Roy. Soc. B* **93**, 306–17.

Ganz, T., Selsted, M. E., Szklarek, D., Harwig, S. S. L., Daher, K., Bainton, D. F., & Lehrer, R. I. (1985). Defensins. *J. Clin. Invest.* **76**, 1427–35.

Gasson, J. C. (1991). Molecular physiology of granulocyte-macrophage colony-stimulating factor. *Blood* **77**, 1131–45.

Grant, S. M., & Heel, R. C. (1992). Recombinant granulocyte-macrophage colony-stimulating factor (rGM-CSF): A review of its pharmacological properties and prospective role in the management of myelosuppression. *Drugs* **43**, 516–60.

Harwig, S. S. L., Park, A. S. K., & Lehrer, R. I. (1992). Characterisation of defensin precursors in mature human neutrophils. *Blood* **79**, 1532–7.

Hasty, K. A., Pourmotabbed, T. F., Goldberg, G. I., Thompson, J. P., Spinella, D. G., Stevens, R. M., & Mainardi, C. L. (1990). Human neutrophil collagenase: A distinct gene product with homology to other matrix metalloproteinases. *J. Biol. Chem.* **265**, 11421–4.

Lehrer, R. I., & Ganz, R. I. (1990). Antimicrobial polypeptides of human neutrophils. *Blood* **76**, 2169–81.

Lehrer, R. I., Ganz, T. & Selsted, M. E. (1991). Defensins: Endogenous antibiotic peptides of animal cells. *Cell* **64**, 229–30.

Lieschke, G. J. & Burgess, A. W. (1992). Granulocyte colony-stimulating factor and granulocyte-macrophage colony-stimulating factor. *New Eng. J. Med.* **327**, 99–106.

Lopez, A. F., Williamson, D. J., Gamble, J. R., Begley, C. G., Harlen, J. M., Klebanoff, S. J., Waltersdorph, A., Wong, G., Clark, S. C., & Vadas, M. A. (1986). Recombinant human granulocyte-macrophage colony-stimulating factor stimulates in vitro mature human neutrophil and eosinophil function, surface receptor expression, and survival. *J. Clin. Invest.* **78**, 1220–8.

Metcalf, D. (1985). The granulocyte-macrophage colony-stimulating factors. *Science* **229**, 16–22.

Metcalf, D., Begley, C. G., Johnson, G. R., Nicola, N. A., Vadas, M. A., Lopez, A. F., Williamson, D. J., Wong, G. G., Clark, S. C., & Wang, E. A. (1986). Biologic properties in vitro of a recombinant human granulocyte-macrophage colony-stimulating factor. *Blood* **67**, 37–45.

Moore, M. A. S. (1991). The clinical use of colony stimulating factors. *Ann. Rev. Immunol.* **9**, 159–91.

Morishita, K., Kubota, N., Asano, S., Kaziro, Y., & Nagata, S. (1987). Molecular cloning and characterization of cDNA for human myeloperoxidase. *J. Biol. Chem.* **262**, 3844–51.

Nagata, S., Tsuchiya, M., Asano, S., Kaziro, Y., Yamazaki, T., Yamamoto, O., Hirata, Y., Kubota, N., Oheda, M., Nomura, H., & Ono, M. (1986). Molecular cloning and expression of cDNA for human granulocyte colony-stimulating factor. *Nature* **319**, 415–19.

Nauseef, W. M. (1986). Myeloperoxidase biosynthesis by a human promyelocytic leukemia cell line: Insight into myeloperoxidase deficiency. *Blood* **67**, 865–72.

Nauseef, W. M. (1987). Posttranslational processing of a human myeloid lysosomal protein, myeloperoxidase. *Blood* **70**, 1143–50.

Nauseef, W. M., & Malech, H. L. (1986). Analysis of the peptide subunits of human neutrophil myeloperoxidase. *Blood* **67**, 1504–7.

Nicola, N. A., Metcalf, D., Matsumoto, M., & Johnson, G. R. (1983). Purification of a factor inducing differentiation in murine myelomonocytic leukaemia cells: Identification as granulocyte colony-stimulating factor. *J. Biol. Chem.* **258**, 9017–23.

Olsen, R. L., & Little, C. (1984). Studies on the subunits of human myeloperoxidase. *Biochem. J.* **222**, 701–9.

Pohl, J., Pereira, A., Martin, N. M., & Spitznagel, J. K. (1990). Amino acid sequence of CAP37, a human neutrophil granule-derived antibacterial and monocyte-specific chemotactic glycoprotein structurally similar to neutrophil elastase. *FEBS Lett.* **272**, 200–4.

Raines, M. A., Liu, L. D., Quan, S. G., Joe, V., DiPersio, J. F., & Golde, D. W. (1991). Identification and molecular cloning of a soluble human granulocyte-macrophage colony-stimulating factor receptor. *Proc. Natl. Acad. Sci. USA* **88**, 8203–7.

Sengeløv, H., Nielson, M. H., & Borregaard, N. (1992). Separation of human neutrophil plasma membrane from vesicles containing alkaline phosphatase and NADPH oxidase activity by free flow electrophoresis. *J. Biol. Chem.* **267**, 14912–17.

Souza, L. M., Boone, T. C., Gabrilove, J. L., Lai, P. H., Zsebo, K. M., Murdock, D. C., Chazin, V. R., Bruszewski, J., Lu, H., Chen, K. K., Barendt, J., Platzer, E., Moore, M. A. S., Mertlesmann, R., & Welte, K. (1986). Recombinant human granulocyte colony-stimulating factor: Effects on normal and leukaemic myeloid cells. *Science* **232**, 61–5.

Spitznagel, J. K., & Okamura, N. (1983). Oxygen independent microbicidal mechanisms of human polymorphonuclear leukocytes. *Adv. Exp. Med. Biol.* **162**, 5–17.

Vadas, M. A., Lopez, A. F., Gamble, J. R., & Elliot, M. J. (1991). Role of colony-stimulating factors in leucocyte responses to inflammation and infection. *Curr. Opin. Immunol.* **3**, 97–104.

Valore, E. V., & Ganz, T. (1992). Posttranslational processing of defensins in immature human myeloid cells. *Blood* **79**, 1538–44.

Weil, S. C., Rosner, G. L., Reid, M. S., Chisholm, R. L., Farber, N. M., Spitznagel, J. K., & Swanson, M. S. (1987). cDNA cloning of human myeloperoxidase: Decrease in myeloperoxidase mRNA upon induction of HL-60 cells. *Proc. Natl. Acad. Sci. USA* **84**, 2057–61.

Weiss, J., Kao, L., Victor, M., & Elsbach, P. (1985). Oxygen-independent intracellular and oxygen-dependent extracellular killing of *Escherichia coli* S15 by human polymorphonuclear leukocytes. *J. Clin. Invest.* **76**, 206–12.

Whetton, A. D., & Vallance, S. J. (1991). Regulation of neutrophil and macrophage production by growth factors. *Int. J. Biochem.* **23**, 1361–7.

Zeya, H., & Spitznagel, J. K. (1966). Cationic proteins of polymorphonuclear leukocytes: I. Resolution of antibacterial and enzymatic activities. *J. Bacteriol.* **91**, 750–4.

3
The generation and recognition of neutrophil-activating factors: Structure and function of neutrophil receptors

Most if not all of the naturally-occurring neutrophil-activating factors elicit their effects on neutrophils after binding to specific receptors on the plasma membrane. Therefore, this chapter describes how many of these factors are generated (many may be generated by the neutrophils themselves) and how they are thought to mediate their effects; where possible, details of the corresponding receptor will be given. In addition, this chapter includes descriptions of the structure and function of the complement and immunoglobulin receptors involved in the regulation of many neutrophil functions, such as adhesion and opsonophagocytosis.

3.1 Leukotriene B_4

3.1.1 Properties

Leukotrienes are generated via the activities of lipoxygenases on arachidonic acid. Arachidonic acid itself is generated largely via the activity of phospholipase A_2 on membrane polyunsaturated fatty acids, although it may also be formed via the activity of diacylglycerol lipase on *sn*-1,2-diacylglycerol (see §§6.3.1.1, 6.3.1.5). Different cells possess lipoxygenases that oxidise arachidonic acid at different C atoms on the molecule. For example, platelets possess 12-lipoxygenase, mast cells have 11-lipoxygenase, but neutrophils, eosinophils, basophils, monocytes and macrophages have 5-lipoxygenase. The initial product of 5-lipoxygenase on arachidonic acid is the short-lived molecule 5-hydroxyperoxy-eicosatetraenoic acid (5-HPETE), which is converted into the unstable peroxide LTA_4. In neutrophils, monocytes and pulmonary macrophages, this LTA_4 is then reduced to LTB_4 (Fig. 3.1) via epoxide hydrolase. Other products formed from 5-HPETE (§6.3.3.2) include the following:

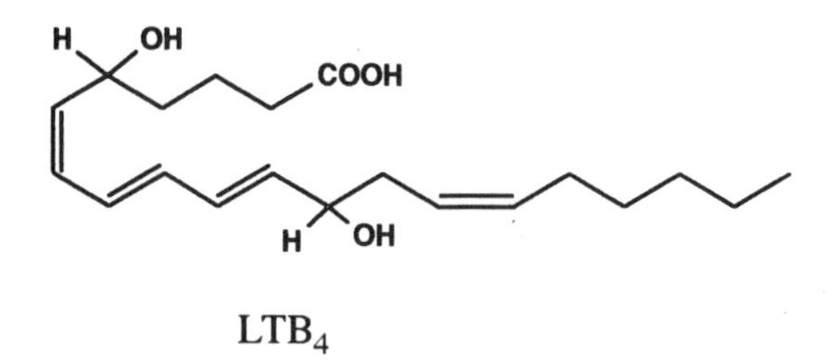

Figure 3.1. Structure of leukotriene B_4 (LTB_4).

i. 5-HETE, which may be reacylated into lysophospholipids or exported (and which may be taken up by platelets and converted by 12-lipoxygenase into 12-*epi*-6-*trans*-8-*cis*-LTB_4) and
ii. LTC_4, which is generated from LTA_4 via LTC_4 synthetase. This requires glutathione *S*-transferase (K_m for glutathione = 3–6 mM) and is produced by stimulated eosinophils, mast cells and monocytes.

The 5-lipoxygenase of neutrophils is membrane-associated and has a Michaelis constant for arachidonic acid of K_m = 10–20 μM, similar to that for cyclooxygenase (which generates prostaglandins) but lower than that for 15-lipoxygenase. Thus, in the neutrophil the enzymes cyclooxygenase, 5-lipoxygenase and 15-lipoxygenase will all compete for arachidonic acid, which is released from membranes via phospholipase A_2 activity. However, activity of 15-lipoxygenase in neutrophil preparations may be due to eosinophil contamination (see §6.3.3.2). 5-lipoxygenase activity is Ca^{2+}-dependent, although the concentrations of this cation required in vitro to activate the enzyme appear to be in excess of those actually found in vivo. It may thus be that, during activation of the enzyme in vivo, the Ca^{2+} affinity required for activation is decreased via conformational changes, covalent modifications and/or G-protein interaction. The rate of 5-lipoxygenase activity is subject to feedback inhibition by the hydroperoxide product, which thus regulates the utilisation of arachidonic acid if the products are not further metabolised or exported.

In neutrophils, LTB_4 may be metabolised by ω-oxidation, which then leads to its inactivation. This reaction is catalysed by LTB_4-20-hydroxylase (K_m for LTB_4 of 0.2–1 μM), an enzyme that is a member of the cytochrome P-450 family and is responsible for the conversion of the C20-methyl into the C20-alcohol-derivative ω-hydroxyLTB_4. The Ca^{2+} ionophore A23187 is a good stimulus for LTB_4 production in neutrophils. This agent stimulates the influx of extracellular Ca^{2+} to intracellular sites in the absence of receptor occupancy. Thus, whilst this agent is a useful laboratory tool for the study of some Ca^{2+}-mediated events, interpretation of its physiological importance is questionable. Zymosan-activated neutrophils also generate large quantities of LTB_4, but cells stimulated by fMet-Leu-Phe or PMA generate only low amounts.

3.1.2 *Effects on neutrophil function*

LTB_4 is a powerful chemoattractant for neutrophils, acting at <nM concentrations. Additionally, it can promote adherence of neutrophils to endothelial cells, induce neutrophil aggregation and up-regulate expression of CR3. It can also activate low levels of degranulation and NADPH oxidase activity, but generally these events only occur after addition of much higher concentrations than are required for chemotaxis. Cytochalasin B augments this LTB_4 stimulated degranulation and oxidant production.

3.1.3 *The LTB_4 receptor*

The LTB_4 receptor may exist in a high-affinity state ($K_d = 4–5 \times 10^{-10}$ M), occupancy of which leads to the initiation of chemotaxis, and there are about $3–20 \times 10^3$ high-affinity binding sites per neutrophil. A low-affinity receptor state also exists ($K_d = 0.6–5 \times 10^{-7}$ M), and occupancy of this receptor ($10–50 \times 10^4$ sites per neutrophil) probably results in the activation of secretion and NADPH oxidase activity. The LTB_4 receptor has a relative molecular mass of 60–65 kDa and may be down-regulated (i.e. the number of receptors expressed on the cell surface may be decreased) by protein kinase C activity. The LTB_4 receptor may be associated with the cytoskeleton, since if radiolabelled LTB_4 is added to neutrophils, it rapidly becomes associated with the Triton X-100-insoluble fraction of the cell homogenate. It is also G-protein linked because such LTB_4-induced functions as chemotaxis, degranulation and actin polymerisation are blocked by the addition of pertussis toxin (§6.2.1). This G-protein is linked to phospholipase C activity because LTB_4 addition causes increases in the concentrations of intracellular Ca^{2+}. Occupancy of the receptor stimulates the activity of an amiloride-sensitive Na^+/H^+ antiport, which causes the influx of Na^+ and alkalisation of the cytoplasm. LTB_4 addition also causes a small increase in the intracellular concentration of cAMP; this is curious because increases in the concentration of this molecule can actually limit some neutrophil functions.

3.2 Complement fragment C5a

3.2.1 Properties

The complement fragment C5a was first discovered by Boyden in 1962 as a heat-stable chemotactic substance produced when antigen–antibody complexes were incubated with rabbit serum, a process resulting in the activation of complement. Snyderman, Gewurz & Mergenhagen (1968) showed that the major chemoattractant activity generated in this way had a relative molecular mass of about 15 kDa and was a cleavage product of C5. Fragments of C5, generated either by treatment with proteases such as trypsin, or else by addition to serum of sheep erythrocytes coated with complement components C1–C4, produced this neutrophil chemoattractant activity.

C5-derived products are chemoattractants for neutrophils, eosinophils, basophils and monocytes and may be generated by either the classical or alternative pathways of complement activation (see §§1.3.2.1–2). They are also generated by protease activity on native C5, C5a being the most potent chemoattractant of the number of products generated: it is chemotactic for human neutrophils at 0.1 nM, whereas at higher concentrations (1–5 nM) migration is decreased. Other complement fragments are biologically active in the immune response and function as anaphylatoxins (e.g. C3a, C4a and C5a all cause contraction of smooth muscle, increased vascular permeability and stimulation of histamine release from mast cells), but only C5a is chemotactic for neutrophils and monocytes.

Human C5a is a cationic glycoprotein with a relative molecular mass of 16 kDa, as determined by gel filtration and SDS-polyacrylamide gel electrophoresis (SDS-PAGE). The molecule comprises 74 amino acids, accounting for 8.2 kDa of the molecule, and also carbohydrate, which adds a further 3 kDa to the structure. These two components total 11.2 kDa, and the anomaly between this value and the physical measurements is due to the presence of the carbohydrates: proteins containing large quantities of carbohydrates do not run strictly according to their size in SDS-PAGE or gel filtration. The single carbohydrate complex of human C5a contains 4 moles of glucosamine, 3–4 moles of sialic acid, 4 moles of mannose and 2 moles of galactose, attached to aspargine 64. Amino-acid-sequence analysis reveals a carboxy-terminal arginine residue essential for anaphylatoxin activity.

Serum contains an anaphylatoxin inactivator that is a carboxypeptidase N and cleaves the carboxy-terminal arginine residue to generate C5a des Arg. Thus, following complement activation, C5a des Arg is the predominant

C5-derived peptide in serum. This molecule lacks anaphylatoxin activity (i.e. it cannot cause smooth muscle contraction), and its ability to cause chemotaxis in neutrophils is about 10–20 times lower than that of C5a. However, human serum also contains a heat-stable, anionic protein termed *cochemotaxin* (relative molecular mass = 60 kDa), which acts in a concentration-dependent manner to permit C5a des Arg to act as a chemoattractant for neutrophils. Thus, C5a des Arg plus cochemotaxin working together probably account for most of the neutrophil chemoattractant activity in vivo following complement activation. The mechanism of action of cochemotaxin is unknown, but it may form a physical complex by attaching to a sialic acid residue on the oligosaccharide chain of C5a des Arg. Deglycosylation of C5a des Arg increases its chemoattractant activity more than 10-fold, and its dependency upon cochemotaxin is decreased.

C5a is inactivated by the myeloperoxidase–H_2O_2 system, which oxidises a methionine residue (Met 70) on the molecule; group A streptococcal endoproteinases also abolish chemotactic activity of C5a and related compounds. Neutrophil lysosomal enzymes (e.g. elastase and cathepsin G) also destroy C5a chemotactic activity, but as these proteases are inhibited by the serum antiproteinases, α_1-antiproteinase and α_2-macroglobulin, the physiological role of neutrophilic proteases in the inactivation of C5a is questionable. Two chemotactic factor inactivators have been found in human serum: an α-globulin that specifically and irreversibly inactivates C5-derived chemotactic factors, and a β-globulin that inactivates bacterial chemotactic factors. These activities are heat labile (destroyed by treatment at 56 °C for 30 min) and are distinct from those attributable to anaphylatoxin inactivator. An apparently specific inhibitor of C5-derived chemotactic activity has also been described in human synovial fluid and peritoneal fluid. This factor (molecular mass of 40 kDa) is heat stable and acts directly on C5a.

3.2.2 Effects on neutrophil function

3.2.2.1 Degranulation

In addition to their chemotactic activity, C5-derived peptides can also induce degranulation in human neutrophils. This process, readily detected in cytochalasin-B-treated neutrophils, occurs rapidly (within 1 min) upon the addition of peptides. The half-maximum responses in cytochalasin-treated human neutrophils are observed at 1–5 nM C5a and 100–800 nM C5a des Arg. In the absence of cytochalasin B, degranulation can still be detected (albeit at lower rates), especially if the neutrophils are adherent prior to the addition of peptides. It may be that C5a can also stimulate microtubule assembly.

3.2.2.2 Oxidative metabolism

C5a and C5a des Arg stimulate aerobic glycolysis, hexose monophosphate shunt activity, glucose uptake and the respiratory burst of human neutrophils. All of these processes are stimulated in neutrophil suspensions incubated in the absence of cytochalasin B, but the responses are considerably enhanced if this inhibitor of microtubule assembly is present. Stimulated rates of oxidative metabolism are maximal within 2 min of addition of peptides, with half-maximal responses obtained at 30–60 nM C5a and 1–3 μM C5a des Arg.

3.2.2.3 Adherence

Low (0.1-nM) concentrations of C5a decrease adhesion of human neutrophils to aortic endothelial cells or human umbilical-vein endothelial cells in vitro, but promote chemotaxis. In contrast, higher concentrations (>0.1 nM) stimulate adhesion but decrease chemotaxis. This concentration-dependent switching from chemotactic activity to adhesion is also observed with the chemotactic peptide fMet-Leu-Phe: low concentrations (0.1–1 nM) stimulate chemotaxis and inhibit adhesion, whereas higher concentrations (>10 nM) stimulate adhesion and decrease chemotaxis. This may be important in regulating neutrophil function during acute inflammation, where the stimulation of adherence to endothelial cells prior to diapedesis and subsequent chemotaxis are both required. It is thus advantageous for an effective inflammatory reponse if the same factor can, depending upon its concentration, elicit different responses from the target cell.

3.2.3 C5a receptors

C5a receptors (which also bind C5a des Arg) have half-saturation binding of 3–7 nM C5a. There are about 100000–300 000 receptors per neutrophil, and the receptor has a relative molecular mass of 42–48 kDa. The C5a receptor has recently been cloned. The U-937 cell line is a rich source of C5a receptor, and because the effects of C5a can be blocked by pertussis toxin, the receptor is G-protein linked. Many receptors of this type belong to the rhodopsin superfamily of genes, which have seven hydrophobic membrane-spanning domains (Fig. 3.2); thus, it was predicted that the C5a receptor would have homology to regions of this superfamily, particularly in one of the membrane-spanning domains. Gerard and Gerard (1991) produced a cDNA library from mRNA isolated from U-937 cells and screened this li-

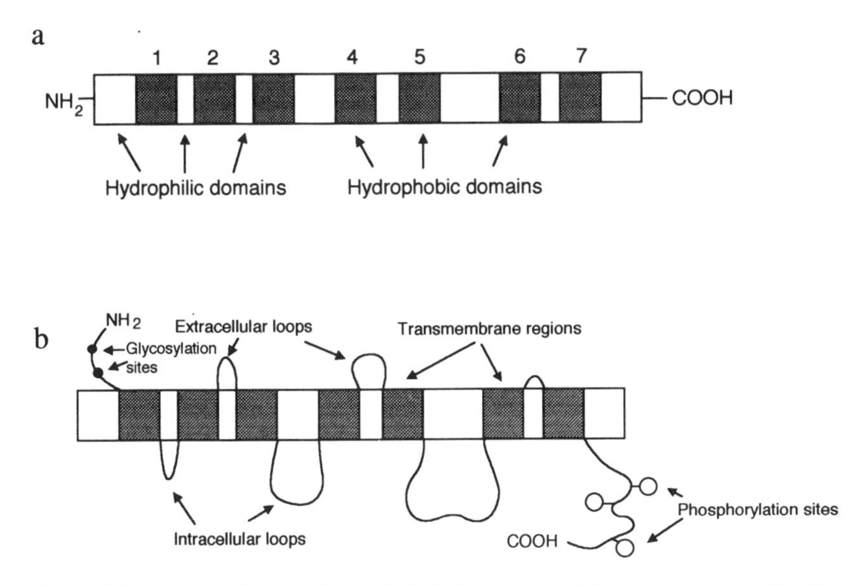

Figure 3.2. Structure of some G-protein-linked receptors of the rhodopsin superfamily: (a) the structural organisation of the predicted hydrophilic and hydrophobic domains of the receptor; (b) how the hydrophilic regions form extracellular and intracellular loops, being anchored by the seven hydrophobic transmembrane domains.

brary with an oligonucleotide probe based on a region of the M7 hydrophobic domain of this rhodopsin superfamily. Twenty clones were isolated, and many of these hybridised to a 2.2-kb mRNA species of U-937 mRNA. One of these clones (NPIIY-18) was selected for further studies and was (i) used to screen a cDNA library constructed from differentiated HL-60 cells and (ii) transfected into COS-7 cells. This clone conferred upon the COS-7 cells the ability to bind C5a on their cell surface (about 10^6 binding sites per cell). The clone was then sequenced and shown to possess a 1050-bp open reading frame, which predicted a protein of 350 amino acids with a relative molecular mass of 39320 da. The size of the receptor measured from binding studies is 42–48 kDa. Sequence analysis has revealed that this receptor does indeed belong to the rhodopsin superfamily because hydrophobicity plots reveal seven membrane-spanning domains.

3.3 Platelet activating factor (PAF)

3.3.1 Properties

Platelet activating factor (PAF) is the term used to describe a family of structurally-related lipids that are all acetylated phosphoglycerides (Fig.

3.3). The major components of this family that are synthesised and released by human neutrophils are 1-*O*-alkyl-2-acetyl-*sn*-glycero-3-phosphocholines, with C16:0 and C18:0 being the major components of the alkyl chain. As with the eicosanoids, which are another family of lipid molecules with the ability to regulate local inflammatory responses, PAF is not stored in large quantities by cells, but instead is rapidly synthesised and released upon cell stimulation. The term 'platelet activating factor' is a historical one, illustrating the first identifiable biological process that was regulated by these compounds, but in fact many cell types are affected by these molecules. Indeed, as again is observed with the eicosanoids, PAFs are intimately involved in the stimulation of acute and chronic inflammation, but also probably play important roles in the regulation of normal tissue processes including heart, liver and kidney function. Its effects during inflammation are to induce hypotension, cause bronchoconstriction and increase vascular permeability.

PAF is produced by neutrophils, platelets, basophils, natural killer (NK) cells, monocytes, macrophages, eosinophils, mast cells and vascular endothelium during inflammation. Identification of PAF production during cell stimulation is not always based upon precise determination of its chemical structure, but rather on the characterisation of molecules by comparison of their relative mobilities on TLC and reverse phase HPLC, and by susceptibility to phospholipases (PAF is inactivated by phospholipases A_2, C and D). In neutrophils, PAF is synthesised and released upon activation by a variety of pathophysiological stimuli, including A23187, aggregated IgG, C5a, C5a des Arg, fMet-Leu-Phe, IgG-coated particles, IgG- and IgA-containing immune complexes, opsonised zymosan and PMA. It may be taken up and catabolised by the same cell that produces it. Thus, during cell activation PAF-synthesising and -utilising enzymes may both be stimulated; hence, the amount of PAF detected is a balance between the amount generated and the amount catabolised.

Biosynthesis of PAF requires the stimulation of phospholipase A_2 to hydrolyse intracellular pools of 1-*O*-alkyl-2-acyl-*sn*-glycero-3-phosphocholine (reaction 1, Fig. 3.4). The resulting lysophospholipid is then acetylated by a specific acetyltransferase to form the biologically-active molecule 1-*O*-alkyl-2-acetyl-*sn*-glycero-3-phosphocholine (reaction 2). Specific degradation of PAF occurs via the activity of an acetylhydrolase (reaction 3) resynthesising lysoPAF, which is then reacylated via acyltransferase (reaction 4) in a reaction requiring long-chain fatty acylCoA. These biosynthetic routes have been determined from experiments showing that radiolabelled precursors are incorporated into these molecules: [^{3}H]-acetate and [^{3}H]lysoPAF are incorporated into PAF (reaction 2); [^{3}H]lysoPAF is incorporated and

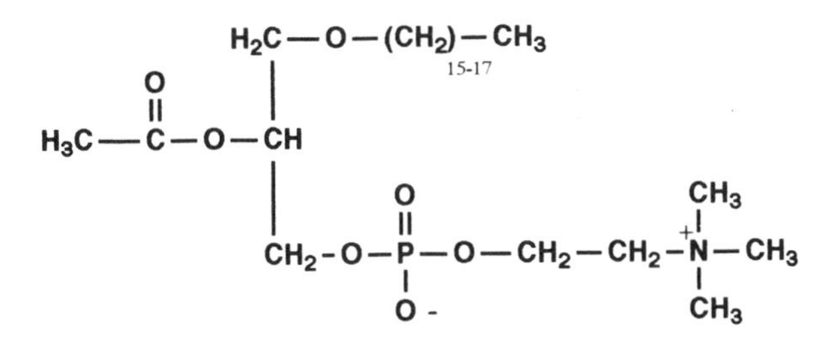

Figure 3.3. Structure of platelet activating factor (PAF). The structure shown is 1-*O*-alkyl-2-acetyl-*sn*-glycero-3-phosphocholine.

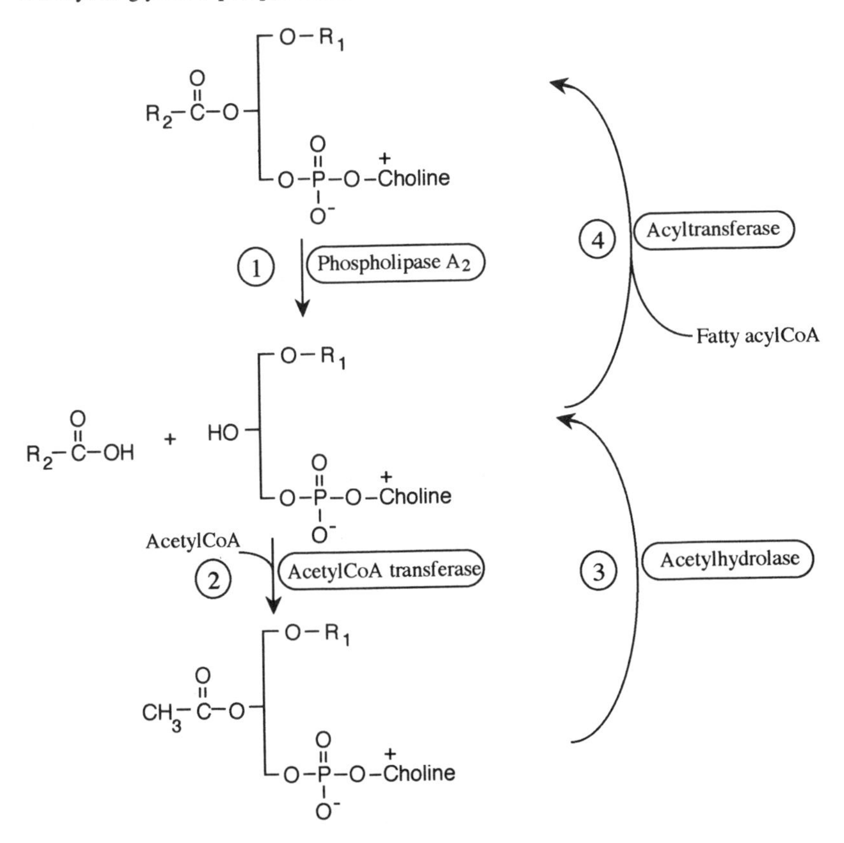

Figure 3.4. Biosynthesis of PAF. See text for details.

converted into 1-*O*-alkyl-2-acyl-*sn*-glycero-3-phosphocholine (reaction 4) and then into [^{3}H]PAF (reactions 1 and 2) upon stimulation.

If the R_2 group of the the substrate 1-*O*-alkyl-2-acyl-*sn*-glycero-3-phosphocholine is arachidonic acid (i.e. if the PAF molecule is 1-*O*-alkyl-2-

arachidonyl-*sn*-glycero-3-phosphocholine), then this polyunsaturated fatty acid will also be released upon stimulation of phospholipase A_2 activity. Arachidonic acid is a precursor of prostaglandin and leukotriene biosynthesis (see §6.3.2.2) and is also required for activation of the NADPH oxidase (see §5.3.2). Indeed, many of the biological activities of PAF are modulated by interactions with eicosanoids. Experimental evidence shows that about 80% of exogenously-added lysoPAF becomes acylated with arachidonic acid. Thus, it is likely that PAF formation and arachidonic acid metabolism are intimately linked during cell activation.

Not all of the newly-synthesised PAF is released by neutrophils: it is thought that release of PAF only occurs after a critical intracellular concentration has been reached; even then, some PAF remains cell associated and does not act as a true extracellular effector molecule. The cell-associated PAF may thus play a role in cell–cell communication (e.g. neutrophil–neutrophil or neutrophil–endothelium interactions). Intra- and extracellular Ca^{2+} levels regulate both the biosynthesis and release of PAF. Increases in intracellular Ca^{2+} may allow PAF synthesis (but not release), whereas extracellular Ca^{2+} increases both synthesis and release in a dose-dependent manner.

The levels of PAF synthesis and release are also modulated by levels of extracellular albumin. In the absence of albumin, neutrophils (stimulated with fMet-Leu-Phe) synthesise only low levels of PAF within 1–2 min of stimulation. In the presence of 0.25% albumin, PAF synthesis is increased, and up to half of this may be released; with 5% albumin, rates of synthesis and release are increased further and sustained over a 30-min period. Newly-synthesised PAF is reincorporated by neutrophils into membrane lipids and is therefore poorly soluble in aqueous media. Thus, extracellularly added albumin will bind to cell-associated PAF and effectively solubilise it at concentrations below its critical micellar concentration (CMC). This will effectively enhance the PAF release rate, which will decrease the concentration of cell-associated PAF; thus, the rate of biosynthesis will be sustained.

3.3.2 Effects on neutrophil function

An intravenous infusion of PAF causes rapid (within 60 s) intravascular platelet aggregation, thrombocytopenia and platelet factor 4 release, as well as a profound and reversible neutropenia, due to enhanced aggregation and adherence of these cells. In vitro, PAF effects on neutrophils are dependent upon extracellular Ca^{2+} and Mg^{2+} and occur within 60 s of addition. The addition of inhibitors of 5-lipoxygenase activity (e.g. ETYA, 5,8,11,14-eicosatetraenoic acid and NDGA, nordihydroguaiaretic acid) – but not those of cyclooxygenase activity – greatly decrease but do not fully inhibit the effects of PAF, thus implicating the lipoxygenase pathway in PAF function. PAF also stimulates the release of arachidonic acid and its metabolism to 5-, 11- and 15-HETE and to LTB_4, this latter compound being 10–100 times more effective in inducing neutrophil adherence. Thus, when determining the effects of PAF on neutrophil function it is necessary to distinguish between its direct effects (i.e. those due to PAF itself) and those that may be secondary and due to biological effector molecules (e.g. LTB_4) whose production is stimulated by PAF.

In addition to chemotaxis, chemokinesis, aggregation and adherence, many other neutrophil functions are stimulated by PAF, depending upon the concentration used. Concentrations within the range 1–1000 nM stimulate degranulation of azurophilic, specific and gelatinase-containing granules (within 30 s of addition) in the presence of cytochalasin B. In the absence of cytochalasin B degranulation is still detected, albeit at considerably lower levels, and higher concentrations (1 μM) of PAF are required. Extracellular Ca^{2+} is not essential for secretion, but lipoxygenase inhibitors again decrease the ability of PAF to cause degranulation. PAF also stimulates the respiratory burst, but concentrations of 0.1–10 μM are required to elicit this response, even in the presence of cytochalasin B. Pre-incubation of neutrophils with 5×10^{-9} M PAF primes the ability of these cells to generate reactive oxidants following the addition of stimuli such as fMet-Leu-Phe, PMA, opsonised zymosan and A23187. PAF at a concentration of 10 pM also primes tumour necrosis factor (TNF)-induced reactive oxidant production. PAF agonists have been reported, such as CV-6209 (2-[*N*-acetyl-*N*-(2-methoxy-3-octadecylcarbamoyloxy proxycarbonyl) aminomethyl]-1-ethylpyridium chloride), BN52021 (ginkgolide B, a 20-carbon-cage terpene) and BN52111 (2-heptadecyl-2-methyl-4-[5-pyridinium pentacarbonyloxymethyl]-1,3-dioxolane bromide). These molecules may thus prove useful in the determination of both the in vivo and in vitro effects of PAF.

3.3.3 PAF receptors

Activation of neutrophils with PAF occurs through a G-protein-linked receptor, and the subsequent transmembrane signalling involves the stimulation of inositol phosphate metabolism. Within 30 s of addition of PAF (0.01–100 nM), intracellular Ca^{2+} levels increase and Ca^{2+} transport from the external medium is enhanced. It seems that phospholipase C-dependent and -independent activation pathways are involved in Ca^{2+} mobilisation. This indirectly suggests that two receptors may be involved in PAF activation. The first of these is pertussis-toxin-insensitive and may be linked to a pre-existing Ca^{2+} channel; the second may be linked to a pertussis-toxin-sensitive G-protein, and its occupancy results in PIP_2 turnover and mobilisation of intracellular Ca^{2+} stores.

The PAF receptor of guinea pig lung cells has recently been cloned by Honda and colleagues (1991). Messenger RNA was isolated from these cells, size fractionated and injected into *Xenopus laevis* oocytes. (Often, mRNA injected into these oocytes is translated and assembled into functional proteins.) The expression of the PAF receptor was then determined electrophysiologically by measuring inward Cl^- current in response to PAF. Once the size fraction of mRNA was identified, a cDNA library was constructed and batches of 6×10^4 clones were analysed. DNA from these clones was transcribed in vitro and injected into oocytes prior to screening for PAF expression. The clones were then subdivided and the process repeated, until eventually a single clone (Z74) was obtained. This clone was sequenced and contained a 3020-bp insert, which encoded a predicted 342-amino-acid protein of relative molecular mass 38 982 da. A hydrophobicity plot revealed that this receptor again had seven hydrophobic membrane-spanning domains, characteristic of the rhodopsin superfamily of genes.

3.4 Histamine

3.4.1 Properties

Histamine (5-β-amino-ethylimidazole, Fig. 3.5) is secreted from mast cells and basophils upon stimulation by agonists such as IgE, C3a, C5a, substance P, ATP and IL-1. It is produced by the decarboxylation of histidine by the enzyme L-histidine decarboxylase, which is dependent upon the cofactor pyridoxal phosphate for activity. Histamine is stored in the granules of mast cells and basophils in association with the anionic side chains of proteoglycans such as heparin and chondroitin sulphates. Mast cells are found within connective tissue of most organs (e.g. around blood vessels, nerves, lymphatics) and are abundant in skin and mucosa of the respiratory, gastrointestinal and reproductive tracts. They are also abundant in the lungs (up to 2% of all alveolar cells). Basophils generally remain in the circulation, only rarely entering tissues, and have a relatively short half-life (1–2 weeks).

3.4.2 Effects on neutrophils

Concentrations of 10^{-4}–10^{-6} M histamine enhance neutrophil chemokinesis in response to zymosan-activated serum, casein and fMet-Leu-Phe, whereas

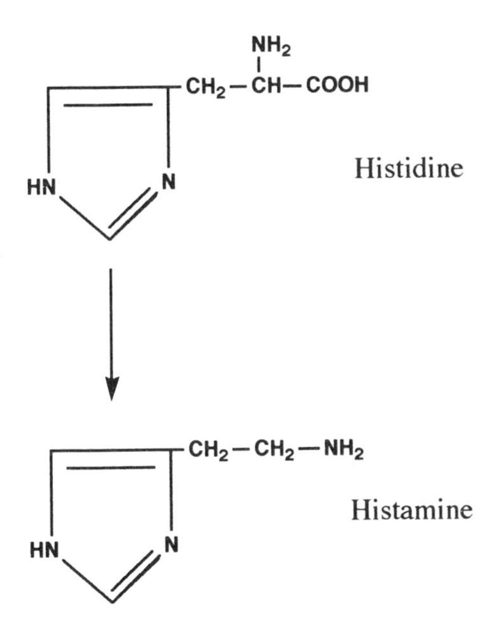

Figure 3.5. Biosynthesis of histamine. Histamine is generated from histidine by the action of L-histidine decarboxylase.

concentrations of 10^{-3}–10^{-6} M inhibit chemotaxis in response to these agonists. These concentrations of histamine alone have no effect on chemokinesis or chemotaxis. The decrease in chemotaxis is associated with increased levels of cAMP, and the inhibitory effect of histamine is prevented by pretreatment of cells with levamisol, a compound that increases cGMP levels. Concentrations of histamine in the range 10^{-2}–10^{-7} M decrease fMet-Leu-Phe-stimulated (but not PMA- nor A23187-induced) degranulation under circumstances that do not affect fMet-Leu-Phe binding or internalisation.

3.4.3 Histamine receptors

There are two distinct types of histamine receptors, H-1 and H-2, with distinct pharmacological properties. There are about 2.5×10^5 H-1 receptors per neutrophil, and these have a single dissociation constant of around 50 nM. Occupancy of these H-1 receptors increases chemokinesis; occupancy of H-2 receptors increases both chemokinesis and intracellular cAMP levels, but decreases chemotaxis and degranulation.

3.5 Cytokines

A large number of cytokines generated during an inflammatory response can affect neutrophil function. Some of these cytokines, such as G-CSF and GM-CSF, can affect the rate of biosynthesis of mature neutrophils in the bone marrow; they can also affect the function of mature cells by priming certain functions (such as the respiratory burst, degranulation and expression of some plasma membrane receptors). These effects are described in detail in Chapters 2 and 7, respectively; the present chapter briefly describes some the properties of cytokines known to affect the function of mature neutrophils.

3.5.1 Interleukin-1 (IL-1)

IL-1 was first discovered in the 1940s when it was found that a substance present in acute exudate fluid could induce a fever when injected into animals. This was shown to be a small (10–20 kDa) protein synthesised de novo either when macrophages were stimulated by endotoxin or during phagocytosis of bacteria. Attempts to characterise this pyrogen were hampered by inadequate purification procedures, and even some apparently-homogenous preparations yielded confusing experimental data. More complete characterisation of the molecular structure and biological effects has come from the molecular cloning of IL-1, which has shown there to be two forms of this cytokine, IL-1α and IL-1β (with relative molecular masses of 17.5 and 17.3 kDa, respectively). These molecules are synthesised as 31-kDa precursor molecules that have only about 25% homology with each other (in the human versions) and possess no detectable signal sequence. Amino-acid-sequence analysis of isolated cDNA clones has reported molecules with predicted NH_2-terminal residues of alanine 117 (for IL-1β) and serine 113 (for IL-1α). IL-1β is secreted by blood monocytes and macrophages; IL-1α is secreted by macrophage cell lines such as P388D.

cDNA sequencing and cloning of the IL-1 genes have led to two major breakthroughs. Firstly, they have helped to elucidate the amino acid composition and sequence of the proteins; secondly, the expression of these cDNA clones in suitable cell lines has provided large quantities of homogenous, recombinant proteins for experimental and clinical use. Monoclonal antibodies have also been raised to these recombinant proteins, and such antibodies may be utilised for a number of purposes: the specific blockage of IL-1 activity, the accurate measurement of IL-1 concentrations in biological samples (in immunoassays) and the measurement of IL-1 production in cells and tissues (in immunohistochemical or immunocytochemical assays).

IL-1 is produced by stimulated monocytes and macrophages, blood-vessel endothelial cells and smooth muscle cells. It has also recently been shown to be generated by human neutrophils (§7.3.4.1), particularly when they have been exposed to other cytokines such as granulocyte–macrophage colony-stimulating factor (GM-CSF). IL-1β is not stored by these cells but is rapidly synthesised de novo by processes requiring transcription and translation. Increases in mRNA levels for IL-1 can be detected within 1 h of exposure of human neutrophils to GM-CSF. Its biological effects include stimulation of synthesis and release of acute-phase proteins by the liver, T- and B-cell stimulation, histamine release from basophils, release of prostaglandin E_2 from fibroblasts and synovial cells, release of neutrophils from bone marrow, increased thromboxane A_2 release from neutrophils, enhanced macrophage cytotoxicity and enhanced degranulation from neutrophils after stimulation with other agonists.

3.5.2 *Interferon-γ*

Interferon-γ was the first secretory product of T cells to be discovered when it was found that supernatants derived from suspensions of T cells that had been treated with mitogenic agents could activate macrophages. This 'macrophage activating factor', subsequently found to interfere with the replication of viruses, was thus named *interferon.* The production of this compound, associated with delayed-type hypersensitivity and cell-mediated immunity, was termed *immune interferon* or *type II interferon.* With the discovery of other lymphokines with interferon-like activity (interferon-α and -β), the compound was finally designated interferon-γ.

Sequence analysis of cDNA clones for interferon-γ predict a mature protein of 17 347 da. Analysis of the native, purified protein on SDS gels (relative molecular masses of 20 and 25 kDa) and by gel filtration (40–70 kDa), indicate that the mature protein is heavily glycosylated. The 25-kDa form is glycosylated at asparagine residues 28 and 100 whereas the 20-kDa form is glycosylated at asparagine 28; both forms have heterogeneous carbohydrate content. There is also some molecular heterogeneity in that there are six possible COOH-termini that may result from variable proteolytic processing. Furthermore, the natural protein does not contain the three NH_2-terminal amino acids (Cys-Tyr-Cys) predicted from cDNA sequence analysis. There are three introns in the gene, which is located on chromosome 6.

Interferon-γ is produced by helper, suppressor and cytotoxic T cells during cell activation, and also by natural killer (NK) cells in response to Interleukin-2 and H_2O_2. In view of the fact that interferon-γ can enhance the ability of phagocytes to produce reactive oxidants such as H_2O_2, this response of NK cells may perform a positive regulatory mechanism to enhance phagocyte function before T cells have time to respond to an immune challenge.

In macrophages, interferon-γ enhances their ability to kill a variety of intracellular pathogens (e.g. *Toxoplasma, Listeria, Leishmania, Legionella* and *Salmonella*), increases the secretion of cytokines (e.g. IL-1, TNF-α), prostanoids and plasminogen activator, and augments the stimulated secretion of reactive oxygen metabolites and reactive nitrogen metabolites. The expression of surface antigens (e.g. Fc receptors, Ia antigens and IL-2 receptor) is increased, whereas expression of the transferrin receptor and C3b receptor is decreased by this cytokine.

In mature neutrophils, interferon-γ induces the expression of FcγR1, increases antibody-dependent cytotoxicity, primes the ability to generate reactive oxidants and selectively stimulates protein biosynthesis. These effects are described in detail in Chapter 7. Additionally, this cytokine has been used clinically for the treatment of chronic granulomatous disease (CGD), which is associated with an increased susceptibility to infections due to an impairment of NADPH oxidase function (§8.2).

Receptors for interferon-γ on the neutrophil have not been characterised in detail. Mature human macrophages possess two types of receptor (K_{d1} = 4.3×10^{-10} M; $K_{d2} = 6.4 \times 10^{-9}$ M) whereas on monocytes there is only one class of receptor (1600 sites per cell) with an affinity of $K_d = 1 \times 10^{-9}$ M.

3.5.3 Colony-stimulating factors (CSFs)

The biosynthesis, molecular properties and receptors for the major colony-stimulating factors, as well as their role in haematopoiesis, is described in Chapter 2. The following section describes the role of G- and GM-CSF in the modulation of the function of mature, circulating neutrophils.

3.5.3.1 Granulocyte-macrophage colony-stimulating factor (GM-CSF)

The properties of GM-CSF and the GM-CSF receptor are given in section 2.2.4. GM-CSF acts on mature neutrophils over the concentration range 1–100 pM. It does not stimulate the generation of oxidant secretion in cells in suspension, but rather 'primes' their ability to generate oxidants when the neutrophils are exposed to a second stimulus: primed neutrophils generate greater levels of oxidants compared to unprimed neutrophils (see §7.2 for details). However, if neutrophils are adhered to surfaces (e.g. plastic surfaces), then GM-CSF on its own can activate a respiratory burst. GM-CSF is initially chemotactic for neutrophils, but after about 30 min exposure to this CSF chemotaxis is inhibited. This dual effect on neutrophil mobility may first attract cells to the site of CSF production and then immobilise them at the site of infection. The mechanisms by which GM-CSF activates neutrophils are only partly explained. GM-CSF receptors are coupled to G-proteins, and the activities of phospholipases A_2 and D are necessary for the up-regulation of receptor expression and the priming of the oxidase.

3.5.3.2 Granulocyte colony-stimulating factor (G-CSF)

Details of the properties of G-CSF and the G-CSF receptor are given in section 2.2.3. G-CSF is a neutrophil-specific cytokine and as such is proving to be more useful than GM-CSF in clinical applications that require a specific enhancement in numbers or function of neutrophils. Many of the effects of GM-CSF on mature neutrophils are also observed upon exposure to G-CSF. G-CSF enhances phagocytosis, primes the respiratory burst and membrane depolarisation upon subsequent stimulation and also augments antibody-dependent cellular cytotoxicity (ADCC) of neutrophils towards tumour cells. By itself, it does not activate a respiratory burst in suspended neutrophils but primes oxidant production in response to subsequent stimulation by fMet-Leu-Phe: in contrast, G-CSF alone stimulates reactive oxidant production in adherent cells. It increases the expression of CD11b (the C3bi receptor) with a concomitant increase in the adhesion of neutrophils to surfaces. It can also up-regulate the affinity of the neutrophil–endothelium 'homing' receptor LAM (leukocyte adhesion molecule)-1 without changing the expression of LAM-1 on the neutrophil surface. These changes in adhesion molecules may account for the rapid, transient neutropenia observed following G-CSF administration, because the neutrophils become associated with the marginating pool.

G-CSF does not induce degranulation of azurophilic granules. After in vivo administration, circulating neutrophils exhibit enhanced levels of alkaline phosphatase due to alterations in secondary-granule formation. In neutrophils of patients with Pelger–Huet anomaly, G-CSF induces changes in their nuclear morphology within 24 h and enhances nuclear segmentation. ADCC is augmented by G-CSF, and protein biosynthesis is selectively up-regulated. It induces the expression and secretion of α-interferon (§7.3.4.5), which inhibits the formation of granulocyte colonies from progenitor cells. This phenomenon may thus represent a feedback mechanism to regulate neutrophil formation at high levels of G-CSF.

3.5.4 Tumour necrosis factor (TNF)

Tumour necrosis factor (TNF) was originally described as a factor produced following exposure of Bacille–Calmette–Guerin-treated animals to bacterial endotoxin. It was so named because it possessed the ability to necrotise tumours. This factor is now named TNF-α to distinguish it from another, related cytokine: lymphotoxin, which is sometimes referred to as TNF-β. Alternative names for TNF-α include cachectin and cytotoxin. Its primary cellular source in the body is the activated macrophage, but some other cell types (e.g. NK cells, astrocytes, some lymphocytes, fibroblasts, many tumour cells, endothelial cells and neutrophils) have also been shown to synthesise this cytokine.

TNF-α was first purified from conditioned medium from HL-60 cells. It has a relative molecular mass of 17 kDa when analysed by SDS-PAGE, but 45 kDa when analysed by gel filtration. Thus, the molecule exists as a nonglycosylated trimer with a pI of 5.3. Each monomer comprises 157 amino acids and contains two cysteine residues that form a disulphide bridge. Trimer formation appears to be due to noncovalent interactions between the monomers. Human TNF-α is synthesised as a 233-amino-acid protein that is proteolytically cleaved during processing. Whilst the 17-kDa form is readily secreted (and hence can function as an extracellular mediator), a 26-kDa transmembrane form has also been identified. This form of TNF-α may thus function in cytotoxicity resulting from cell–cell contact.

cDNA for human TNF-α has been isolated and sequenced, recombinant protein has been synthesised and monoclonal antibodies have been raised to both the purified and recombinant proteins. The gene is located on chromosome 6, close to the genes for the major histocompatability complex (MHC) and lymphotoxin. The gene comprises four exons spread over 3 kb of the genome. The regulation of expression of TNF-α can occur at several steps. Unstimulated macrophages possess mRNA for this cytokine as well as unprocessed protein. Within minutes of stimulation (e.g. by LPS), this precursor protein is processed into biologically-active molecules. Secretion of active protein is inhibited by PGE_2, IL-4, IL-6 and dexamethasone. TNF-α has also been shown to be expressed by human neutrophils (§7.3.4.2).

TNF-α elicits a wide range of responses in cells and tissues. Apart from causing the lysis of certain tumours (by mechanisms probably related to the ability of the target to induce the synthesis of the mitochondrial manganese-dependent superoxide dismutase, Mn-SOD), it can also kill normal cells. These effects on normal cells are more apparent when biosynthesis is blocked – for example, by the addition of inhibitors of macromolecular biosynthesis. Many cells, including some tumour cells, are resistant to the effects of TNF-α.

TNF-α is chemotactic for monocytes and neutrophils. Its effects on neutrophils are numerous: for example, it can prime degranulation and reactive oxidant production, enhance phagocytosis and ADCC and up-regulate the expression of some surface receptors, such as CR3. Whilst low concentrations of TNF-α are required to prime the cells subsequent to stimulation by other agonists, such as fMet-Leu-Phe, higher concentrations of TNF-α alone can activate low levels of oxidant production. This activity is even more pronounced if the neutrophils are adhered to surfaces.

Two types of receptors for TNF-α have been identified. The type A receptor has a relative molecular mass of 75 kDa and binds the ligand with high affinity. The type B receptor is 55 kDa and has an affinity for TNF-α that is between five- and sevenfold lower than that of the type A receptor. The level of expression of the type A receptor can be regulated independently of changes in expression of the type B receptor.

3.5.5 Interleukin 8 (IL-8)

Interleukin-8 (IL-8), or neutrophil attractant protein-1 (NAP-1), is a small (8-kDa) protein that was originally identified in culture supernatants of LPS-stimulated monocytes or macrophages. It is an extremely potent chemoattractant for neutrophils. It has been purified from monocyte culture supernatants and been cloned, and its amino acid sequence has been predicted from DNA sequencing and confirmed by amino acid sequencing. It is also produced by epithelial and endothelial cells, lymphocytes, fibroblasts and keratinocytes. It has recently been shown to be expressed by human neutrophils after they have been exposed to priming agents such as GM-CSF (§7.3.4.4). In addition to LPS, inflammatory signals such as zymosan, IL-1 and TNF-α can induce its production by certain cells types. It is now appreciated that IL-8 is just one member of a family of related proteins that play important roles in inflammatory processes. Its importance in neutrophil function in vivo is confirmed by its presence in inflammatory sites such as psoriatic scales, synovial fluid of patients with rheumatoid arthritis and the bronchial lavage of patients with cystic fibrosis.

Apart from being a potent activator of neutrophil chemotaxis, IL-8 can also induce degranulation and reactive oxygen metabolite production in human neutrophils. These activities are usually very low or undetectable but are augmented when neutrophil suspensions are pretreated with cytochalasin B. Addition of IL-8 also induces elevations in intracellular free Ca^{2+}.

The IL-8 receptor is expressed on neutrophils, and because its effects are blocked by pertussis toxin, this receptor is coupled to a G-protein. Recently, Thomas and colleagues (Thomas, Pyun & Navarro, 1990; Thomas, Taylor & Navarro, 1991) screened a rabbit neutrophil cDNA library with an antisense oligonucleotide probe deduced from the coding region of the second transmembrane domain of rhodopsin-like G-protein-coupled receptors. They isolated a clone, F3R, which they originally believed to encode the fMet-Leu-Phe receptor; however, this clone showed little sequence homology to another putative fMet-Leu-Phe receptor clone isolated by Boulay and colleagues (1990a,b, 1991; see §3.6.3). More recently, the F3R clone has been shown actually to encode the IL-8 receptor. Sequence analysis thus reveals that this receptor also belongs to the superfamily of receptor molecules possessing seven membrane-spanning hydrophobic domains.

3.6 fMet-Leu-Phe

3.6.1 Properties

It has long been known that peptides of bacterial origin, such as *N*-formylated oligopeptides, are potent activators of neutrophils. Bacterial protein biosynthesis is initiated by the codon AUG, which codes for polypeptide chains at the NH_2 terminus to start with *N*-formylmethionine. However, very few mature bacterial proteins actually have this amino acid at the NH_2 terminus because *N*-formylmethionine is cleaved off by proteolytic processing. Sometimes just this amino acid is cleaved, but often several adjacent residues are also removed with it. These observations formed the basis for the chemical synthesis of a variety of *N*-formylated oligopeptides and an assessment of their ability to activate neutrophils in vitro. The most potent of these formylated peptides is *N*-formylmethionyl-leucyl-phenylalanine (fMet-Leu-Phe).

3.6.2 Effects on neutrophil function

The formylated peptide fMet-Leu-Phe is probably the most commonly-used activator of neutrophils in vitro. It is used as a model agonist to study receptor-mediated processes, generating intracellular signalling molecules that then activate cell functions. This compound can, depending upon the concentration used, activate many varied functions, such as chemotaxis, aggregation, reactive oxidant production, cytoskeletal changes and (particularly in combination with cytochalasin B) degranulation.

3.6.3 fMet-Leu-Phe receptor

The formylated peptide fMet-Leu-Phe binds rapidly to specific receptors on the plasma membrane of neutrophils. Because this receptor also binds analogues of fMet-Leu-Phe, such as *N*-formyl-Nle-Leu-Phe-Leu-[^{125}I]iodo-Tyr-Lys (where Nle is norleucine) and [^{3}H]-labelled fMet-Leu-Phe, compounds such as these have thus been used to characterise it. Such experiments have shown that the receptor is a glycoprotein of relative molecular mass 55–70 kDa and with two isoforms having pI values of 5.8 and 6.2. It possesses two oligosaccharide chains, which may be (sequentially) removed to yield proteins of 40–50 kDa and 33 kDa by treatment with pronase or Endo-*b*-*N*-acetylglucosaminidase F (Endo F). The deglycosylated receptor binds fMet-Leu-Phe as efficiently as the native receptor, but still exists as two isoforms distinguished by their pI values. The two oligosaccharide chains are *N*-linked on the distal 1–3-kDa portion of the receptor. The receptor is also present on monocytes and differentiated HL-60 cells, and the nonglycosylated form appears identical to that present on neutrophils.

Unstimulated neutrophils are spherical. Upon addition of fMet-Leu-Phe, neutrophils become polarised even in the absence of a gradient of chemoattractant. This observation suggests that the ability to polarise may be an intrinsic response of the cells to orientate themselves prior to movement in a particular direction. This may arise from the cell having an asymmetrical distribution of fMet-Leu-Phe receptors on the plasma membrane or else from an ability to redistribute these receptors rapidly upon stimulation. The polarised cell is characterised by a ruffled region at the front, a midregion containing the nucleus and a trailing tail possessing retraction fibres as the cell moves along a surface. The front end of the cell is more sensitive to stimulation than the rear, possibly because of an asymmetrical distribution of receptors. Bound receptors are rapidly capped to the tail and are internalised by endocytosis. In order for the neutrophil to be able to sustain chemotaxis along a gradient by sensing changes in concentrations of chemoattractant, there must be a continuous re-expression of receptors at the front of the cell. This may occur via recycling of receptors or else via the mobilisation of new receptors from internal pools.

Experiments using fluorescein-labelled fMet-Leu-Phe indicate that the majority of re-expressed receptors that appear on the plasma membrane within 4–10 min after stimulation arise from the mobilisation of internal pools. Subcellular fractionation studies indicate that the pools of these receptors are the membranes of specific granules, although it is possible that these are on other membranes (e.g. on gelatinase-containing granules or secretory vesicles). In unstimulated cells there are about 15 000 receptors per cell, and this number can be rapidly up-regulated to 30 000–50 000 per cell via the mobilisation of these subcellular stores, at rates of up to 10 000/min via a Ca^{2+}-dependent process. Recycling of internalised, ligand-bound receptors to the plasma membrane may also occur, but only more slowly. Up-regulation of fMet-Leu-Phe receptors can also occur during activation by agents such as PMA or during priming by cytokines. The internal pool of receptors has a low affinity for ligand (5–10 nM), but the affinity of the receptor can be up-regulated (0.5–1 nM) perhaps via interaction with G-proteins or else by covalent modifications such as phosphorylation.

The involvement of G-proteins with the fMet-Leu-Phe receptor has been appreciated for some time. For example, non-hydrolysable analogues of GTP have been found to increase the affinity of binding of fMet-Leu-Phe to isolated membranes, and the addition of fMet-Leu-Phe to neutrophil homogenates stimulates GTPase activity. The G-protein coupled to the receptor is of the G_i type (§6.2.1), as defined by the ability of pertussis toxin to block fMet-Leu-Phe stimulated neutrophil activation. This G-protein activates phospholipases C (to generate inositol phosphates and diacylglycerol), A_2 (to generate arachidonic acid) and D (to release phosphatidic acid and diacylglycerols). Changes in intracellular Ca^{2+} levels are required to mediate some fMet-Leu-Phe responses, and these may arise from mobilisation of intracellular pools and also from stimulated Ca^{2+} influx from the outside of the cell. For example, in cells in which intracellular Ca^{2+} transients are blocked by loading with high concentrations of Quin-2, chemotaxis and secretion of reactive oxygen metabolites are inhibited. Further details of the signal transduction pathways activated during fMet-Leu-Phe stimulation are given in Chapter 6.

Because the fMet-Leu-Phe receptor is present only at low levels in neutrophils ($\sim 12 \times 10^{-15}$ g of receptor per cell), it has proved difficult to purify and characterise. Researchers have therefore turned to molecular cloning techniques to gain insight into the molecular structure of this receptor. This approach itself has not been easy because, in the absence of an antibody that specifically binds to the receptor, or else without some amino acid sequence data that can be used to synthesise oligonucleotide probes, cDNA libraries cannot be screened to isolate relevant clones. Therefore, experimental systems in which functional fMet-Leu-Phe receptors are expressed on the surfaces of transfected cells have been used. Two main systems have been utilised: expression of mRNA injected into *Xenopus laevis* oocytes and cDNA cloning into the COS-cell expression vector.

Cells of the line HL-60 can be induced to differentiate into more mature cells, and as they do so they begin to express the fMet-Leu-Phe receptor.

Messenger RNA isolated from differentiated HL-60 cells is thus a convenient source of transcripts for the receptor, and this mRNA (50–100 ng) can be injected into *X. laevis* oocytes. These oocytes actively translate foreign mRNA molecules, and often the expressed protein is functional: 2–5 days later, functional fMet-Leu-Phe receptors can be detected on the cell surface. The identity of a functional fMet-Leu-Phe receptor was confirmed because:

i. addition of fMet-Leu-Phe resulted in an increase in intracellular Ca^{2+} levels in the injected oocytes, and this increase was prevented by the fMet-Leu-Phe receptor inhibitor, *t*-butoxycarbonyl-Met-Leu-Phe;
ii. fMet-Leu-Phe addition could cause changes in the membrane potential of the oocytes;
iii. fMet-Leu-Phe addition to injected oocytes stimulated Ca^{2+} efflux.

When the mRNA was size-fractionated on a sucrose gradient prior to injection into oocytes, it was found that the size of the fMet-Leu-Phe receptor message was 1.5–2.0 kb.

A more specific approach to clone the fMet-Leu-Phe receptor was reported by Boulay et al. (1990a,b, 1991), who constructed a cDNA library from differentiated HL-60 cells in the COS expression cell line CDM8. This library was screened for its ability to bind the hydrophilic derivative *N*-formyl-Met-Leu-Phe-Lys, which was labelled with ^{125}I. A clone (fMLP-R98) was isolated of 1.9 kb that contained a 1050-bp open reading frame encoding 350 amino acids. The predicted protein had two potential glycosylation sites (Asn-Ser-Ser and Asn-Ile-Ser) in agreement with the observations of the glycosylated protein in neutrophil membranes. A hydrophobicity plot revealed seven hydrophobic domains – a pattern seen in rhodopsin-type receptors, which predicts that the receptor is arranged in the plasma membrane with seven transmembrane helices (see Fig. 3.2). COS cells transformed with fMLP-R98 possessed both high- (0.5–1.0 nM) and low- (5–10 nM) affinity binding sites.

An alternative approach to clone the receptor was taken by Thomas and colleagues (1990, 1991). They screened a cDNA library from peritoneal rabbit cells with an antisense oligonucleotide probe deduced from the coding region of the second transmembrane domain, which is common to G-protein-coupled receptors (fMet-Leu-Phe is a G-protein-linked receptor). Screening of this library yielded a putative receptor clone F3R, characterised as follows:

i. DNA sequencing and prediction of its amino acid sequence indicated that it belonged to the G-protein-linked receptor family;
ii. mRNA hybridising to this clone was only detected in neutrophils (and not uterine smooth muscle, skeletal muscle, lung, liver or brain);

iii. in vitro translation of the message generated a protein of the same size (30–32 kDa) as the non-glycosylated receptor;
iv. *X. laevis* oocytes injected with F3R message bound [^{125}I]-labelled fNle-Leu-Phe-Nle-Tyr-Lys (where Nle is norleucine) and generated Ca^{2+} transients in response to fMet-Leu-Phe.

This clone, however, had only 56% homology at the amino acid level to the clone described by Boulay et al., and it has since been shown actually to encode the IL-8 receptor, which is also a G-protein-linked receptor with seven transmembrane loops (see Fig. 3.2).

3.7 Purinergic receptors

For many years, it has been recognised that incubation of platelets with neutrophils enhances some fMet-leu-Phe-stimulated functions, such as aggregation and the respiratory burst. The platelet-derived factor responsible for this effect does not stimulate neutrophils in the absence of external stimuli (e.g. fMet-Leu-Phe), and the effects may be mimicked by the addition of ATP or ADP to neutrophil suspensions.

These effects of ATP are blocked by pertussis toxin, and so the putative ATP receptor is G-protein linked. ATP addition results in phospholipase C activation, which may be detected as increased inositol phosphate metabolism and subsequent elevations in cytosolic free Ca^{2+}. Purinergic receptors on many types of cells are classified as type P_1 or P_2. Neutrophils possess P_2-type receptors, which are activated by ATP and ADP, and also P_1-type receptors, which are activated by adenosine. Occupancy of P_2-type receptors enhances fMet-Leu-Phe-mediated effects, whilst occupancy of P_1-type receptors has the opposing effect. Some pharmacological evidence suggests that the P_2-type receptor on neutrophils is distinct from the P_{2x} and P_{2y} subtypes that have been described in other cell types.

3.8 Adhesion molecules involved in inflammation

3.8.1 *Properties*

The recruitment of neutrophils from the bloodstream into infected tissues depends upon them being able to respond to chemoattractants generated at the site of infection, but also requires physical interactions with the endothelial cells that line the capillary walls. Firstly, neutrophils, which may be moving at speeds of up to 1000–2000 μm/s in the circulation, must attach to the endothelial walls and then squeeze through the spaces between these

Table 3.1. *General features of the major selectins*

Name	Old name	Occurrence
L-selectin	LECAM-1, LAM-1	Neutrophils, lymphocytes, many leukocytes
E-selectin	ELAM-1	Endothelium
P-selectin	gmp140, CD62	Endothelium and platelets

cells to begin their migration up the chemoattractant concentration gradient towards the infective locus. The first feat to accomplish is thus to leave the fast-moving circulation and attach to the endothelial cell lining. Secondly, the attached neutrophils must then begin the process of *diapedesis,* which involves changes in cell shape and deformability via modifications in the cytoskeletal network. These two processes require two different sets of adhesion receptors on the neutrophil and endothelium, and the expression of these during inflammatory activation is differentially regulated by cytokines.

There are three groups of adhesion molecules responsible for these processes: the selectins, the integrins and the immunoglobulin (Ig) superfamily.

3.8.2 Selectins

There are three types of selectins responsible for the attachment of leukocytes to the endothelium, as shown in Table 3.1. Selectins possess three types of domains: a C-type lectin domain, an epidermal growth-factor-like domain and between two and nine regulatory domains.

On the neutrophil, the major selectin expressed is *L-selectin.* This molecule is constitutively expressed on mature neutrophils but may be expressed at low levels (50% of adult) in neonates. Stimulation of endothelial cells with thrombin, histamine, IL-1 and some other agents induces neutrophils (and other leukocytes) to leave the circulation and adhere to the endothelium. They do this by 'rolling' onto the surface of the endothelium, to which they attach via *P-selectin* translocated from storage sites in Weibel–Palade bodies to the surface of the endothelium upon activation. The expression of P-selectin is short-lived and is replaced on the endothelial surface by *E-selectin* (whose expression is also regulated by some cytokines), which continues the endothelial–leukocyte interaction.

Rolling involves sequential attachment and detachment of the leukocyte with the endothelium. The selectin receptors bind very quickly and tether via recognition of their carbohydrate moieties with the ends of flexible pro-

Table 3.2. *General features of the major integrins*

Name	Old name	Occurrence
CD11a/CD18	LFA-1	All human leukocytes
CD11b/CD18	Mac-1, CR3	Neutrophils, monocytes, NK cells
CD11c/CD18	p150,95	Monocytes, macrophages, neutrophils
CD49d/CD29	VLA-4	Monocytes, eosinophils, lymphocytes

Note: LFA-1 = lymphocyte function-associated molecule, VLA-4 = very late antigen 4; see text for details.

tein counter-structures. Attachment is probably via capping of leukocyte selectins at the tips of uropods, so that only small regions of the leukocyte plasma membrane are actually in contact with the endothelial surface.

3.8.3 Integrins

After the initial attachment of neutrophils to the endothelium via the selectins, the next stage in the process is performed via the integrins. L-selectin on the neutrophil surface is shed (the shed molecule being some 8–10 kDa smaller than the bound molecule) via a process that probably involves proteolytic cleavage. The major groups of integrins present on leukocytes are shown in Table 3.2; further properties of these molecules are given later.

Attachment of neutrophils to the endothelium via the integrins results in a much tighter cell–cell interaction than that achieved by the selectins. Expression of integrins on resting neutrophils is low and thought to be independent of selectin–ligand binding, but rather controlled by the secretion of substances such as PAF by the endothelial cells. Up-regulation of integrin expression on neutrophils occurs via the translocation of preformed receptors from the membranes of specific granules or other subcellular structures (e.g. secretory vesicles and gelatinase-containing granules) to the plasma membrane; it is stimulated by exposure to cytokines or other neutrophil-activating factors (see §7.2.1). The major neutrophil integrins are CR3 (CD11b/CD18) and LFA-1 (CD11a/CD18). Ligand–integrin binding appears to be dependent upon cations: Ca^{2+} binds to inactive integrins and must be replaced by Mg^{2+} for binding to occur. Expression is transient and may be localised to the leading edges of the cell, with clustering being dependent upon changes in the cytoskeletal network. Anti-CD11b/anti-CD11a antibodies block migration of neutrophils through the endothelium without affecting their initial adhesion, thus supporting the role of the selectins in initial attachment but of integrins in the later stages of the attachment process, which leads to diapedesis.

Table 3.3. *General features of the immunoglobulin superfamily of adhesion molecules*

Name	Other name	Function
ICAM-1	adhesion mol-1	binds CD11a/CD18, CD11b/CD18
ICAM-2		binds CD11a/CD18
VCAM-1	vascular cell adhesion molecule	binds VLA-4

3.8.4 Immunoglobulin superfamily

These molecules interact with the leukocyte integrins and thus are important during the later stages of the inflammatory response, which are independent of selectin function. The major molecules present on endothelial cells are shown in Table 3.3.

ICAM-1 and -2 are constitutively expressed on endothelial cells; ICAM-1 may be further up-regulated by exposure to cytokines. ICAM-3 has recently been described (its identity based on the unique specificity of a monoclonal antibody) and is a 124-kDa glycoprotein present on the surfaces of T cells, monocytes and neutrophils; its expression may be up-regulated by stimulants such as mitogens. VCAM-1, which is expressed on the luminal surface of cytokine-exposed endothelial cells, binds T cells via VLA-4. It is also expressed on the surface of some leukaemic cell lines, on rheumatoid synovial cells and on some tumours.On the other hand, ICAM-1 is found on all endothelial surfaces, and its interaction with neutrophil integrins is the major mechanism that results in the stimulation of transendothelial migration.

3.9 Neutrophil integrins

3.9.1 General properties

The integrins comprise a family of cell-surface proteins that are involved in adhesion, a process vital for many processes, such as anchorage, migration, growth and differentiation. Cells may adhere to other cells (cell–cell adhesion) or may interact with soluble molecules that constitute the extracellular matrix (cell–extracellular matrix). The integrins are linked to elements of the cytoskeleton, and so they provide a bridge between the external cellular environment and intracellular activation processes.

The structure of the integrin family has been deduced largely from the cDNA sequences of the cloned genes. All integrins comprise α- and β-subunits, and the ligand binding that occurs at the binding site comprises an interaction between the two subunits. The α-subunits can combine with different β-subunits to form integrin molecules with different ligand-binding properties. To date 11 distinct α-subunits and 6 distinct β-subunits have been identified, and different combinations of these result in the formation of 16 different integrin molecules with unique properties. No doubt different integrin molecules will soon be identified based upon different combinations of these α- and β-subunits. Some integrins are cell specific: for example, gp IIb/IIIa is expressed exclusively on megakaryocytes; LFA-1 (CD11a/CD18; $\alpha_L\beta_2$), Mac-1 (CD11b/CD18, $\alpha_M\beta_2$) and p150,95 (CD11c/CD18, $\alpha_X\beta_2$) are found only on leukocytes; $\alpha_6\beta_4$ is specific to epithelial cells and tumour cells derived from them. The leukocyte integrins are thus an example of three separate molecules having different α-subunits (α_L, α_M or α_X), but a common (β_2) β-subunit (Fig. 3.6).

3.9.2 The CD11/CD18 adhesion molecules

CR3, LFA-1 and p150,95 are a family of leukocyte proteins with distinct α-subunits but a common β-subunit (Fig. 3.6). They are involved in cell–cell contact as well as in cell–substrate interactions. They thus function in a variety of adhesive processes.

i. *CR3 (Mac-1, CD11b/CD18).* This is involved in the binding of C3bi-coated particles to phagocytes, in neutrophil aggregation and in the binding of neutrophils to endothelial cells. The receptor also binds fibrinogen, and its activation results in the stimulation of phagocytosis, the respiratory burst, granule enzyme release, adhesion and spreading onto surfaces such as glass, plastic, endothelial cells and epithelial cells. Receptor occupancy can also activate chemotaxis and aggregation. This is the major adhesion glycoprotein on neutrophils.
ii. *LFA-1 (CD11a/CD18).* This receptor is involved in the adhesion between cytotoxic T cells and their targets before activation of cytotoxicity (i.e. it has been shown that this activation is prevented by anti-CD11a antibodies). It is involved in antibody-independent cell lysis by NK cells and in ADCC by monocytes and neutrophils. It also functions in lymphocyte–endothelium interactions and is possibly involved in neutrophil chemotaxis. Anti-CD11a antibodies also prevent T-lymphocyte blastogenesis in response to alloantigen or lectin stimulation. ICAM-1 binds to LFA-1 and is the receptor for rhinoviruses such as the common cold; ICAM-1 fragments have been shown to prevent the entry of such viruses into tissues.
iii. *p150,95 (CD11c/CD18).* This receptor is thought to promote binding of neutrophils and macrophages to substrates such as endothelial cells; it may also be involved in C3bi-binding.

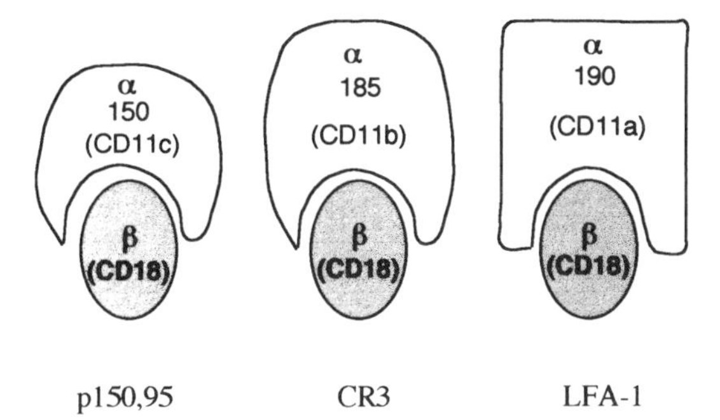

Figure 3.6. Structure of leukocyte integrins. These molecules possess a common β-subunit (CD18) but distinct α-subunits.

Many integrins bind extracellular matrix components such as fibronectin, fibrinogen, laminin, collagen, entactin, tenascin, thrombospondin, von Willebrand factor and vitronectin. Others, however, bind to membrane proteins on other cells (Fig. 3.7). For example, LFA-1 binds ICAM-1 and ICAM-2 (members of the Ig superfamily) and so mediates cell–cell contact; the $\alpha_4\beta_1$ integrin (VLA-4), however, can either bind VCAM-1 and thereby mediate cell–cell contact, or bind fibronectin and so mediate cell–extracellular contact. During platelet aggregation, platelets bind to each other via the soluble factors, fibrinogen and von Willebrand factor. The integrin gp IIb/IIIa also binds fibronectin and vitronectin, and this interaction may assist in the binding of activated platelets to the subendothelial matrix.

Much progress has been made in recent years in the identification of the recognition site for integrin binding. On the extracellular matrix proteins such as fibronectin, this is the peptide sequence Arg-Gly-Asp (RGD). However, the conformation of this RGD site determines the specificity with which the extracellular matrix molecule binds to different integrins. This conclusion has been reached by experiments in which short synthetic peptides carrying this RGD motif have been designed and in which the conformation of this motif has been varied by altering the nature of the amino acids flanking it. Thus, by designing synthetic peptides that form cyclic structures to modify the conformation at the RGD site, the specificity of different peptides to bind different integrins can be determined in vitro. Such peptides may thus have therapeutic applications. For example, cyclic peptides containing the RGD motif can bind gp IIb/IIIa and thus may function as inhibitors of platelet aggregation. Furthermore, there is some evidence from experiments with experimental tumours that synthetic peptides containing RGD sequences may interfere with the adhesion of tumour cells and thus prevent their anchorage.

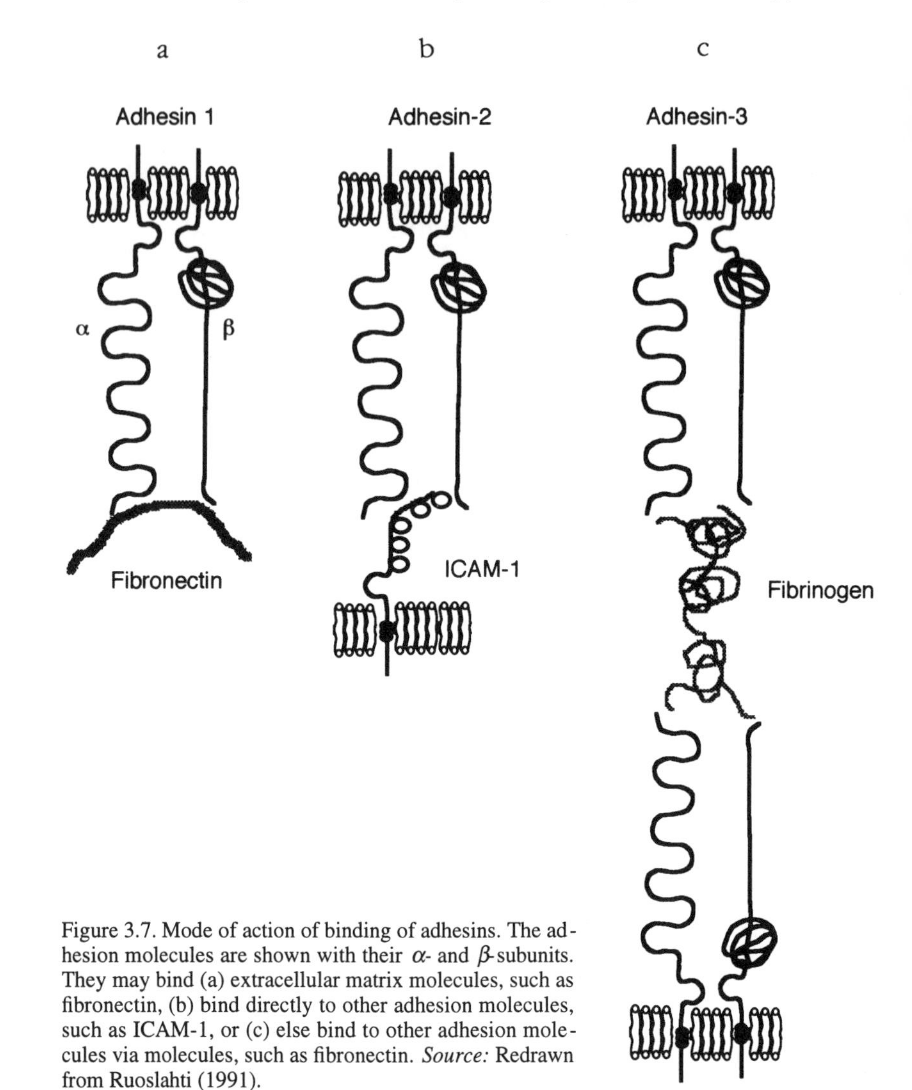

Figure 3.7. Mode of action of binding of adhesins. The adhesion molecules are shown with their α- and β-subunits. They may bind (a) extracellular matrix molecules, such as fibronectin, (b) bind directly to other adhesion molecules, such as ICAM-1, or (c) else bind to other adhesion molecules via molecules, such as fibronectin. *Source:* Redrawn from Ruoslahti (1991).

3.9.3 Neutrophil adhesins

3.9.3.1 Complement receptors

During complement activation, several components are generated that serve as ligands for leukocytes (see §§1.3.2.1–2) and that bind to specific receptors on the plasma membrane. Those complement components recognised by neutrophils may be soluble (e.g. C3a, C5a) and initiate chemotaxis and

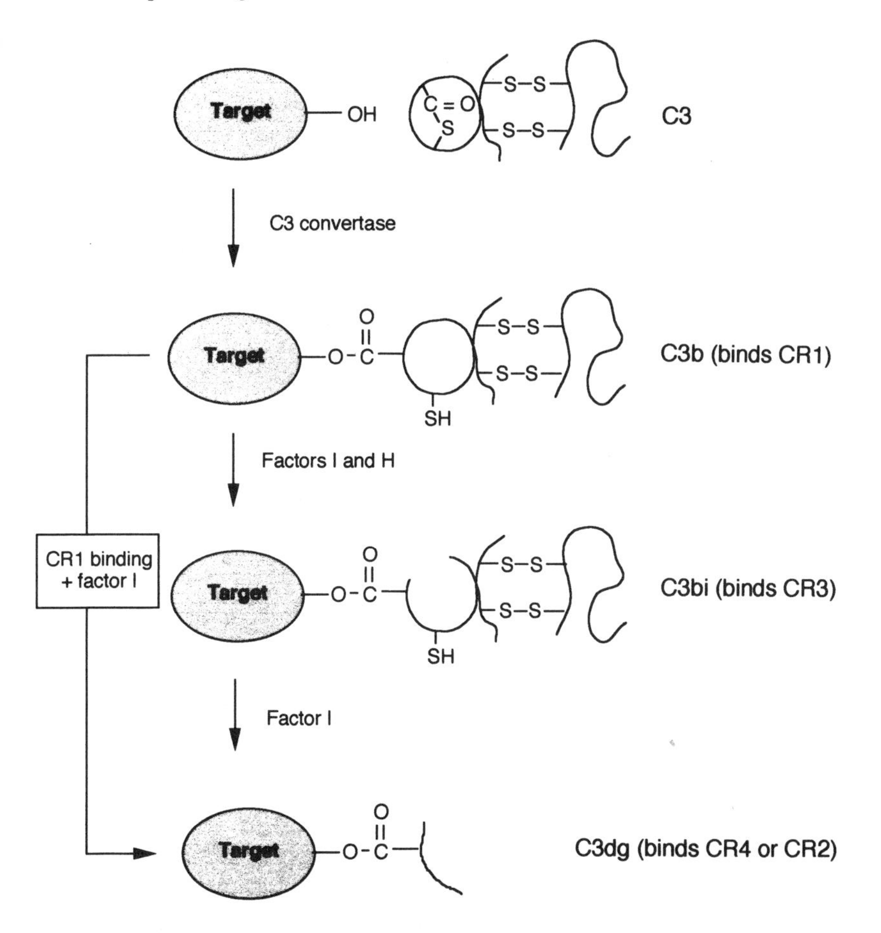

Figure 3.8. Conversions of C3. Binding of C3 to the target causes a rearrangement in an internal thioester bond, yielding a sulphydryl group on C3 and an ester link between C3 and a target hydroxyl group. The covalently bound C3 is now termed C3b. Factors I and H may then cleave the α-subunit of C3b to generate C3bi. If C3bi-coated particles are not immediately phagocytosed, then further activity of Factor I yields C3dg.

secretion, or else they may serve as opsonins (e.g. surface-bound fragments of C3 such as C3b, C3bi and C3dg) for adhesion or phagocytosis. This section describes the neutrophil receptors that recognise C3 fragments bound to surfaces.

Upon activation of complement (by either the classical or alternative pathways), C3 binds covalently to the target, and an internal thioester bond in C3 is rearranged to yield a free sulphydryl group on C3 and an ester link between C3 and a hydroxyl group on the target (Fig. 3.8). The resulting covalently bound C3 molecule is termed *C3b* and is recognised by the receptor CR1 (i.e. complement receptor type 1). However, if C3b is deposited on the target surface following activation of the classical pathway, then it is relatively unstable and within minutes its α-chain is cleaved by factor H and factor I (C3b inactivator). This cleavage generates *C3bi,* which is relatively stable and is recognised by the receptor CR3 (also known as Mac-1). Alternatively, if C3b is deposited via activation of the alternative pathway, then some of it is more stable to factor I activity. Thus, some foreign particles will be coated with C3b if they activate the alternative pathway, whereas some will be coated with C3bi if they activate the classical pathway. Particles coated primarily with C3b will thus be recognised by CR1 whilst those coated with C3bi will be recognised by CR3. If such C3b- or C3bi-coated particles are not immediately phagocytosed, then the further activity of factor I converts C3 to *C3dg.* Whilst neutrophils are reported to possess C3dg receptors (CR4), the role of these receptors in initiating phagocytosis is uncertain: C3dg may also be recognised by CR2 on B lymphocytes. Complement fragment C4b is also reported to be recognised by CR1.

3.9.3.2 Properties of complement receptors

(i) *CR1.* This is a glycosylated protein of which four allelic forms have been identified in humans, with relative molecular masses of 160, 190, 220 and 250 kDa. Individuals may be *homozygous* (and so express only one form of CR1) or else *heterozygous* and express two forms of the receptor. CR1 binds C3b with an affinity of 5×10^7 M^{-1}, but native C3 is not recognised and C3bi binds only very weakly. Although the affinity of CR1 for C3b is not very high, during complement activation many hundreds or thousands of molecules of C3b will be bound to the target; hence, in practical terms, many of the CR1 receptors will be bound to ligand.

The gene for CR1 has been cloned and partly characterised. The predicted structure of the CR1 protein comprises a single 25-amino-acid domain that spans the membrane and a short (43-amino-acid) COOH-terminal cytoplasmic domain. The extracellular, ligand-binding domain is comprised of many (>30) repeated units arranged in tandem, with each of these single consensus repeats (SCR) of 60–70 amino acids. Whilst these SCRs have variable length and sequence, 29 of the amino acids in these are conserved, and the position of four cysteine residues is invariant. These cysteine residues within a SCR bind each other rather than form disulphide bonds with cysteines in other SCRs. An intriguing observation is that many other proteins, such as factor H, C4-binding protein, CR2, clotting factor XIIIb and the IL-2 receptor, also contain 12–20 copies of an SCR, but the functional significance of this is unknown.

The extracellular domain also possesses at least four tandem repeats of 450 amino acids, each comprised of seven SCRs. These structures are 70–99% homologous with each other and form the basis for the allelic differences in CR1 structure. For example, it has been proposed that the 160-kDa form possesses four tandem repeats, the 190-kDa form five, the 220-kDa form six and the 250-kDa form seven of these repeats.

Whilst CR1 is found on the surfaces all phagocytes, it is also present on the membranes of erythrocytes and also on some lymphocytes. Binding of CR1 to C3b accelerates its cleavage, but the product is C3dg rather than C3bi. Thus, binding of C3b to non-phagocytic cells bearing CR1 may prevent the formation of C3bi (via factor H activity) and so prevent an acceleration of the alternative pathway of complement activation, which could ultimately damage host cells.

(ii) *CR3 (Mac-1).* This receptor recognises C3bi but neither C3 nor C3dg. It binds monomeric C3bi with only a very low affinity, but binds avidly to C3bi-coated particles that contain 10^3–10^5 molecules of C3bi. Unlike CR1 and Fc receptors, CR3 binding with its ligand requires high concentrations (0.5 mM) of Ca^{2+} and Mg^{2+}, and binding is strictly temperature dependent. Studies with synthetic peptides have shown that CR3 recognises a 21-amino-acid stretch between residues 1383 and 1403 of C3. This region contains the triplet RGD, which also serves as the recognition sequence for many adhesins (see §3.9.2).

CR3 comprises two polypeptides, an α-subunit of 185 kDa and a β-subunit of 95 kDa, which are covalently linked into an $\alpha_1\beta_1$ structure. Both chains are exposed at the cell surface, whilst the α-chain forms the major portion of the ligand-binding site. The β-subunit is common to the other adhesins, LFA-1 and p150,95. These have an antigenically-distinct α-chain but a common β-chain (see Fig. 3.6). Of these three receptors with common β-chains, only CR3 binds significant amounts of C3bi.

Whilst CR3 avidly binds C3bi-coated particles, it also binds zymosan, *Leishmania, E. coli* and *Staphylococcus* that have not been exposed to complement. Such observations suggest that microbes may be able to bind CR3 (and possibly LFA-1 and p150,95) in the absence of complement via sugar phosphates on their cell surfaces. CR3 may also bind lipopolysaccharide at the lipid region, which comprises fatty acylated diglucosamine biphosphate. Because the fatty acids of LPS are not exposed at the bacterial surface (as they are buried in the outer membrane), the diglucosamine phosphate is probably the ligand that binds CR3.

CR3 is present on neutrophils, monocytes and macrophages. Macrophages possess more CR3 than do monocytes, and macrophages from all sources so far examined (e.g. spleen, peritoneum, lung) express this receptor. Neutrophils express low amounts on the plasma membrane, but expression can be rapidly up-regulated via the mobilisation of preformed pools present on the membranes of cytoplasmic granules (specific or gelatinase-containing granules, and possibly secretory vesicles) during activation and priming (see §7.2.1).

3.9.3.3 Regulation of CR1 and CR3 function

The role of CR3 in neutrophil function can be identified in patients who have a defect in either C3 or in CR3 because such patients have a greatly increased susceptibility to infections (see §8.7). Similarly, experiments that block monoclonal antibodies to these receptors, or in patients with CR3 deficiency, show that this receptor plays an important role in phagocytosis. The extracellular domain binds the ligand whilst the cytoplasmic domain activates the intracellular signalling systems that initiate the phagocytic processes. Blood neutrophils have a low capacity to bind C3bi-coated particles, but this ability is rapidly increased (within 15 min) after addition of activating agents such as fMet-Leu-Phe or PMA, or else priming agents such as GM-CSF. In unstimulated blood cells there are about 10 000–20 000 molecules of CR3 per cell, and the up-regulation due to mobilisation of preformed receptors from the granules is independent of de novo protein biosynthesis. Pretreatment of neutrophils with agents that up-regulate CR3 can result in a 2–10-fold increase in binding to C3bi-coated particles.

There is, however, a wide variation in the reported increase in the expression or binding properties of these receptors after activation or priming, because many procedures used to purify neutrophils from blood also cause such an up-regulation in receptor expression (see Fig. 7.7). However, this increase in the number of receptors appearing on the plasma membrane does not, in itself, account for the enhanced ability of activated or primed neutrophils to bind more avidly and then phagocytose C3bi-coated particles. Thus, it is believed that in resting neutrophils the expression of CR3 is low, but the expressed receptors are not fully functional. Priming or activation of the cells then increases the activity of CR3 at two stages: (i) receptor number on the plasma membrane is up-regulated via the mobilisation of preformed subcellular pools and (ii) receptor function is enhanced, allowing them to bind more avidly to the ligand.

Similarly, activation of neutrophils with PMA can also increase CR1 expression by a factor of 2–3, whereas CR1-dependent phagocytosis is enhanced 20–30-fold. This, again, may be due to activation of receptor function as well as to increasing receptor number on the cell surface. In addition to altering the ability of CR1 and CR3 to bind their respective ligands, priming or activation may also alter the ways in which the cytoplasmic portions of these receptors interact with their intracellular signalling pathways.

Deactivation of CR1 and CR3 can also occur and, again, does not appear to be directly related to decreases in receptor number. Prolonged (>1h) exposure of neutrophils to PMA results in a two- to threefold decrease in the number of CR1 on the cell surface with little change in CR3 expression, but a large (10–20-fold) decrease in their ability to phagocytose C3b- and C3bi-coated particles. Hence, down-regulation of CR1 and CR3 function may also occur. This phenomenon may be important in decreasing the binding of activated neutrophils to surfaces such as endothelial cells, and thus allow them either to migrate along a surface or else to squeeze through cell spaces following the initial attachment.

In some circumstances CR1 and CR3 may efficiently bind their ligands, but phagocytosis may not be initiated. As described above, pretreatment with PMA may increase the ability of neutrophils to ingest these bound particles. Several naturally-occurring molecules can also promote the ingestion of coated particles; presumably, this is achieved via alterations in the efficiency of coupling of receptor occupancy to intracellular signalling systems.

Whereas molecules such as albumin or collagen do not promote C3b- and C3bi-mediated phagocytosis, the extracellular matrix substances, fibronectin, serum amyloid P and laminin can all stimulate CR1- and CR3-activated phagocytosis. Fibronectin and serum amyloid P do not act as opsonins per se but rather regulate the activity of opsonin receptors. In order for *fibronectin* to exert this effect it must be bound to a surface. It is an elongated molecule with binding sites for heparin, collagen and fibrin, and can also avidly bind to bacteria. *Serum amyloid P* is a discoid decamer that binds amyloid to portions of the galactose molecule (present in the cell walls of yeast and some bacteria) and also to components of the basement membrane. *Laminin* is a T-shaped molecule that binds to type-IV collagen, the collagen of basement membranes. All three of these extracellular matrix molecules are thus present in the basal lamina, attached to elastin fibres in loose connective tissue. Phagocytes present within such connective tissue may thus have their CR1 and CR3 function up-regulated via interactions with the extracellular matrix proteins present in these tissues. Fibronectin is also present in serum and may attach to bacteria, a process that will also affect the ability of phagocytes to bind fibronectin-coated bacteria. Fibronectin receptors of phagocytes recognise the RGD motif of fibronectin, and addition of synthetic peptides with this sequence to phagocytes activates their

CR1 and CR3. Thus, it is envisaged that occupancy of fibronectin receptors somehow also activates CR1 and CR3.

The mechanisms responsible for activation of CR1 and CR3 are not fully defined. This process does not require protein biosynthesis, is not regulated by proteolysis and, as mentioned above, is not related to the number of receptors expressed. Because PMA (which activates protein kinase C) can cause this receptor activation, it has been proposed that phosphorylation reactions directly or indirectly regulate receptor activation. Phosphorylations may thus cause a conformational change in the receptors but may also affect their distribution on the cell surface. Thus, it is possible that in resting neutrophils CR1 and CR3 are randomly distributed as monomers on the plasma membrane, but upon activation they may cluster into an arrangement of 6–10 receptors. Whereas monomeric receptors may be inactive, the clusters may be active, and the clustering process itself may be controlled by phosphorylation. It may also be that clustering of ligand is important for efficient phagocytosis.

Observations of patients with CD11/CD18 deficiency (§8.7) has clarified the role of these receptors in immune function. These patients suffer life-threatening infections, and their phagocytes fail to bind C3bi-coated particles. Their neutrophils do not migrate from the bloodstream through the endothelium towards infected loci, nor do they adhere to protein-coated surfaces or to endothelial cells in vitro. Hence, CD11/CD18 molecules are required for adhesion to either vascular endothelium or C3-opsonised targets.

Other adhesion receptors that are structurally and functionally related include the receptors for fibronectin, vitronectin, platelet glycoproteins IIb and IIIa and the VLA (very-late antigen) series. All molecules involved in adhesion recognise the RGD motif and require the divalent cations Ca^{2+} and Mg^{2+} for binding. All are dimers of glycosylated proteins with relative molecular masses 95–190 kDa. There is also some sequence homology between the β-chain (CD18) and one chain of the fibronectin receptor.

3.10 Immunoglobulin receptors

3.10.1 General properties

The interaction of antibody–antigen complexes with cells of the immune system results in the activation of a variety of responses ranging from ADCC, mast-cell degranulation, lymphocyte proliferation, antibody secretion and phagocytosis. All these processes are activated via the binding of the Fc domain of the antibody molecule, which is exposed during antibody–

antigen interaction (see Fig. 1.4), to specialised receptors on the surface of the immune cells. These Fc receptors (FcR) are both structurally and functionally diverse. They may recognise the Fc regions of different types of antibody molecules, and may possess distinct transmembrane and cytoplasmic domains that are coupled to different signal transduction systems in order to elicit the appropriate response. Thus, the receptors are characterised by the nature of the immunoglobulin molecules they recognise, and different receptors exist on immune cells for the major immunoglobulin classes (IgG, IgM, IgA, IgE, IgD). Because neutrophils possess receptors for the Fc regions of IgG (FcγR) and IgA (FcαR), these are discussed in detail. (The properties of IgG and IgA are given in §1.3.1.1.)

Human neutrophils express three types of receptor for IgG: FcγRI, FcγRII and FcγRIII. Whilst these appear to have overlapping activities, it is clear that they may mediate different functions appropriate under different physiological or pathological conditions. Furthermore, the expression of these three receptors can vary considerably depending upon the past history of the neutrophils. For example, FcγRI is only expressed on neutrophils after exposure to certain cytokines such as γ-interferon, whereas the expression of FcγRIII can be regulated up (via the translocation of preformed intracellular pools to the plasma membrane) or down (via shedding from the cell surface). Furthermore, there is considerable evidence to suggest that, as with the complement receptors, the function of these receptors can be modulated (in the absence of a change in receptor number) so that the affinity of binding to the ligand is altered.

Analysis of the structure and function of the neutrophil immunoglobulin receptors has been greatly facilitated by molecular cloning techniques (which allow for a structural comparison of different receptors and hence a prediction of their function) and the availability of monoclonal antibodies that specifically bind to individual receptors. The use of such monoclonal antibodies has allowed the nature of the ligand binding to be elucidated (by blocking the interaction of the receptor with the natural ligand) and has permitted the molecular events following receptor occupancy to be determined (by cross-linking antibodies on the cell surface in order to activate the receptor). The use of fluorescently-linked antibodies is also useful to determine the number of receptors on the cell surface (by FACS analysis) and to assess which are located intracellularly (after making the cell permeable by treatment with mild detergents). The receptors share many common features: both FcγRI and FcγRII are integral membrane glycoproteins; all three types contain extracellular domains related to a C2-set of immunoglobulin domains and possess an intracytoplasmic region of variable length. These structurally-related molecules arise from gene duplications and alternative splicing.

3.10.2 Structure of Fcγ receptors

3.10.2.1 FcγRI

This is a heavily-glycosylated, 72-kDa transmembrane protein. cDNA sequence analysis predicts a mature protein of 40 kDa and six potential glycosylation sites. It is present on the cell surface of monocytes and macrophages, but is absent from bloodstream neutrophils, NK cells, mast cells, basophils and lymphocytes. The expression of this receptor on neutrophils is up-regulated by prior exposure to γ-interferon (up to 400 U/ml for 24 h). This treatment results in the expression of about 10^4 receptors/cell (a value comparable to that found on freshly-isolated blood monocytes), but has no effect on the expression of FcγRII or FcγRIII. γ-interferon treatment also up-regulates the expression of FcγRI on monocytes. FcγRI expression has also been reported on neutrophils of patients undergoing G-CSF therapy, but it is unknown if this up-regulation is due to the G-CSF itself or else to the generation of some other cytokine (e.g. γ-interferon) generated secondarily to the G-CSF. This receptor is also present at low levels on neutrophils isolated from the synovial fluid of patients with rheumatoid arthritis (see §8.9).

The increase in expression of FcγRI on neutrophils results not from the mobilisation of preformed subcellular pools but rather from the enhanced transcription and translation of the gene for this receptor. Whilst three transcripts for this gene have been cloned, Southern analyses reveal a single gene for this receptor, which is present on chromosome 1q. Two of the transcripts arise from polymorphisms, whilst the third represents diversity in the predicted cytoplasmic domain and may be either genuine or due to a cloning error. The transcript size is 1.7 kb. This increase in FcγRI expression on neutrophils and monocytes is associated with an enhancement of ADCC and phagocytosis. Cotreatment of neutrophils or monocytes with γ-interferon and dexamethasone (200 nM) largely prevents the up-regulation in FcγRI expression (by 40–75%, dependent upon the donor) and also decreases ADCC and phagocytosis.

FcγRI (unlike FcγRII and FcγRIII) binds monomeric IgG with high affinity ($K_d = 10^{-8}$–10^{-9} M); bloodstream neutrophils, which lack this receptor, thus do not bind monomeric IgG. The affinity of binding of murine monomeric IgG is $IgG_{2a} = IgG_3 \gg IgG_1 = IgG_{2b}$, whilst that for human monomer-

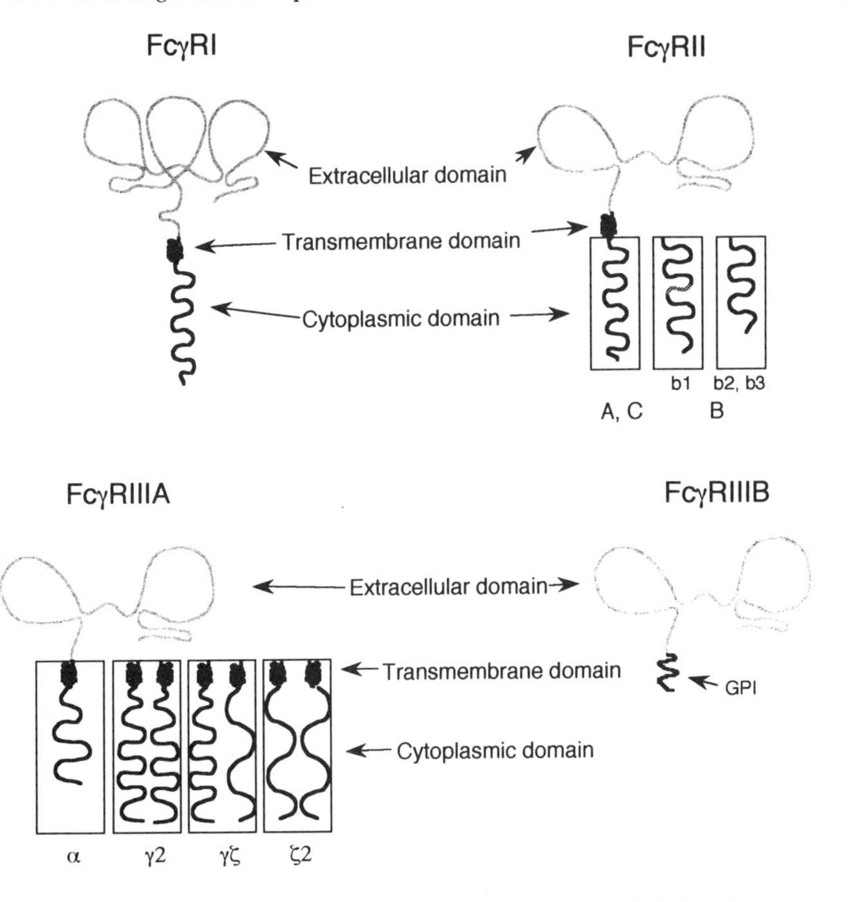

Figure 3.9. Structure of immunoglobulin G receptors. The large ovals indicate immunoglobulin-like domains that constitute the extracellular ligand-binding regions of the molecules. FcγRI possesses three of these domains, whilst FcγRII and FcγRIII possess only two. Three separate genes for FcγRII exist (A, B and C), and these code for different types of transcripts. Furthermore, alternative splicing of the FcγRIIB gene product can yield three separate transcripts, b1, b2 and b3. Two distinct forms of FcγRIII exist and are designated A and B. In FcγRIIIA, membrane spanning and cytoplasmic domains are present, and these are associated with cytoplasmic subunits. The α-chain may be associated with a disulphide-linked dimer of γ-chains, a homodimer of ζ-chains, or else a heterodimer of γζ. In FcγRIIIB (which is expressed on neutrophils) the membrane protein anchor is replaced by a glycosyl-phosphatidylinositol (GPI) anchor.

ic IgG is $IgG_1 = IgG_3 > IgG_4 \gg IgG_2$. The molecule comprises an extracellular domain consisting of three immunoglobulin-like domains (Fig. 3.9) whose structures are maintained by disulphide bridges. Two of these domains share homology with the two extracellular, immunoglobulin-like domains of FcγRII and FcγRIII; the third does not, however, and it is likely that this extra domain confers upon FcγRI the ability to bind monomeric IgG with high affinity. The murine gene is 65–70% homologous with the human gene in the extracellular and transmembrane domains, but the human cytoplasmic domain is 25 amino acids shorter and is only 25% homologous with its murine counterpart.

3.10.2.2 FcγRII

This is a 40-kDa transmembrane glycoprotein present on a variety of immune cells, such as neutrophils, B lymphocytes, platelets, eosinophils, monocytes and macrophages; however, it is not expressed on NK cells. It binds monomeric IgG with very low affinity but binds well to dimers, trimers or aggregated IgG ($K_d > 10^{-7}$ M). There are about 10^3 receptors per platelet, 10^5 per monocyte and $1–2 \times 10^4$ per neutrophil.

In mice, FcγRII is encoded by a single gene giving rise to a low-affinity receptor that binds IgG_1, IgG_{2a} and IgG_{2b}. This gene gives rise to two transcripts, b1 and b2, both of which have been cloned. These transcripts arise from alternative splicing, with b1 having a 47-amino-acid insertion into the cytoplasmic domain. Transcript b1 is preferentially expressed on lymphocytes whilst b2 is preferentially expressed on macrophages.

However, in humans, there are three different forms of the receptor (FcγRIIA, FcγRIIB and FcγRIIC), and these are encoded by three separate genes located on chromosome 1q23. These genes give rise to six separate transcripts:

FcγRIIA yields two transcripts (HR, LR) of 1.8 and 2.5 kb, due to alternative polyadenylation sites.

FcγRIIB generates three transcripts from alternative splicing of exons encoding either the cytoplasmic region (transcripts b1, b2) or the signal sequence (b3).

FcγRIIC gives rise to a single (1.8-kb) transcript.

The mature proteins of FcγRIIA and FcγRIIC are 95% homologous in the extracellular and transmembrane domains, and 100% identical in the intracytoplasmic domain. The FcγRIIB proteins b1, b2 and b3 are 95% homologous to FcγRIIC in the extracellular and transmembrane domains, but the FcγRIIB intracytoplasmic domain is unrelated to FcγRIIA and FcγRIIC. Furthermore, there is a 19-amino-acid insertion into the cytoplasmic domains of b1 and b3 (but not b2), due to an alternative splicing event.

These forms of human FcγRII are differentially expressed in immune cells. For example, FcγRIIB transcripts are detectable in monocytes, macrophages and lymphocytes, but not in neutrophils, NK cells or T-cell lines. On the other hand, FcγRIIA and FcγRIIC are expressed on monocytes, macrophages and neutrophils but not on NK cells nor lymphocytes. FcγRIIA has two potential glycosylation sites whilst FcγRIIB and FcγRIIC possess three and the murine receptors have four potential sites. The specificity of binding of IgG for the murine receptor is $IgG_{2b} > IgG_{2a} > IgG_1 \gg IgG_3$, whilst that of the human receptor is $IgG_1 = IgG_3 \gg IgG_2 = IgG_4$.

3.10.2.3 FcγRIII

This receptor, a glycoprotein appearing as a broad band of 50–70 kDa in SDS-PAGE, is expressed on neutrophils, eosinophils, macrophages and NK cells. It is not expressed on blood monocytes, but macrophages derived from cultured monocytes, and macrophages isolated from the peritoneum, lung and liver, are found to express FcγRIII. The protein backbone of the molecule (predicted from cDNA analysis or after removal of the sugars) is 29 or 33 kDa. The murine receptor has four potential glycosylation sites; the two human forms have five or six. FcγRIII binds IgG complexes with low affinity ($K_d < 10^{-7}$ M in humans, 10^{-6} M in murine cells) but is present on the cell surface in large numbers. In human neutrophils there are 100 000–200 000 receptors per cell, and so it is by far the most abundant Fcγ receptor on these cells. It is additionally present on the plasma membrane, but subcellular pools of this receptor also exist and may be mobilised to the plasma membrane in order to maintain or augment expression. On human neutrophils FcγRIII binds IgG molecules with the specificity $IgG_1 = IgG_3 \gg IgG_2 = IgG_4$; on murine cells it binds IgG with the specificity $IgG_3 > IgG_{2a} > IgG_{2b} \gg IgG_1$.

The murine and human genes for FcγRIII are homologous in terms of sequence, genomic organisation, cellular distribution and function. In murine cells, the single gene gives rise to a single transcript of 1.6 kb, and the extracellular domain is 95% homologous with that of FcγRII; the transmembrane and cytoplasmic domains are unrelated, however. Transfection of COS cells or L cells with the murine cDNA for FcγRIII results in the appearance of low-affinity binding sites for IgG complexes, but these receptors are only expressed at low levels. It was found that the coexpression of another molecule, the γ-chain of FcεR1, resulted in a 50–100-fold increase in expression of FcγRIII. Interestingly, transcripts for this γ-chain of FcεR1 are found in NK cells and macrophages, even though these cells do not express FcεR; they do, however, express FcγRIII. Thus, it appears that the expression of this γ-chain somehow affects FcγRIII's expression, structure or stability.

The two human-cell forms of FcγRIII are encoded for by two separate genes, FcγRIIIA and FcγRIIIB; a single mRNA size of 2.2 kb is detected in Northern analyses. The genes are mapped to chromosome 1, and FcγRII and FcγRIII are located within 200 kb of the genome. The existence of two forms of FcγRIII was predicted from studies of patients with *paroxysmal nocturnal haemoglobinuria (PNH).* These patients have a greatly decreased expression of FcγRIII on their neutrophils (only 10–15% of that on normal neutrophils), but the expression of this receptor on NK cells from these patients is normal. It has been shown that FcγRIIIA, which is present on the surface of NK cells, macrophages and cultured monocytes, comprises extracellular, transmembrane and cytoplasmic domains, whereas that present on neutrophils comprises an extracellular domain anchored to the membrane via a glycosyl-phosphatidylinositol (GPI) linkage. This linkage is cleaved by proteases such as elastase and pronase or by phosphatidylinositol-specific phospholipase C. Thus, the receptor may be shed from the neutrophil membrane upon activation (Fig. 3.10). This switch from the receptor having a transmembrane domain or a GPI linkage results from a change in amino acid 203 of the sequence: in FcγRIIIB, residue 203 is Ser (which is followed by a termination codon), resulting in a GPI linkage; in FcγRIIIA, residue 203 is phenylalanine, resulting in a transmembrane linkage. The cytoplasmic domain of FcγRIIIA is a 25-amino-acid sequence that arises from a T $\rightarrow$ C transition in the TGA stop sequence present in FcγRIIIB (Figs. 3.9, 3.11). Transfection experiments of cell lines with cDNA molecules for FcγRIII indicate that, as for the murine gene, co-infection with either the γ-chain of FcεR1 or the ζ-chain of the CD3/TcR (T-cell receptor) is required for maximal expression of FcγRIIIA. Co-infection with these molecules is not required, however, for expression of FcγRIIIB. Thus, FcγRIIIA molecules are designated FcγRIIIAα, FcγRIIIAγ or FcγRIIIAζ depending upon the nature of the subunit associated with them.

Allogeneic forms of FcγRIIIB also exist. These are recognised by an antibody present in the serum of patients with a form of autoimmune neutropenia, and separate monoclonal antibodies recognise these allotypic forms (NA1, NA2) of the receptor. The two allotypic forms can also be recognised by different mobilities on gels. Whilst cDNA for both NA1 and NA2 have been cloned and sequenced, both molecules predict primary translation products of equal size; however, when the immunoprecipitated NA-1 and NA-2 molecules are deglycosylated, the apparent relative molecular masses are 29 and 33 kDa. Similarly, in vitro translation of mRNA for these two allotypic forms yields proteins of 24 and 26 kDa for NA1 and NA2, respectively. Such a translation system would not be expected to glycosylate the newly-made proteins, so these differences in apparent size cannot be taken into account by differential glycosylation. These cDNA molecules differ in only five nucleotides; these predict four amino acid substitutions, which account for the fact that there are six potential glycosylations sites on NA2 but only four in NA1 (Fig. 3.11). The anomalous mobility of these two non- or

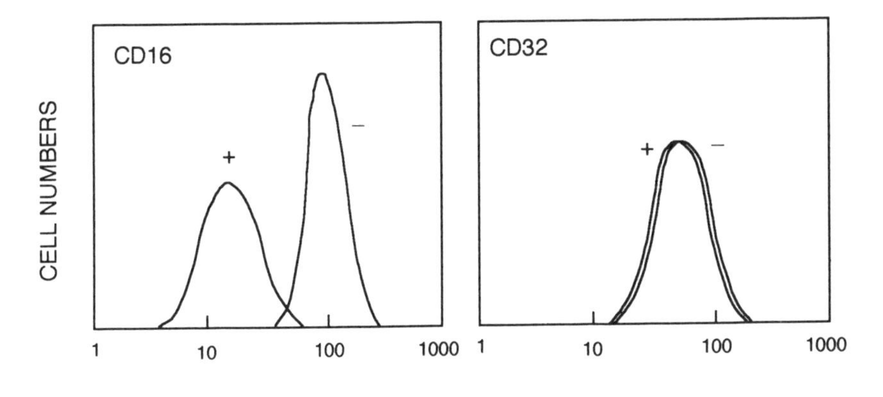

Figure 3.10. Effect of pronase on FcγRII (CD32) and FcγRIII (CD16) expression on neutrophils. Neutrophils were incubated in the absence (–) or presence (+) of pronase (50 μg/ml) for 30 min at 37 °C. After this incubation, expression of FcγRII and FcγRIII was determined by FACS analysis.

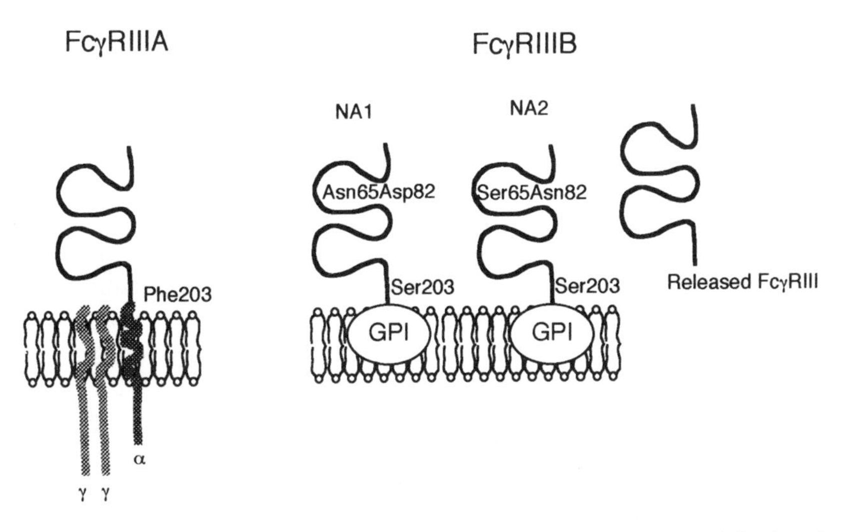

Figure 3.11. Structure of FcγRIIIA and FcγRIIIB. In FcγRIIIA, residue 203 is phenylalanine (Phe), which is followed by a hydrophobic transmembrane domain and a cytoplasmic domain. In FcγRIIIB, residue 203 is serine (Ser), which is followed by a stop codon. See text for details.

deglycosylated proteins on SDS-PAGE, yielding different apparent molecular masses (even though amino acid prediction of cDNA molecules suggest that this should not be the case) may thus result from conformational differences in these two proteins due to the small differences in their amino acid composition. It is also reported that NA2-type FcγRIII receptors have a lower capacity to mediate phagocytosis than do their NA1 counterparts.

3.10.3 Function of Fcγ receptors

Several approaches can be taken to determine the function of the different neutrophil Fcγ receptors during binding to immune complexes:

i. It is possible to use monoclonal antibodies to block binding of the ligand to a specific receptor.
ii. Fab fragments of different monoclonal antibodies that recognise specific receptors may be used to cross-link with second antibodies in order to achieve receptor activation.
iii. Cell lines that do not normally express Fcγ receptors can be transfected with cDNA clones for these molecules; the consequences of ligand binding may then be observed.
iv. The expression of some receptors may be experimentally manipulated either by inducing expression (e.g. of FcγRI by incubation with γ-interferon) or by decreasing expression (e.g. of FcγRIII by incubation with pronase).
v. Experiments may involve patients, such as those with PNH, who have defective expression of FcγRIII.

These kinds of research have not fully resolved the role of Fcγ receptors in immunoglobulin binding. What appears to be the case is that these receptors interact with each other so that, for example, whilst one receptor may be primarily required for binding the ligand, another may actually transduce the signal, though this event requires occupancy of the first receptor. Because of such interactions, experiments using the approaches described above do not provide definitive answers. Furthermore, it is likely that occupancy of one type of receptor may alter the binding/signalling properties of another type.

The physiological role of FcγRI on neutrophils is difficult to assess because this receptor is expressed only when cells have been exposed to γ-interferon. As is observed in monocytes, glucocorticoids such as dexamethasone can prevent the γ-interferon-dependent up-regulation of this receptor. Hence, it is predicted that the FcγRI gene contains both γ-interferon- and glucocorticoid-response elements. Because the increase in expression of FcγRI by γ-interferon and inhibition of this phenomenon by dexamthasone is paralleled by corresponding changes in ADCC and associated phagocytosis, it is assumed that expression of this receptor is required for these processes. However, because expression of this receptor requires fairly long (10–24 h) exposure to this cytokine, it is difficult to imagine how this may be important for neutrophil function during bacterial infections. The neutrophils are generally required to respond fairly quickly to such a microbial challenge, before the microbe has the opportunity to multiply and migrate throughout the body. However, the observation of γ-interferon-dependent up-regulation of FcγRI expression raises the possibility that, under some circumstances, neutrophils may respond to an inflammatory challenge by a slow adaptation of their cytotoxic armory that requires active de novo biosynthesis. This has two implications:

i. It argues against the idea that neutrophils have only a short functional lifespan.
ii. It suggests that blood neutrophils are not terminally differentiated and that, at some sites in the body, they may possess a different molecular constitution, which confers upon them different cellular functions, such as ADCC.

In view of the fact that neutrophils isolated from some inflammatory sites have been exposed to agents that selectively activate their gene expression (which may result in the expression of FcγRI; see Fig. 7.9 and §8.9), such molecular adaptation of neutrophils to their environment may indeed occur.

The role of FcγRIII is controversial. In neutrophils this receptor does not possess a transmembrane or cytoplasmic domain, being anchored instead to the cells by an easily-cleaved GPI linkage (see Fig. 3.11). This would indicate that occupancy of this receptor cannot lead directly to the generation of intracellular signals via coupling to phospholipases or other signalling enzymes. The receptor is present on the neutrophil plasma membrane in quantities ten times greater than FcγRII, and numbers can be up-regulated via the mobilisation of preformed subcellular pools of receptors present on intracellular stores. These stores are believed to be a granule population cosedimenting with the specific granules, including a fraction cosedimenting with 'latent' alkaline phosphatase activity. Whilst alkaline phosphatase activity is generally thought to be a marker for the plasma membrane, 'latent' activity (i.e. activity enhanced by agents such as Triton, which disrupt membranes) may be a marker for secretory vesicles. Thus, intracellular FcγRIII stores may be specific granules and/or the secretory vesicles that contain albumin and possess latent alkaline phosphatase activity.

FcγRIII is actively synthesised by blood neutrophils, and its rate of expression can be increased up to twofold after exposure of neutrophils to GM-CSF (§7.3.3). It is also shed from the membrane during activation by, for example, PMA or fMet-Leu-Phe, as neutrophils age or when treated with agents such as proteases (see Fig. 3.10). Intracellular pools of FcγRIII are present on the membranes of granules, and these granules also contain CR1 (C3b receptor), CR3 (C3bi receptor), fMet-Leu-Phe receptors, decay-accelerating factor and the cytochrome b of the NADPH oxidase. However, priming or stimulation of neutrophils results in an increase in expression on the plasma membrane of all of these molecules except for FcγRIII, whose expression is largely unchanged. This observation was believed to represent that FcγRIII was present in a subcellular pool separate from the other molecules and that the translocation of this pool was under separate control. The intracellular pool of FcγRIII can be detected by immunoelectron microscopy or by permealising the cells (in a 0.1% BSA/PBS / 0.1% glycine / 0.04% saponin buffer) and FACS analysis. Shed FcγRIII can be detected by analysis of culture medium by immunoblotting.

In resting neutrophils, about 50% of the total cellular FcγRIII pool is expressed on the cell surface. There is considerable variation in this value because many methods used to isolate neutrophils can also inadvertently mobilise these subcellular receptors. The remainder of the total cellular FcγRIII that is not expressed on the plasma membrane is present in the subcellular pool. However, if the FcγRIII normally present on the plasma membrane is cleaved (e.g. via the action of elastase or pronase) and the cells subsequently activated, then FcγRIII reappears on the cell surface via the mobilisation of these pools. Thus, the expression can be restored to up to 70% of the resting level within 15 min via such a translocation. During activation (and presumably priming), FcγRIII (together with other plasma membrane markers) is also translocated to the plasma membrane; however, because the receptor is also shed from the cell, the total number of receptors on the cell surface remains largely unchanged. There is also some evidence that continued expression of FcγRIII on the cell surface requires de novo biosynthesis of this receptor (see Fig. 7.8).

In spite of the fact that FcγRIII possesses neither a transmembrane nor a cytoplasmic domain, its occupancy leads to the generation of intracellular signals such as Ca^{2+} transients. Indeed, some reports suggest that occupancy of both FcγRII and FcγRIII can lead to the generation of Ca^{2+} fluxes independently of each other. However, it is generally believed that FcγRIII functions primarily for the binding of immune complexes to the neutrophil plasma membrane, whereas occupancy of FcγRII leads to the activation of the cells to generate reactive oxidants, cause degranulation and induce phagocytosis. Observations of patients with PNH appear to confirm these roles: in these patients the expression of FcγRIII on neutrophils is only 10–15%

of normal whereas expression of FcγRII is normal. The binding of IgG dimers to these neutrophils is greatly decreased, but their ability to activate a respiratory burst in response to IgG-coated latex particles is largely unaltered.

Human neutrophils bind IgG_3 complexes up to three times faster than IgG_1 complexes, but the same number of both types of complex will eventually become bound. Blocking the binding of these complexes with monoclonal antibodies against these receptors has shown that anti-FcγRIII antibodies block the binding of IgG_3 or IgG_1, whereas anti-FcγRII antibodies decrease the affinity of binding but do not affect the total number of complexes bound. Thus, it seems apparent that FcγRII and FcγRIII function cooperatively to bind immune complexes, but that only occupancy of FcγRII results in the generation of intracellular signals that activate the respiratory burst and degranulation. FcγRIII appears to function primarily to bind small complexes (dimers and trimers); large complexes (e.g. opsonised *S. aureus*) are bound by FcγRII.

3.10.4 IgA receptor

Neutrophils, monocytes, T and B lymphocytes and NK cells possess plasma membrane receptors that specifically bind IgA. The first evidence for Fcα-receptors in neutrophils came from experiments in which rosettes were formed when neutrophils were incubated with erythrocytes that had been coated with IgA. Furthermore, neutrophils were then shown to possess IgA-containing immune complexes, indicating that these had been taken up by the cells in vitro. Aggregated IgA can inhibit chemotaxis and chemokinesis but can stimulate the respiratory burst. The role of IgA molecules as opsonins has been controversial, however, because in most bodily secretions IgA exists as a dimer, with the two molecules joined at their Fc regions via the J chain (see Fig. 1.13). Thus, these Fc regions of IgA are not fully exposed during ligand binding (as are the Fc regions of IgG molecules) and so may not be accessible to IgA receptors on the neutrophil surface.

The IgA receptor on neutrophils is a glycosylated protein with a relative molecular mass of 50–70 kDa. Heat aggregates containing IgA_1 or IgA_2 are both capable of triggering a respiratory burst (as detected by chemiluminescence) and degranulation. Both monomeric and dimeric IgA can cause these effects, so the Fc regions of IgA must be accessible to the Fcα-receptor. The affinity of binding of IgA to the receptor is about 5×10^{-7} M, and there are about 7-15 $\times 10^3$ receptors per neutrophil, a value comparable to the level of expression of FcγRII. Thus, these findings indicate that, in bodily fluids (where levels of complement may be too low to act as an efficient opsonin), the presence of large quantities of IgA may serve to opsonise microorganisms prior to their phagocytosis by neutrophils.

3.11 Bibliography

Akerley, W. L. I., Guyre, P. M., & Davis, B. H. (1991). Neutrophil activation through high-affinity Fcγreceptor using a monomeric antibody with unique properties. *Blood* **77**, 607–15.

Akira, S., Hirano, T., Taga, T., & Kishimoto, T. (1990). Biology of multifunctional cytokines: IL-6 and related molecules (IL-1 and TNF). *FASEB J.* **4**, 2860–7.

Albelda, S. M., & Buck, C. A. (1990). Integrins and other cell adhesion molecules. *FASEB J.* **4**, 2868–80.

Anderson, C. L., & Abraham, G. N. (1980). Characterisation of the Fc receptor for IgG on a human macrophage cell line, U937. *J. Immunol.* **125**, 2735–41.

Becker, E. L. (1987). The formylpeptide receptor of the neutrophil. *Am. J. Pathol.* **129**, 16–24.

Berger, M., Wetzler, E. M., Welter, E., Turner, J. R., & Tartakoff, A. M. (1991). Intracellular sites for storage and recycling of C3b receptors in human neutrophils. *Proc. Natl. Acad. Sci USA* **88**, 3019–23.

Boulay, F., Mery, L., Tardif, M., Brouchon, L., & Vignais, P. (1991). Expression cloning of a receptor for C5a anaphylatoxin on differentiated HL-60 Cells. *Biochemistry* **30**, 2993–9.

Boulay, F., Tardif, M., Brouchon, L., & Vignais, P. (1990a). Synthesis and use of a novel *N*-formyl peptide derivative to isolate a human *N*-formyl peptide receptor cDNA. *Biochem. Biophys. Res. Commun.* **168**, 1103–9.

Boulay, F., Tardif, M., Brouchon, L., & Vignais, P. (1990b). The human *N*-formylpeptide receptor: Characterisation of two cDNA isolates and evidence for a new subfamily of G-protein-coupled receptors. *Biochemistry* **29**, 11123–33.

Boyden, S. (1962). The chemotactic effect of mixtures of antibody and antigen on polymorphonuclear leukocytes. *J. Exp. Med.* **115**, 453–66.

Coats, W. D., & Navarro, J. (1990). Functional reconstitution of fMet-Leu-Phe receptor in *Xenopus laevis* oocytes. *J. Biol. Chem.* **265**, 5964–6.

Cockcroft, S., & Stutchfield, J. (1989). ATP stimulates secretion in human neutrophils and HL60 cells via a pertussis toxin-sensitive guanine nucleotide-binding protein coupled to phospholipase C. *FEBS Lett.* **245**, 25–9.

Djeu, J. Y., Matsushima, K., Oppenheim, J. J., Shiotsuki, K., & Blanchard, D. K. (1990). Functional activation of human neutrophils by recombinant monocyte-derived neutrophil chemotactic factor/IL-8. *J. Immunol.* **144**, 2205–10.

Fanger, M. W., Shen, L., Graziano, R. F., & Guyre, P. M. (1989). Cytotoxicity mediated by human Fc receptors for IgG. *Immunol. Today* **10**, 92–9.

Gerard, N. P., & Gerard, C. (1991). The chemotactic receptor for human C5a anaphylatoxin. *Nature* **349**, 614–17.

Greenblatt, M. C., & Elias, L. (1992). The type B receptor for tumor necrosis factor-α mediates DNA fragmentation in HL-60 and U937 cells and differentiation in HL-60 cells. *Blood* **80**, 1339–46.

Hickstein, D. D., Hickey, M. J., Ozols, J., Baker, D. M., Back, A. L., & Roth, G. J. (1989). cDNA sequence for the αM subunit of the human neutrophil adherence receptor indicates homology to integrin α subunits. *Proc. Natl. Acad. Sci. USA* **86**, 257–61.

Honda, Z.-I., Nakamura, M., Miki, I., Minami, M., Watanabe, T., Seyama, Y., Okado, H., Toh, H., Ito, K., Miyamoto, T., & Shimizu, T. (1991). Cloning by functional expression of platelet-activating factor receptor from guinea pig lung. *Nature* **349**, 342–6.

Huizinga, T. W. J., Kerst, M., Nuyens, J. H., Vlug, A., von dem Borne, A. E. G. K., Roos, D., & Tetteroo, P. A. T. (1989). Binding characteristics of dimeric IgG subclass complexes to human neutrophils. *J. Immunol.* **142**, 2359–64.

Huizinga, T. W. J., Roos, D., & von dem Borne, A. E. G. K. (1990). Neutrophil Fc-γ receptors: A two-way bridge in the immune system. *Blood* **75**, 1211–14.

Huizinga, T. W. J., van der Schoot, C. E., Jost, C., Klaassen, R., Kleijer, M., von dem Borne, A. E. G. K., Roos, D., & Tetteroo, P. A. T. (1988). The PI-linked receptor FcRIII is released on stimulation of neutrophils. *Nature* **333**, 667–9.

Huizinga, T. W. J., van Kemenade, F., Koenderman, L., Dolman, K. M., von dem Borne, A. E. G. K., Tetteroo, P. A. T., & Roos, D. (1989). The 40-kDa Fc γ receptor (FcRII) on human neutrophils is essential for the IgG-induced respiratory burst and IgG-induced phagocytosis. *J. Immunol.* **142**, 2365–9.

Hundt, M., & Schmidt, R.E. (1992). The glycosylphosphatidylinositol-linked Fcγ receptor III represents the dominant receptor structure for immune complex activation of neutrophils. *Eur. J. Immunol.* **22**, 811–16.

Indik, Z., Kelley, C., Chien, P., Levinson, A. I., & Schreiber, A. D. (1991). Human FcγRII, in the absence of other Fcγ receptors, mediates a phagocytic signal. *J. Clin. Invest.* **88**, 1766–71.

Jones, D. H., Looney, R. J., & Anderson, C. L. (1985). Two distinct classes of IgG Fc receptors on a human monocyte line (U937) defined by differences in binding of murine IgG subclasses at low ionic strength. *J. Immunol.* **135**, 3348–53.

Kerr, M. A. (1990). The structure and function of human IgA. *Biochem. J.* **271**, 285–96.

Kerr, M. A., Mazegera, R. L., & Stewart, W. W. (1990). Structure and function of immunoglobulin A receptors on phagocytic cells. *Biochem. Soc. Trans.* **18**, 215–17.

Kurzinger, K., & Springer, T. A. (1982). Purification and structural characterization of LFA-1, a lymphocyte function-associated antigen, and Mac-1, a related macrophage differentiation antigen associated with the type three complement receptor. *J. Biol. Chem.* **257**, 12412–18.

Mazengera, R. L., & Kerr, M. A. (1990). The specificity of the IgA receptor purified from human neutrophils. *Biochem. J.* **272**, 159–65.

Merritt, J. E., & Moores, K. E. (1991). Human neutrophils have a novel purinergic P_2-type receptor linked to calcium mobilization. *Cell. Signalling* **3**, 243–9.

Micklem, K. J., Stross, W. P., Willis, A. C., Cordell, J. L., Jones, M., & Mason, D. Y. (1990). Different isoforms of human FcRII distinguished by CDw32 antibodies. *J. Immunol.* **144**, 2295–303.

Murphy, P. M., Gallin, E. K., Tiffany, H. L., & Malech, H. L. (1990). The formyl peptide chemoattractant receptor is encoded by a 2 kilobase messenger RNA. Expression in *Xenopus* oocytes. *FEBS Lett.* **261**, 353–7.

Omann, G. M., Traynor, A. E., Harris, A. L., & Sklar, L. A. (1987). LTB_4-induced activation signals and responses in neutrophils are short-lived compared to formylpeptide. *J. Immunol.* **138**, 2626–32.

Ory, P. A., Clark, M. R., Talhouk, A. S., & Goldstein, I. M. (1991). Transfected NA1 and NA2 forms of human neutrophil Fc receptor III exhibit antigenic and structural heterogeneity. *Blood* **77**, 2682–7.

O'Shea, J. J., Brown, E. J., Seligman, B. E., Metcalf, J. A., Frank, M. M., & Gallin, J. I. (1985). Evidence for distinct intracellular pools of receptors for C3b and C3bi in human neutrophils. *J. Immunol.* **134**, 2580–7.

Peltz, G. A., Grundy, H. O., Lebo, R. V., Yssel, H., Barsh, G. S., & Moore, K. W. (1989). Human FcγRIII: Cloning, expression, and identification of the chromosomal locus of two Fc receptors for IgG. *Proc. Natl. Acad. Sci. USA* **86**, 1013–17.

Petrequin, P. R., Todd, R. F., III, Devall, L. J., Boxer, L. A., & Curnutte, J. T., III. (1987). Association between gelatinase release and increased plasma membrane expression of the Mo1 glycoprotein. *Blood* **69**, 605–10.

Ravetch, J. V., & Kinet, J.-P. (1991). Fc receptors. *Ann. Rev. Immunol.* **9**, 457–92.

Rosales, C., & Brown, E. J. (1991). Two mechanisms for IgG Fc-receptor-mediated phagocytosis by human neutrophils. *J. Immunol.* **146**, 3937–44.

Ruoslahti, E. (1991). Integrins. *J. Clin. Invest.* **87**, 1–5.

Schreiber, A. D., Rossman, M. D., & Levinson, A. I. (1992). The immunobiology of human Fc γ receptors on hematopoietic cells and tissue macrophages. *Clin. Immunol. Immunopathol.* **62**, S66–72.

Seifert, R., Hoer, A., Offermans, S., Buschauer, A., & Schunack, W. (1992). Histamine increases cytosolic Ca^{2+} in dibutyryl-cAMP-differentiated HL-60 cells via H_1 receptors and is an incomplete secretagogue. *Am. Soc. Pharmacol. Exp. Therapeut.* **42**, 227–34.

Seifert, R., Hoer, A., Schwaner, I., & Buschauer, A. (1992). Histamine increases cytosolic Ca^{2+} in HL-60 promyelocytes predominantly via H_2 receptors with an unique agonist/antagonist profile and induces functional differentiation. *Am. Soc. Pharmacol. Exp. Therapeut.* **42**, 235–41.

Selvaraj, P., Rosse, W. F., Silber, R., & Springer, T. A. (1988). The major Fc receptor in blood has a phosphatidylinositol anchor and is deficient in paroxysmal nocturnal haemoglobinuria. *Nature* **333**, 565–7.

Shen, L. (1992). Receptors for IgA on phagocytic cells. *Immunol. Res.* **11**, 273–82.

Simmons, D., & Seed, B. (1988). The Fc γ receptor of natural killer cells is a phospholipid-linked membrane protein. *Nature* **333**, 568–70.

Snyderman, R., Gewurz, H., & Mergenhagen, S.E. (1968). Interactions of the complement system with endotoxic lipopolysaccharide: Generation of a factor chemotactic for polymorphonuclear leukocytes. *J. Exp. Med.* **128**, 259–75.

Stewart, W. W., & Kerr, M. A. (1990). The specificity of the human neutrophil IgA receptor (Fc αR) determined by measurement of chemiluminescence induced by serum or secretory IgA1 or IgA2. *Immunol.* **71**, 328–34.

Stutchfield, J., & Cockcroft, S. (1990). Undifferentiated HL60 cells respond to extracellular ATP and UTP by stimulating phospholipase C activation and exocytosis. *FEBS Lett.* **262**, 256–8.

Suchard, S. J., Burton, M. J., & Stoehr, S. J. (1992). Thrombospondin receptor expression in human neutrophils coincides with the release of a subpopulation of specific granules. *Biochem. J.* **284**, 513–20.

Takahashi, S., Yoshikawa, T., Naito, Y., Tanigawa, T., Yoshida, N., & Kondo, M. (1991). Role of platelet-activating factor (PAF) in superoxide production by human polymorphonuclear leukocytes. *Lipids* **26**, 1227–30.

Thomas, K. M., Pyun, H. Y., & Navarro, J. (1990). Molecular cloning of the fMet-Leu-Phe receptor from neutrophils. *J. Biol. Chem.* **265**, 20061–4.

Thomas, K. M., Taylor, L., & Navarro, J. (1991). The Interleukin-8 receptor is encoded by a neutrophil-specific cDNA clone, F3R. *J. Biol. Chem.* **266**, 14839–41.

Van Damme, J., Rampart, M., Conings, R., Decock, B., Van Osselaaer, N., Willems, J., & Billiau, A. (1990). The neutrophil-activating proteins Interleukin 8 and β-thromboglobulin: In vitro and in vivo comparison of NH_2-terminally processed forms. *Eur. J. Immunol.* **20**, 2113–18.

Vedder, N. B., & Harlan, J. M. (1988). Increased surface expression of CD11b/ CD18 (Mac-1) is not required for stimulated neutrophil adherence to cultured endothelium. *J. Clin. Invest.* **81**, 676–82.

Veys, P. A., Wilkes, S., & Hoffbrand, A. V. (1991). Deficiency of neutrophil FcRIII. *Blood* **78**, 852–3.

4
The cytoskeleton: The molecular framework regulating cell shape and the traffic of intracellular components

The cytoplasm of the neutrophil is highly structured and is organised into four distinct components that constitute the fibrillar meshwork: the microfilaments, microtubules, intermediate filaments and the microtrabecular lattice. These components form the *cytoskeleton,* the supporting framework for the cell, within which the intracellular components are embedded. Thus, the translocation of cytoplasmic organelles and granules, the movement of cytosolic proteins, the recycling of receptors – as well as key neutrophil functions such as cell movement, phagocytosis, NADPH oxidase activation, degranulation and receptor regulation – are all intimately associated with changes in the organisation of the cytoskeleton. Therefore, in order for the neutrophil to be able to mount the appropriate response upon exposure to stimuli, plasma membrane occupancy must be linked to changes in cytoskeletal organisation via the generation of second messengers. Furthermore, because activation of some neutrophil functions or changes in morphology may be observed within seconds of an agonist binding to its receptor, the cytoskeleton must be capable of responding very rapidly to the production of such second-messenger molecules. It is also now becoming accepted that cytoskeletal reorganisations may also terminate some responses – as in the termination of second-messenger production by interaction of cytoskeletal elements with receptors to down-regulate their function, or by receptor internalisation. Thus, the cytoskeleton is a highly-complex network comprising numerous proteins, and these respond to second messengers to undergo rapid changes in molecular reorganisation upon appropriate stimulation.

4.1 The microfilaments

Perhaps the most dynamic components of the cytoskeleton, the microfilaments are directly involved in cell movement and phagocytosis. They are

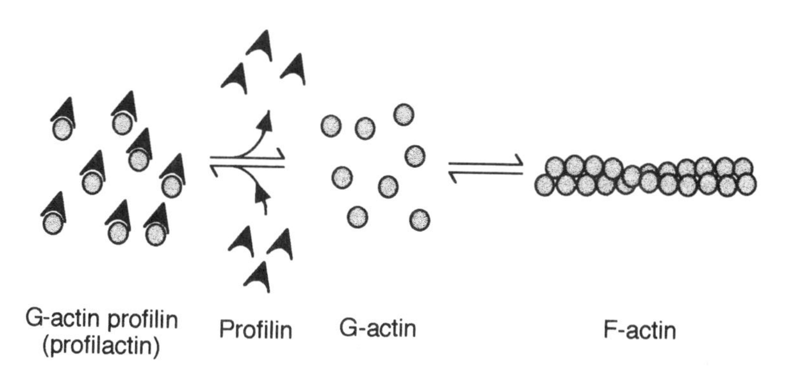

Figure 4.1. Actin polymerisation. Actin monomers (G-actin) may reversibly assemble into actin filaments (F-actin). Profilin binds to G actin (to form profilactin) and thus prevents its polymerisation.

found mainly in regions beneath the plasma membrane and also in regions of the cell that are either undergoing phagocytosis or forming pseudopodia. They comprise actin subunits arranged into highly-organised structures.

4.1.1 Actin

The actin molecule is a bilobed globular protein whose concentration may be as high as 24 mg/ml in the neutrophil cytoplasm. It may exist in a monomeric form, termed *G-actin,* or else in a filamentous form, termed *F-actin* (Fig. 4.1). G-actin is a single polypeptide with a relative molecular mass of 42 kDa. It can bind Ca^{2+} and ATP, and be cleaved by proteases to give fragments of 33 and 9 kDa. The amino acid sequence of actin is highly conserved in cells from different species, there being only 6% difference in residues between molecules found in the protozoan *Acanthamoeba castellanii* and in rabbit skeletal muscle. All actin monomers comprise 375 amino acids and are NH_2-terminal blocked with an acetyl group. Amino acid residue 73 is an unusual amino acid, *N*-methylhistidine, whose function is unknown, and actins from different tissues can be distinguished by amino acid differences in the NH_2-terminal 18 amino acids. Sequences 20–75, 299–356 and 365–375 are highly conserved in actin molecules from different tissues, implying that these regions are involved in the binding properties, and hence function, of actin.

Actin molecules possess four possible binding sites: two binding sites for adjacent molecules to form strands (*intrastrand binding*) and two sites for binding of actin molecules in neighbouring strands (*interstrand binding*).

Each molecule contains five cysteine residues, but because none of these forms disulphide bonds they do not contribute to the tertiary structure of the molecule. G-actin molecules also possess high-affinity binding sites for divalent cations and nucleotides. For example, the dissociation constants for cation binding are 1.9×10^{-9} M and 1×10^{-8} M for Ca^{2+} and Mg^{2+}, respectively. Because physiological concentrations of Ca^{2+} and Mg^{2+} are 0.1–1.0 μM and 1 mM, respectively, most of these binding sites will be occupied by Mg^{2+}. ATP and ADP also bind G-actin with K_d values of 1.3×10^{-10} M and 2.2×10^{-8} M, respectively; thus, at the respective ATP and ADP physiological concentrations of 3–5 mM and 0.5–1 mM, the nucleotide that will be bound is ATP. Interactions between the cation- and nucleotide-binding sites exist. For example, exchanging Ca^{2+} for Mg^{2+} accelerates the release of ATP from actin, whilst ATP hydrolysis lowers the affinity of actin for Ca^{2+} and increases the affinity for Mg^{2+}.

4.1.2 Actin polymerisation

Assembly and disassembly of the actin network is controlled at two major steps: (i) the length of individual actin filaments and (ii) the extent to which these filaments are cross-linked to each other (Fig. 4.2). Filament length itself is controlled by the rates of:

i. addition of G-actin to existing filaments (but note that one end of the actin filament grows at a faster rate than the other);
ii. initiation of new sites for polymerisation (nucleation);
iii. joining of short filaments end to end;
iv. filament breakage.

Several experimental approaches can be employed to determine the pool sizes of polymerised and non-polymerised actin. Firstly, the enzyme DNAse I is inhibited by monomeric (G) actin, but not by polymerised (F) actin. Secondly, polymerised actin can be directly visualised by use of fluorescent derivatives of phalloidin, a cyclic peptide isolated from the toadstool *Amanita phalloides* that selectively binds to polymerised actin with high affinity.

Actin comprises about 5–8% of the total protein of the neutrophil, and in resting cells about 50–70% of the actin pool exists as a monomer. This proportion of monomeric G-actin is far in excess of what would be predicted from the critical concentration for actin assembly in vitro. Thus, in vivo actin polymerisation and depolymerisation is regulated by the activities of a number of binding proteins, cations and other regulatory molecules, which are in turn regulated by the activation status of the cell.

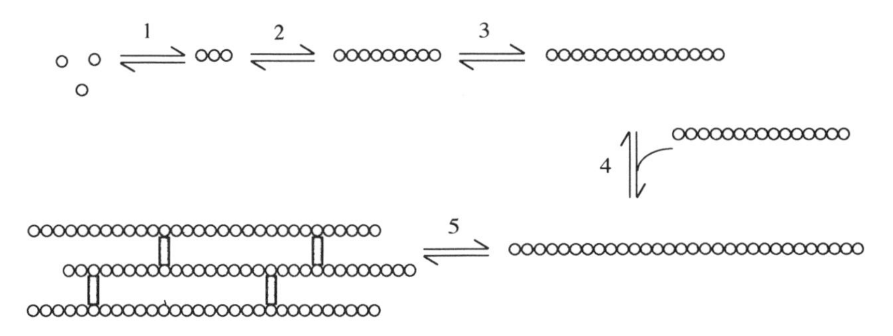

Figure 4.2. Control of the assembly of the actin network. The actin network may be regulated by: (1) the rate of formation of new nucleation centres (composed of three actin monomers); (2) filament formation from these nucleation centres; (3) extension of filaments; (4) joining of filaments end to end; (5) cross-linking of filaments. All of these steps are reversible.

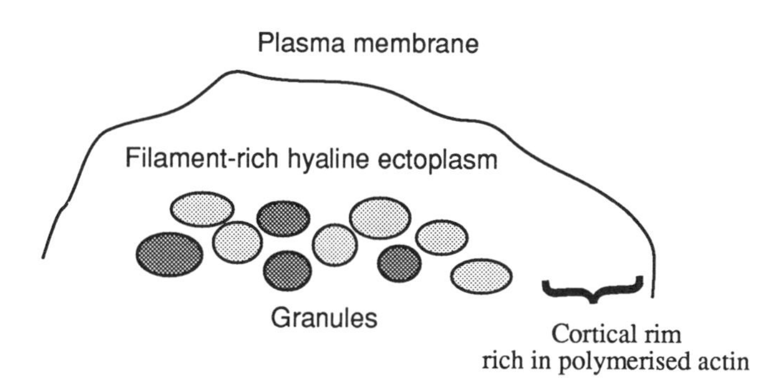

Figure 4.3. Location of polymerised actin in resting neutrophils. In resting neutrophils, the majority of the polymerised actin is present in the *cortical rim,* a region of cytoplasm just beneath the plasma membrane that is devoid of cytoplasmic granules.

In resting cells the large, unpolymerised pool is spread throughout the cytoplasm. This has the advantage that there may be rapid polymerisation of actin at any portion of the cell where it is required (e.g. for directional movement or for phagocytosis) without the need for depolymerisation and redistribution of monomers to occur first. In resting cells almost all of the polymerised actin is present in the cortical, submembrane layer, with a few actin filaments present between the cortex and the nucleus (Fig. 4.3). This cortical meshwork is 0.1–0.2 μm and excludes organelles but contains short actin fibres.

The initial step in actin polymerisation is the activation of actin monomers by binding to divalent cations, a process that causes a conformational change in the monomer. Such activated actin monomers may then either form nucleation oligomers or else add on to the ends of existing actin filaments. The smallest actin oligomer that can grow into a filament is a trimer, which is formed in a two-step process via the combination of three activated monomers. Dimers and trimers are more stable in the presence of Mg^{2+} rather than Ca^{2+}, and ATP-actin forms nuclei at a faster rate than ADP-actin.

Elongation of actin filaments is dependent upon the rates of association and disassociation of actin monomers with existing filaments. The actin filament is polarised with either 'barbed' or 'pointed' ends, because the actin monomers have a polarity and can only polymerise head to tail. Thus, the assembled filaments have a defined polarity, with subunits aligned or 'pointed' in the same direction (Fig. 4.4). The filaments may also be decorated with molecules of meromyosin, giving an arrowhead appearance to the filaments. At the 'barbed' ends monomers are exchanged 30 times more quickly than at the 'pointed' ends. The concentration of G-actin in dynamic equilibrium with actin filaments (the critical monomer concentration) is 15 times lower for the 'barbed' end compared to the 'pointed' end. The critical concentration of actin at the two ends depends upon the concentrations of ions in the microenvironment. For example, in the presence of Mg^{2+} the critical concentration is different at each end, but in the presence of Ca^{2+} it is the same at each end of the filament. Because the exchange is faster at the 'barbed' end, the critical concentration of actin is equal to that of the 'barbed' end (i.e. 0.1 μM). Therefore, the growth of an actin filament occurs at the 'barbed' end because here the rate of growth equals:

20 monomers added per sec per μM actin monomer and
2 monomers removed per sec per μM actin monomer.

Monomeric (G) actin binds 1 mol ATP (G-ATP), which is hydrolysed to ADP during assembly into F-actin (F-ADP); hence, actin filaments contain bound ADP. However, dissociation of actin subunits is not associated with concomitant ATP synthesis, indicating that depolymerisation may not simply be a reversal of polymerisation. Interestingly, polymerisation can still occur if the G-actin is bound to a non-hydrolysable ATP analogue. Short filaments (either formed from the elongation of actin nuclei or else from the growth of pre-existing filaments) may then be joined end to end to form longer filaments.

Actin filaments are thought to exist in a double-stranded, right-hand helix with 14 subunits (per strand) per complete turn (Fig. 4.4), and a crossover distance of 38 nm. This 'strings of beads' appearance is 70 Å in diameter and thought to represent the structure of thin filaments. As the new fila-

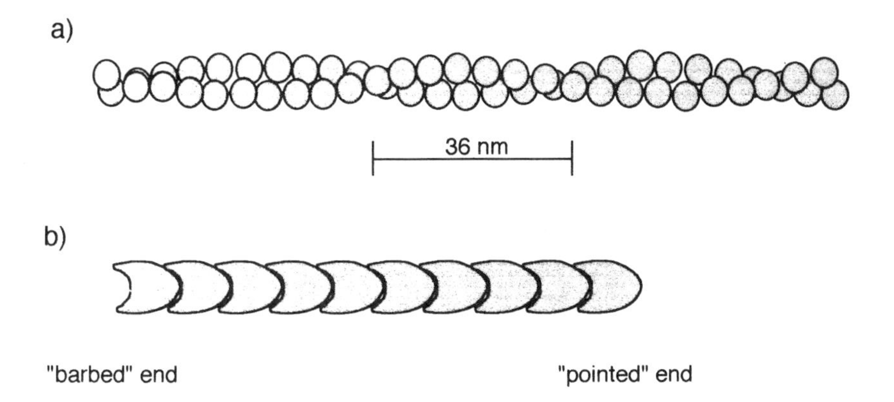

Figure 4.4. Structure of actin filaments: (a) shows the 'beads on a string' appearance of an actin filament. This filament comprises actin monomers, which themselves have a polarity, so that they can only assemble 'head to tail', as shown in (b); thus, the actin filament is polarised, having a 'barbed' and a 'pointed' end.

ments grow, they form straight rods that align at random but tend to become parallel. As the filament sizes increase in length, they overlap and hinder the rotational movement of each other. Some evidence also exists, however, to suggest that F-actin can exist as a linear, single-stranded polymer, with a left-hand helix and an axial rise of 2.73 nm per subunit and 2.16 subunits per turn.

4.1.3 Actin-binding proteins

4.1.3.1 Proteins affecting filament growth

In the resting neutrophil, about 50% of the actin is present in filaments within the cytoskeleton (and hence insoluble in detergents such as Triton X-100), whereas the remainder is detergent soluble and hence is not associated with the cytoskeleton. Data from studies of actin polymerisation in vitro predict that almost all of the actin within the cell should be F-actin (i.e. present in microfilaments). Upon stimulation of neutrophils with agonists such as fMet-Leu-Phe or PMA, actin polymerisation is activated extremely rapidly. There are two important questions: Firstly, how is actin maintained in the unpolymerised state in resting cells? Secondly, how is it rapidly assembled into the cytoskeleton during activation? The answers to these questions lie in understanding the functions of the numerous proteins involved in the assembly and disassembly of actin filaments (Table 4.1).

Table 4.1. *Properties of some actin-binding proteins*

Components	Function	Mass (kDa)	Regulated by
Filamin	branching of actin filaments	540	?
Myosin	binds F-actin and causes contraction	480	Ca^{2+}/calmodulin
α-actinin	bundles F-actin filaments	220	Ca^{2+}
Caldesmon	binds barbed ends	150	Ca^{2+}
Vinculin	bundles and anchors F-actin	130	Ca^{2+}
Gelsolin	blocks barbed ends of actin filaments	84	Ca^{2+}, PIP_2
Tropomyosin	stabilisation of actin filaments	70	?
Acumentin	caps pointed ends	64	?
Profilin	maintains G-actin monomer pool	15	PIP_2

(i) *Profilin.* This is a basic protein with a relative molecular mass of 20 kDa. It binds to G-actin in a 1: 1 configuration (K_d = 3 μM), and the complex (profilactin) cannot form elongation nuclei nor add to existing fibres to elongate them (see Fig. 4.1). Profilin thus buffers the pool of actin monomers and can exist in several different affinity states. In the high-affinity state it binds G-actin so tightly that only denaturing agents can reverse the binding. Apart from inhibiting polymerisation, binding of profilin to actin increases the rate of ATP exchange on the monomer and also inhibits actin monomer ATPase activity. In its low-affinity state profilin can compete only poorly with the fast-growing ends of F-actin for G-actin monomers, and so this form of profilin can promote actin polymerisation. The signalling molecule phosphatidylinositol 4,5-bisphosphate (PIP_2) induces this affinity change and can release actin from profilactin. This may be an important regulatory mechanism for the control of actin polymerisation during cell activation because phospholipid metabolism is one of the first events to be stimulated upon receptor binding (§6.3.1). Thus, profilin may provide a readily-utilisable pool of actin monomers within the cell, which can be used for filament elongation upon cell stimulation. It is also possible that other mechanisms exist for the initiation of affinity changes at different ends of the actin filaments in order to achieve directional filament elongation.

(ii) *Gelsolin.* This 84-kDa protein is one of a number of proteins that can 'cap' the barbed ends of growing actin filaments. When it binds to a barbed end it prevents the exchange of monomers, thus preventing filament extension at this end. Because it does not bind to the pointed ends of the filaments, elongation is still possible at these ends. It also severs actin–actin bonds and so can split existing filaments. Because the concentration of actin monomer may thus increase during these processes, new nuclei of actin elongation may form, but these elongate only slowly. Thus, gelsolin decreases the rate of actin-filament elongation and increases number of short actin filaments. The action of cytochalasins resembles that of gelsolin by inhibiting the exchange of actin monomers with the fast-growing end, thus shortening the length of the filament.

The action of gelsolin is critically regulated by calcium concentrations. It is activated by μM Ca^{2+}, and this activated gelsolin binds firstly to a single actin molecule with fairly-high affinity (10^{-8} M^{-1}). Binding with actin exposes another cryptic binding site, which then binds a second actin molecule with even higher affinity (10^{-10} M^{-1}). Binding to the second actin molecule severs the filament. cDNA sequence analysis reveals that the actin-severing site is located at the NH_2-terminus of the gelsolin, whilst the Ca^{2+} binding site is at the COOH-terminus. EGTA prevents the binding of actin at the first but not the second binding site. On the other hand, PIP_2 inhibits binding of actin at the second binding site, and so prevents the severing of the actin filaments, even in the presence of high (μM) concentrations of Ca^{2+}. Both PIP_2 and PIP can uncap gelsolin-blocked barbed ends, and thus inhibit filament severing.

(iii) *Acumentin.* This protein caps the pointed ends of actin filaments. It has a relative molecular mass of 65 kDa and may constitute up to 6% of the total protein of some leukocytes. It binds to five monomers on F-actin (independently of Ca^{2+}) and may also promote nucleation of new actin filaments.

4.1.3.2 *Proteins affecting cross-linking*

The above processes describe how the growth and depolymerisation of actin filaments (*thin filaments*) is controlled. However, actin filaments are assembled into filamentous networks, and these three-dimensional structures are themselves controlled and also stabilised by a number of proteins:

i. *Filamin (actin-binding protein).* This is a non-covalently-linked dimer of two identical subunits, each of relative molecular mass of 270 kDa. These two polypeptides are joined head to head to produce a flexible, hinged protein that binds actin filaments at 40-nm intervals with a stoichiometry of 1 : 15 actin monomers and with a relatively high affinity ($K_d = 2 \times 10^{-8}$ M). Thus, filamin perpendicularly cross-links adjacent actin filaments to form a branched network.
ii. *Caldesmon.* This is a 150-kDa actin-binding protein that cross-links F-actin chains. Its effects are inhibited by Ca^{2+}/calmodulin.

iii. *α-actinin.* This is a non-covalently linked dimer with a relative molecular mass of 210 kDa. It is a short (100-nm), rodlike protein that joins actin filaments side by side as bundles. Each end of the α-actinin rod binds adjacent fibres, but this protein itself will not cause the Sol → Gel transition (see §4.5). It cross-links actin filaments in the absence of Ca^{2+}, but concentrations of 1 μM Ca^{2+} prevent this activity. Diacylglycerols and palmitic acid promote α-actinin–F-actin interaction.

4.1.3.3 Anchorage proteins

Vinculin is one of a number of proteins that bind the actin network to the plasma membrane. It is a 130-kDa protein that is phosphorylated at a tyrosine residue by a kinase that is controlled by the *src* gene. Vinculin function is Ca^{2+}-dependent, and it is found in close association with α-actinin.

4.1.3.4 Myosin

Myosin accounts for about 1% of the total protein of neutrophils. It comprises a pair of heavy (200-kDa) chains, each of which has a distinctive long, rod-shaped domain and a globular head (Fig. 4.5). Additionally, each myosin molecule possesses two pairs of light (15–20-kDa) chains that are located near the heads of the heavy chains. Thus, the entire myosin complex is 480 kDa. The two heavy chains have very long α-helical tails that interwind into a two-strand coil, and globular heads to which the light chains bind. Between each head and tail domain is a flexible stalk (Fig. 4.6). Trypsin can cleave the tail domain to yield two fragments called *light meromyosin (LMM)* and *heavy meromyosin (HMM)*. Papain cuts the stalks to give fragments termed *S1* (headpieces plus light chains) and *S2*. The S1 regions possess the binding sites for actin filaments and the ATPase activity. Under appropriate conditions, myosin molecules form short (300-nm) bipolar filaments (*thick filaments*), in which the rodlike tails form the backbone of the filament and the globular heads project from the shafts of the filament at each end (Fig. 4.7). The heads of these filaments interact with F-actin, thus forming a characteristic arrowhead configuration, which gives directionality to the actin fibre.

Assembly of the actin network merely by interaction with these binding proteins can itself account for pseudopodia formation and propulsive movement. However, there is some evidence to suggest that F-actin–myosin interactions are required for vectorial movement; hence it has been demonstrated that pseudopodia contain filament networks comprising actin and myosin. Myosin plays a role in the contractile movement of neutrophils in a

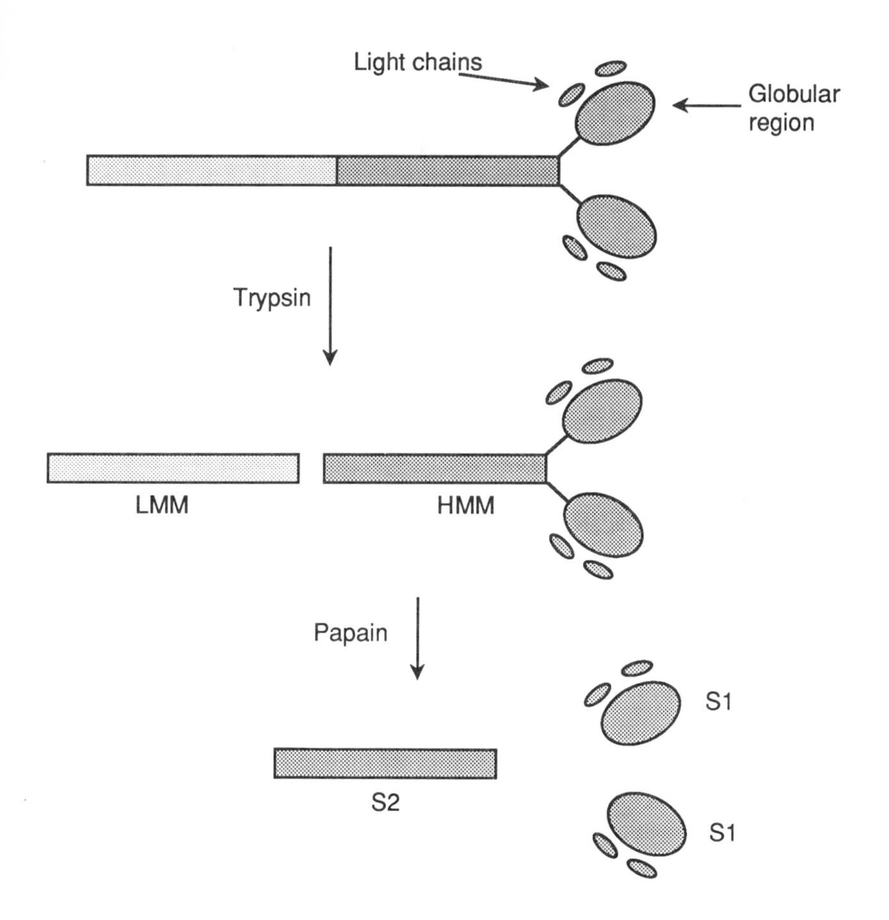

Figure 4.5. Structure of myosin. Myosin comprises both light and heavy chains. The heavy chains may be cleaved by trypsin to generate light meromyosin (LMM) and heavy meromyosin (HMM). Papain digestion of HMM yields subfragments S1 and S2; each S1 fragment contains an ATPase site and an actin-binding site. The light chains modify the activity of the ATPase.

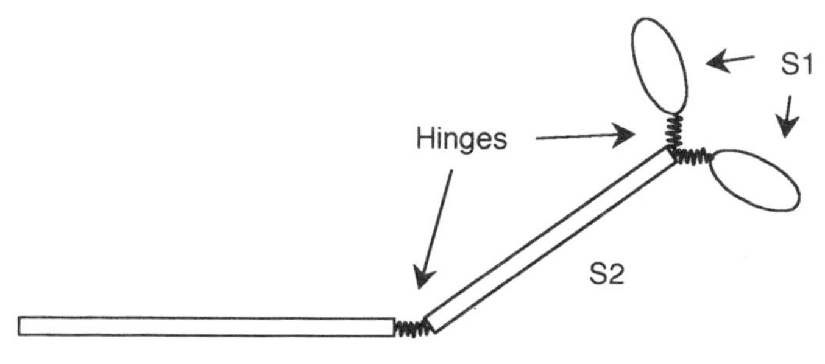

Figure 4.6. Hinge regions of myosin. Two hinge regions are present, one between LMM and S2, the other between S1 and S2.

Figure 4.7. A thick filament of myosin.

way similar to that in which it functions in the movement of muscle cells. In muscle cells the contractile force is generated by the cyclic formation and by dissociation of complexes of the S1 heads of myosin with actin. Thus, the S1 heads bind to the actin molecules of thin filaments, and the heads then tilt to move the thin filament by about 75 Å (Fig. 4.8). The S1 heads then detach from the thin filament and can undergo further cycles of attachment, tilting and detachment. These processes result in ATP hydrolysis.

Myosin S1 headpieces bind both ADP and ATP with high affinity and possess Mg^{2+}-dependent ATPase activity. In the resting state myosin binds ADP and Pi (stage 1, Fig. 4.8); upon stimulation, the S1 headpiece binds actin in a perpendicular fashion (stage 2). The bound ADP and Pi are released, and the S1 head tilts 45° to the actin molecule, causing filament movement (stage 3). The head then binds ATP and then detaches (stage 4). The hydrolysis of ATP (to bound ADP and Pi via the inherent ATPase activity of the myosin) occurs as the head returns to the perpendicular position. This ATPase activity is regulated by the phosphorylation status of the light chain. A myosin light-chain kinase is responsible for this phosphorylation (and thus stimulation of ATPase activity); the activity of this kinase is itself activated by Ca^{2+}/calmodulin. Ca^{2+} levels also play a regulatory role in these processes because rises in the concentration of this cation stimulate the binding of actin to the myosin–ADP–Pi complex.

4.2 Microtubules

In resting neutrophils it is estimated that there are about 11–23 microtubules per cell, with a diameter of approximately 25 nm and a wall width of 5 nm. They are long, tubular structures made by the helical formation of tubulin molecules, which are either α- or β-subunits, each with a relative molecular mass of 55 kDa (Fig. 4.9). Each subunit is present in equimolar amounts in a tubulin molecule, and these subunits exist as dimers of one α- and one β-subunit. Because microtubules are polar, growth of the fibre is biased towards one end, termed the *plus end.* A number of microtubule-associated proteins (MAPs) affect the dynamic shape of the microtubule, and in the resting neutrophil about 35–40% of the tubulin pool is assembled, whilst the remainder can be assembled very rapidly after cell stimulation.

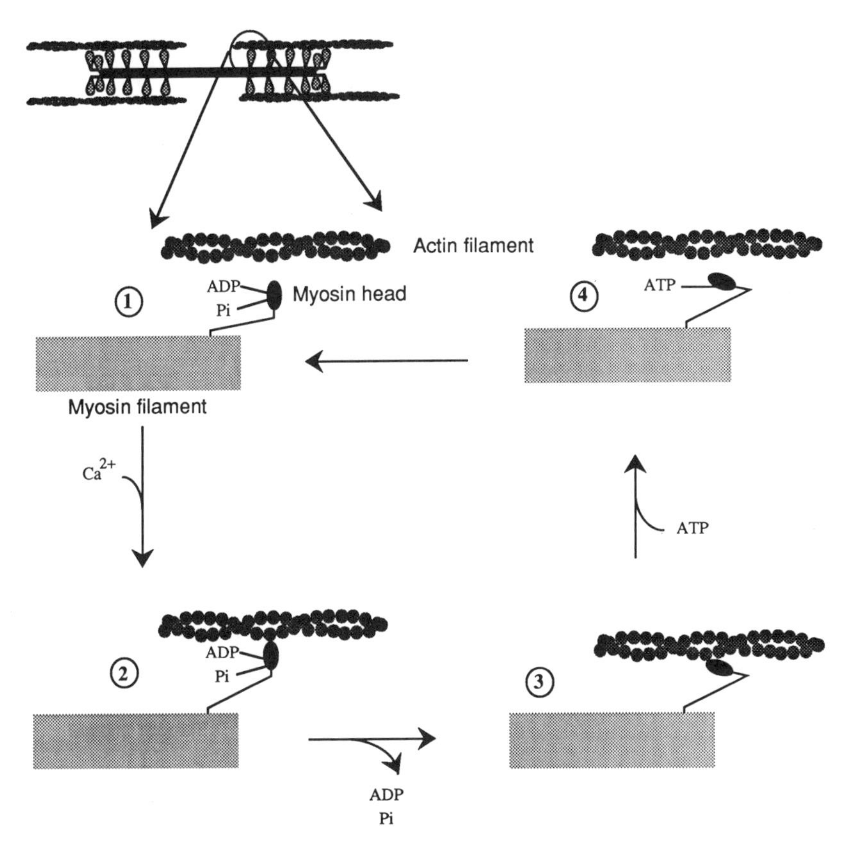

Figure 4.8. Mechanism of force generation by the interaction of actin and myosin. See text for details.

Microtubules may function as a form of 'skeletal support' for microfilaments. Agents that increase intracellular cGMP favour the assembly of microtubules, whereas those that increase intracellular Ca^{2+} and cAMP result in the dissolution of tubulin fibres. Furthermore, the oxidation state of the neutrophil may affect the integrity of the tubulin fibres. Oxidised glutathione (which is increased during oxidative metabolism) regulates tubulin disassembly, and oxidation may increase tubulin tyrosylation, which also promotes disassembly.

The role of microtubules in neutrophil function can be investigated using agents such as colchicine, colcemid, vinblastine and vincristine, which disrupt these structures. Stimulation of neutrophils with chemotactic agents causes a rapid and transient assembly of microtubules, but this assembly does not affect chemotaxis. Similarly, *cytoplasts* (neutrophils devoid of nu-

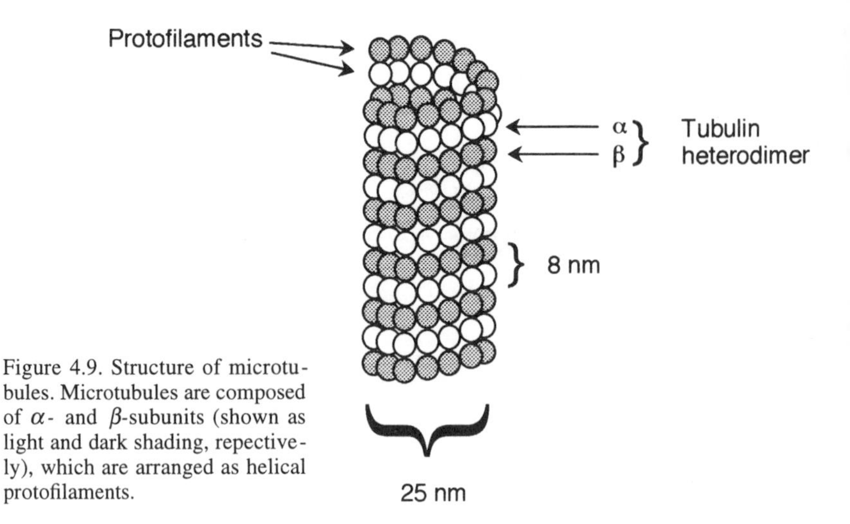

Figure 4.9. Structure of microtubules. Microtubules are composed of α- and β-subunits (shown as light and dark shading, repectively), which are arranged as helical protofilaments.

cleic and cytoplasmic organelles) lack microtubules, but are still capable of directional movement. Most reports also suggest that phagocytosis is unaffected when neutrophils are pretreated with microtubule-disrupting agents.

The role of microtubules in secretion is more clearly defined. Colchicine and vinblastine inhibit secretion, even in cytochalasin-B-treated cells, and D_2O (which promotes tubulin assembly) enhances secretion in cytochalasin-treated cells. Microtubules may also be necessary for the translocation of phagocytic vesicles from the neutrophil periphery into the central region of the cytoplasm. Drugs affecting microtubule assembly may inhibit particle-induced oxidase activation or else increase oxidase activation in response to soluble agents such as fMet-Leu-Phe.

4.3 Intermediate filaments

Although intermediate filaments are not universally associated with the cytoskeleton, neutrophils possess intermediate filaments of the vimentin type. *Vimentin* is a rod-shaped molecule of relative molecular mass 57 kDa that readily polymerises under physiological conditions to produce stable filaments 10–12 nm in diameter. Intermediate filaments are more robust than microfilaments and microtubules, and in neutrophils they form an open network of single filaments in the perinuclear space.

Their role in neutrophil function is not yet established. They may be located, by immunofluorescence and electron microscopy, in areas between the nucleus and the trailing end of the cell, and in uropodal areas of chemoattractant-treated cells. As these filaments are unable to move independently of other cellular structures, it may be that intermediate filaments provide a mechanical framework through which the peripheral and central regions of the cell are connected.

4.4 Microtrabeculae

The microtrabecular lattice is part of the cytoskeleton and is comprised of an intricate network of fine strands of variable (7–18-nm) diameter. This lattice is made up of many (perhaps up to 400) polypeptides, of which actin is the major component. The cytoplasmic granules appear to be distributed at different depths within this lattice, an organisation that may allow phagosome and granule fusion to occur more efficiently.

4.5 Changes in cytoskeletal arrangement during neutrophil activation

Early descriptions of the changes in cytoplasmic structure that occurred during pseudopodia formation referred to a soluble (Sol) and a gelled or semi-solid (Gel) state. Thus, it was assumed that the cytoplasm in the organelle-free peripheral cytoplasm was in the Sol state, whilst in pseudopodia (which resist deformation by mechanical forces) it is in the Gel state. Thus, the control of neutrophil functions via changes in cytoplasmic structure may be explained by understanding this so-called *Sol → Gel transition.* It is perhaps more convenient to think of this transition as the processes that elongate and cross-link actin filaments.

How, then, are these processes controlled during the shape changes that are required for such neutrophil functions as chemotaxis and phagocytosis? Both of these processes require three-dimensional changes in cell shape, for chemotaxis in order to migrate along a chemotactic gradient, and for phagocytosis in order to engulf foreign particles (Fig. 4.10). In many ways, these processes may appear analogous: they both require protrusion of pseudopodia to a site forward of the cell, and then contraction, either to pull the cell along the substratum during chemotactic movement or else to engulf the particle. (It is not certain during phagocytosis whether the neutrophil 'pulls' the particle into the cell body or if the neutrophil cytoplasm moves around the target.) It is evident that within pseudopodia the cytoplasm is in the Gel state because of the abundance of actin filaments and activity of actin-binding proteins that cross-link these filaments. Increases in intracellular Ca^{2+} ($>10^{-7}$ M) may dissolve the gel (i.e. promote a Gel → Sol transition) because this cation activates gelsolin, which shortens actin filaments and prevents elongation. When intracellular Ca^{2+} levels decline, gelsolin is inactive, the filaments reanneal and so a Sol → Gel transition is promoted. Thus, it may be envisaged that localised Ca^{2+} changes within an activated neutrophil (as may occur during directional movement or during phagocytosis) may cause localised changes in actin assembly and disassembly, and so directional extension of pseudopodia. Thus, changes in actin-filament length and the extent of cross-linking may account for protrusion, whilst the activity of myosin may account for contraction. Actin–myosin interactions can therefore lead to force generation (at the expense of ATP hydrolysis) and to the contraction of the (at least partially) solated gel.

4.6 Changes in cytoskeletal structure during chemotaxis and phagocytosis

4.6.1 Proposed models

Two models have been proposed to explain the changes in cytoskeletal arrangements that occur during three-dimensional movement. Stossel (1988) and his colleagues propose a model in which the actin filaments are cross-linked by filamin into an isotropic array associated with gelsolin, myosin and other proteins. Because such an extensively cross-linked gel would be rigid, any myosin activated within such a gel would be working against the gel's elasticity rather than producing net movement or shape change. If the gel is made less rigid (i.e. partially solated), myosin can then generate net relative movement in this loosened filament network. Gelsolin (which is Ca^{2+} activated) cleaves the actin filaments and hence solates the gel in the presence of Ca^{2+}. Thus, shape changes are modulated by Ca^{2+} (plus calmodulin) and myosin light-chain kinase, which then control myosin activity. Thus, intracellular Ca^{2+} rises would decrease the rigidity of the gel and stimulate the interaction of myosin with actin. This model explains shape changes but does not, however, explain protrusion.

Oster (1984) and colleagues have proposed the following model. The actin-filament network is a net negatively-charged, cross-linked (with filamin) polymer network, contained within an ionic aqueous environment This may thus be considered as a charged polymer trapped within a semi-

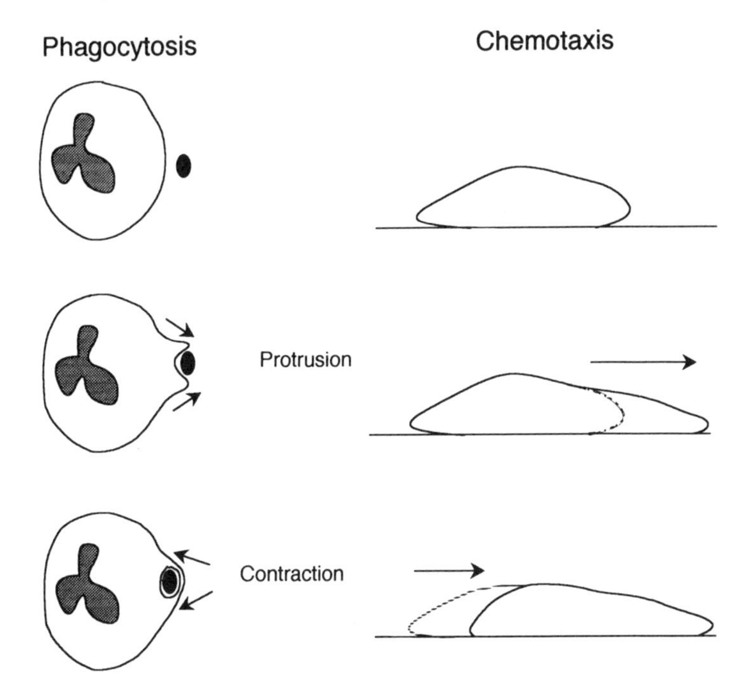

Figure 4.10. Changes in cell shape during phagocytosis and chemotaxis. Both of these neutrophil functions require protrusion of pseudopodia (either to engulf the bacterium or else to move the cell along a chemoattractant gradient) followed by contraction.

permeable membrane. This charged mass is subject to Donnan influx of ions, which raises the osmotic pressure within the vicinity of filaments, resulting in swelling. Net negative charges may also contribute to interfilament repulsion, again causing swelling. The amount of swelling of this meshwork is restricted by its filamin cross-links; it therefore will adopt an equilibrium volume whose dispersive forces are equal and opposite to the elastic energy stored within the cross-linked meshwork.

If the cross-links or polymers are severed, then some elastic energy is released and the system will adopt a new (larger) equilibrium volume where greater distortion is conferred upon a meshwork that has fewer cross-links. Thus, upon increases in Ca^{2+}, gelsolin activity leads to an increase in the volume of the actin-filament network. The additional influence of myosin on such a meshwork is similar to that proposed in the Stossel model. Thus, three-dimensional Ca^{2+} gradients (between a localised region of the cell surface and an external structure) can result in complex shape changes.

Protrusion may be due to growth of new actin filaments, which requires net polymerisation of new filaments, and also by the organisation of actin-binding proteins into higher-order structures. Random movements of flexible membranes away from the filaments may result in gross distortion of actin polymerisation at the barbed ends. Thus, once a critical size is reached, ion pumping (i.e. of Ca^{2+}) may occur at the tip of a pseudopod, which further aids directional changes in the network.

4.6.2 Actin assembly during chemotaxis

Neutrophils may move at speeds of up to 20 μm min^{-1} in response to chemoattractants such as denatured proteins, lipids, peptides or C5a. Movement may be defined either as *chemokinesis,* which is generalised (non-directional) locomotive activity, or as *chemotaxis,* which is orientation and directional migration up a concentration gradient. A concentration difference at opposite ends of the cell of only 1% is sufficient to activate such directional movement. However, neutrophils do not respond chemotactically to static gradients of chemoattractants, and both temporal and directional changes in chemoattractant concentrations are required.

The cytoplasm extends pseudopodia or *lamellipodia* (which are devoid of cytoplasmic granules) in the direction of locomotion (Fig. 4.11). These briefly attach to the substratum to grip, and thus generate a frictional force. The trailing edge of the cell may acquire a 'tail', which is a region of shedding or internalisation of capped membrane structures that become cross-linked after binding to ligands. The tail is attached to the substratum, and the region may become tethered to form thin strings of retraction fibres as the cell moves forward. Movement is thus a combination of protrusion of pseudopods in the direction of the gradient, and contraction of the tail.

Pseudopodia are rich in actin and actin-binding proteins; their extension may thus result from the growth of actin filaments within this region. Alternatively, such extension may result from contraction of the cortical actin, which may exert hydrostatic pressure on the internal fluid cytoplasm, and/or swelling of the cortical actin. Actin polymerisation in pseudopods may thus occur via localised changes in intracellular Ca^{2+}, which at high concentrations shortens the filaments and at low concentrations reanneals the filaments. After the formation of a pseudopod, adhesion sites may be formed; these are patches of integral membrane proteins into which actin filaments are inserted. These filaments may then be stabilised via interactions with vinculin at the insertion sites, and stabilised along the filament via interactions with filamin, α-actinin and other actin-binding proteins.

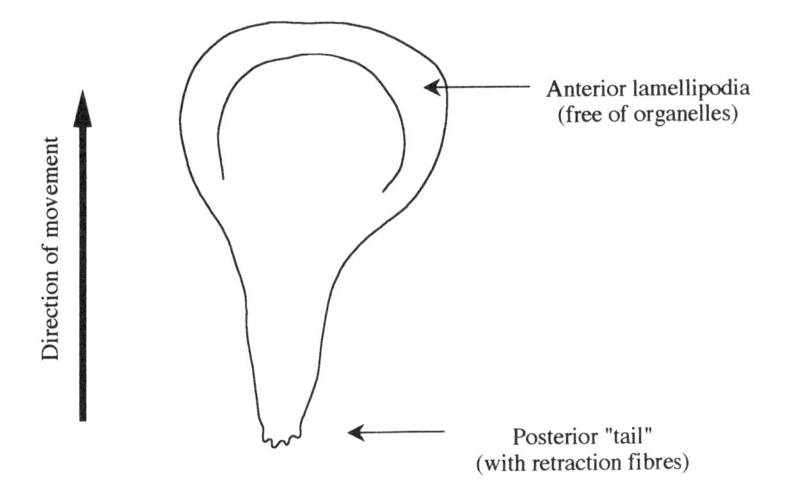

Figure 4.11. Polarisation of a neutrophil during phagocytosis. During directional movement (*chemotaxis*) the neutrophil becomes polarised, assuming an anterior 'head', which is flattened and free of cytoplasmic granules, and a posterior 'tail'.

Propulsion may also be regulated by Ca^{2+} changes: upon activation of gelsolin, the actin matrix will be weaker, thus allowing myosin to interact with F-actin bundles. In the presence of ATP, the myosin will be released from the actin filaments, and if Ca^{2+} is sequestered, then the rigidity of the matrix is restored. During this process, occupied receptors are internalised or capped to the tail of the cell. In order for movement to be continued, new or recycled receptors must then appear on the leading edge of the plasma membrane. Movement then continues in the direction of pseudopod formation until the cell can no longer detect a chemotactic gradient.

4.6.3 Actin assembly during phagocytosis

The initial step in phagocytosis is the binding of a particle to the cell surface. This recognition and subsequent binding is aided by the coating of the particle with opsonins (serum proteins such as complement fragments and immunoglobulins) and the attachment of such opsonised particles to complement and immunoglobulin receptors on the neutrophil plasma membrane (see §§3.8–10). Lamellipodia then extend and progressively migrate over the particle surface until they meet and then fuse to enclose the particle completely in a membrane-bound vesicle known as a *phagosome*.

Cytochalasins bind to the barbed ends of filaments with a K_d of 10^{-7}–10^{-8} M, but with much lower affinity to the sides of filaments or to monomers ($K_d = 10^{-5}$ M). The I_{50} for inhibition of phagocytosis by cytochalasins is closer to the concentration required for fragmentation of filaments rather than for binding to filament ends. Therefore, phagocytosis may not require net actin assembly at the barbed ends, but rather may require the tensile or structural properties of filaments.

Actin filaments are preferentially associated with the sites of particle ingestion, as determined by immunocytochemical analyses. Myosin and filamin are also localised at these sites. The organelle-free cytoplasm that surrounds the engulfed particle (which has the characteristics of the cortical rim of cytoplasm) appears rich in actin filaments. The engulfed particle then passes through the cortical meshwork deep into the cytoplasm. Actin filaments are assembled during engulfment and are subsequently disassembled. This process is not inhibited by cytochalasin D, implying that the pointed ends, but not the barbed ones, are involved in polymerisation. Recent evidence suggests that, in the presence of cytoplasmic concentrations of inorganic phosphate, the critical concentration at opposite ends of the filament are the same.

Thus, assembly of actin filaments may occur at the pointed ends during phagocytosis, but at the barbed ends during locomotion.

4.6.4 Other neutrophil functions

The cytoskeleton also plays an important role in regulating receptor–ligand activation processes, particularly in the termination of second-messenger generation during activation. For example, fMet-Leu-Phe agonist–receptor complexes rapidly become associated with the cytoskeleton after neutrophil activation, and this may coincide with the termination of second-messenger generation. These receptors are coupled to heterotrimeric G-proteins (with α-, β- and γ-subunits; see §6.2.1), and in resting cells the majority of the α-subunits are associated with the cytoskeleton. Binding of fMet-Leu-Phe with its receptor results in the binding of GTP to the α-subunit, which then releases both from the cytoskeleton and from the β- and γ-subunits. It then associates with and activates a phospholipase. The fMet-Leu-Phe–receptor complex may then associate with the cytoskeleton, and this may prevent recombination with the G-proteins, thus terminating the generation of second messengers and also assisting in internalisation. Low-molecular-weight G-proteins (*ras*-related proteins; see §6.2.2) are also associated with the cytoskeleton in resting cells, and during receptor occupancy these may also be released and so perform regulatory functions.

The cytoskeleton also plays an important role in the regulation of the NADPH oxidase (§5.3.2). This oxidase is activated by the assembly of individual components present in the plasma membrane, on the membranes of subcellular organelles (specific and gelatinase-containing granules and secretory vesicles) and in the cytoplasm. Hence, the assembly of these components, and their continuous recruitment from subcellular locations, requires the participation of the cytoskeleton. Indeed, cytochalasin B can affect oxidase function. During stimulation by soluble agonists (such as fMet-Leu-Phe), cytochalasin B enhances oxidase activity (possibly because disruption of the cytoskeleton results in loss of control of movement of the cytoplasmic contents and granules). In contrast, stimulation of oxidant production by particulate stimuli is prevented by cytochalsin B, presumably because the cytoskeleton is required for the physical processes involved in particle uptake.

4.7 Bibliography

Bengtsson, T., Dahlgren, C., Stendahl, O., & Andersson, T. (1991). Actin assembly and regulation of neutrophil function: Effects of cytochalasin B and tetracaine on chemotactic peptide-induced O_2^- production and degranulation. *J. Leuk. Biol.* **49**, 236–44.

Bochsler, P. N., Neilsen, N. R., Dean, D. F., & Slauson, D. O. (1992). Stimulus-dependent actin polymerisation in bovine neutrophils. *Inflammation* **16**, 383–92.

Crawford, N., & Eggleton, P. (1991). Dynamic changes in neutrophil cytoskeleton during priming and subsequent surface stimulated functions. *Biochem. Soc. Trans.* **19**, 1048–55.

Downey, G. P., Elson, E. L., Schwab, B. I., Erzurum, S. C., Young, S. K., & Worthen, G. S. (1991). Biophysical properties and microfilament assembly in neutrophils: Modulation by cyclic AMP. *J. Cell Biol.* **114**, 1179–90.

Edwards, S. W. (1989). Interactions between bacterial surfaces and phagocyte plasma membranes. *Biochem. Soc. Trans.* **17**, 460–2.

Ginis, I., Zaner, K., Wang, J.-S., Pavlotsky, N., & Tauber, A. I. (1992). Comparison of actin changes and calcium metabolism in plastic- and fibronectin-adherent human neutrophils. *J. Immunol.* **149**, 1388–94.

Harvath, L. (1990). Regulation of neutrophil chemotaxis: Correlations with actin polymerization. *Cancer Invest.* **8**, 651–4.

Haston, W. S., & Wilkinson, P. C. (1987). Gradient perception by neutrophil leukocytes. *J. Cell Sci.* **87**, 373–4.

Jesaitis, A. J., Tolley, J. O., & Allen, R. A. (1986). Receptor-cytoskeleton interactions and membrane traffic may regulate chemoattractant-induced superoxide production in human granulocytes. *J. Biol. Chem.* **261**, 13662–9.

Niggli, V., & Keller, H. (1991). On the role of protein kinases in regulating neutrophil actin association with the cytoskeleton. *J. Biol. Chem.* **266**, 7927–32.

Oster, G. F. (1984). On the crawling of cells. *J. Embryol. Exp. Morphol.* **83** (Suppl), 329–64.

Rothwell, S. W., Nath, J., & Wright, D. G. (1989). Interactions of cytoplasmic granules with microtubules in human neutrophils. *J. Cell Biol.* **108**, 2313–26.

Salmon, J. E., Brogle, N. L., Edberg, J. C., & Kimberly, R. P. (1991). Fcγreceptor III induces actin polymerization in human neutrophils and primes phagocytosis mediated by Fcγ receptor II. *J. Immunol.* **146**, 997–1004.

Sarndahl, E., Bokoch, G. M., Spicher, K., Stendahl, O., & Andersson, T. (1991). G-proteins and the association of ligand/receptor complexes with the cytoskeleton in human neutrophils. *Biochem. Soc. Trans.* **19**, 1127–9.

Sheterline, P., & Rickard, J. E. (1989). The cortical actin filament network of neutrophil leucocytes during phagocytosis and chemotaxis. In *The Neutrophil: Cellular Biochemistry and Physiology* (Hallett, M. B., ed.), pp. 141–65, CRC Press, Boca Raton, Fla.

Smith, W. B., Gamble, J. R., Clark-Lewis, I., & Vadas, M. A. (1991). Interleukin-8 induces neutrophil transendothelial migration. *Immunol.* **72**, 65–72.

Stossel, T. P. (1988). The mechanical responses of white blood cells. In *Inflammation: Basic Principles and Clinical Correlates* (Gallin, J. I., Goldstein, I. M., & Snyderman, R., eds.), pp. 325–42, Raven Press, New York.

Wilkinson, P. C. (1990). How do leukocytes perceive chemical gradients? *FEMS Microbiol. Immunol.* **64**, 303–12.

Yuruker, B., & Niggli, V. (1992). α-Actinin and vinculin in human neutrophils: Reorganization during adhesion and relation to the actin network. *J. Cell Sci.* **101**, 403–14.

Zigmond, S. H. (1989). Chemotactic response of neutrophils. *Am. J. Respir. Cell Mol. Biol.* **1**, 451–3.

5

The respiratory burst: The generation of reactive oxygen metabolites and their role in microbial killing

5.1 Oxygen metabolism and neutrophil function

It has been known for many years that the oxygen metabolism of neutrophils is unusual. As long ago as 1933 Baldridge and Gerard recognised the importance of oxygen for bacterial killing when they observed a 'respiratory burst' of increased O_2 uptake that accompanied phagocytosis. It was known that, in most cells, O_2 was needed to generate energy during mitochondrial respiration, and hence the respiratory burst was mistakenly thought to be required to supply the extra energy needed for the physical processes involved in phagocytosis. It was not until 1959 that the unusual nature, and hence the unusual enzyme system, of the respiratory burst was revealed, when it was shown that phagocytosis and bacterial killing could occur in neutrophils where mitochondrial respiration was poisoned by cyanide. This discovery led to the conclusion that the respiratory-burst enzyme has nothing to do with mitochondrial respiration, and so the extra O_2 consumed must serve some other purpose in phagocytosing neutrophils. In fact, mature neutrophils possess very few, if any, active mitochondria, and they obtain their ATP for energy-utilising processes from glycolysis, a process thst does not require O_2 supply. This independence of energy generation on O_2 supply is important because it means that many neutrophil functions can occur efficiently in the O_2-depleted environments that may be found in certain pathological circumstances (e.g. in inflamed tissues). Yet if O_2 was not needed by phagocytosing neutrophils for oxidative phosphorylation, then for what other purpose was it required?

In the early 1960s it was proposed, and then later shown experimentally, that H_2O_2 was generated during phagocytosis; hence, it was suggested that the respiratory burst was required for the generation of toxic oxygen metabolites that were involved in pathogen killing. This link between the respiratory burst and pathogen killing was confirmed when it was discovered that phagocytes from patients who have chronic granulomatous disease (CGD, formerly called 'fatal granulomatous disease of childhood') and who are predisposed to life-threatening infections cannot mount an efficient respiratory burst during phagocytosis. This observation led to the correct conclusion that CGD neutrophils have a defective respiratory-burst enzyme. Around this time Ivan Fridovich discovered the superoxide dismutases (SODs) and made the remarkable finding that all aerobic cells generate O_2^- during oxidative metabolism (Fridovich 1978). In his experiments he devised an assay to measure the production of O_2^- by cells and tissues that was based upon the reduction of ferri-(oxidised) cytochrome c, which was preventable by the addition of SOD to the assay (see §5.4.2.1). Bernard Babior and his colleagues (Babior, Kipnes & Curnutte, 1973) used this type of assay to show that phagocytosing neutrophils generate O_2^-. In fact, it has been calculated that all of the extra O_2 consumed during the respiratory burst is converted to O_2^-; hence, the extra O_2 consumed during the respiratory burst is converted to O_2^-, and this is then converted into H_2O_2. These and other O_2-centred molecules play important roles in pathogen killing. The CGD neutrophil (see §8.2) is thus of great importance in phagocyte research. Firstly, it has provided the link between the generation of reactive oxygen metabolites and pathogen killing. Secondly, the identification of the defect(s) in the respiratory-burst enzyme in these patients has enabled phagocyte researchers to identify, and then to characterise, the molecular structure of the enzyme responsible for the production of O_2^-.

5.2 Oxygen, oxygen metabolites and toxicity

Before we discuss how neutrophils (and other professional phagocytes) generate O_2^- and other oxygen metabolites during the respiratory burst, let us first consider why they produce such compounds. If these metabolites are produced in order to kill phagocytosed microorganisms, why and how are they toxic? The answer comes from an examination of the chemistry of the oxygen molecule itself.

The O_2 molecule is essential to all aerobic forms of life, but many anaerobic organisms (e.g. anaerobic bacteria such as *Clostridia* spp.) are killed after only brief exposures to molecular O_2. However, it is well established that even aerobic organisms, including man and other animals, show signs of oxygen toxicity when exposed to O_2 tensions above those normally found in air (i.e. >21% O_2). Such toxicity does not normally occur because aerobic cells possess protective enzymes that prevent either the formation or the accumulation of oxygen metabolites. It is only when these protective systems be-

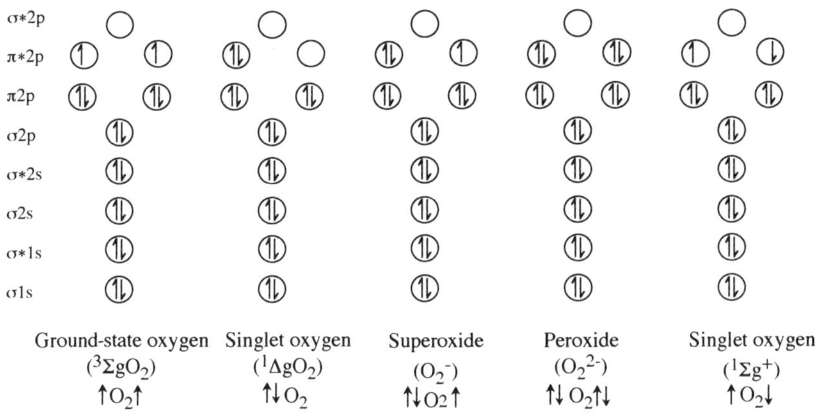

σ*2p
π*2p
π2p
σ2p
σ*2s
σ2s
σ*1s
σ1s

Ground-state oxygen ($^3\Sigma gO_2$) $\uparrow O_2 \uparrow$ | Singlet oxygen ($^1\Delta gO_2$) $\uparrow\downarrow O_2$ | Superoxide (O_2^-) $\uparrow\downarrow O_2 \uparrow$ | Peroxide (O_2^{2-}) $\uparrow\downarrow O_2 \uparrow\downarrow$ | Singlet oxygen ($^1\Sigma g^+$) $\uparrow O_2 \downarrow$

Figure 5.1. Electronic configuration of oxygen and oxygen metabolites. Each electron in a bonding orbital is represented as an arrow whose direction represents the spin quantum number.

come saturated (because of overproduction of oxygen metabolites) or when they are present at reduced levels (e.g. in anaerobic organisms or in diseased states) that the toxic effects of oxygen metabolism become apparent. Toxicity thus arises because of the formation of oxygen-centered free radicals under conditions where the normal protective systems are inadequate. A *free radical* is defined as a species containing one or more unpaired electrons. The O_2 molecule itself (dioxygen, O_2) is thus a free radical (it has two unpaired electrons), but fortunately it is only poorly reactive.

The oxygen molecule's two unpaired electrons are each present in different π^* antibonding orbitals (Fig. 5.1). This electronic structure can be represented as $\uparrow O_2 \uparrow$, with each of the unpaired electrons possessing the same spin quantum number, or each having parallel spin. This is the ground state of oxygen and it is the most stable state. The complete reduction of O_2 requires four electrons (and 4 H^+), and the product is two molecules of water (Fig. 5.2). In aerobic organisms, most of the O_2 consumed by tissues is converted into water by terminal oxidases (such as mitochondrial cytochrome oxidase) during aerobic electron transport, but these four electrons can only be added one at a time. Thus, the complete reduction of O_2 to water is said to occur via the sequential, univalent addition of electrons. No partially-reduced forms of oxygen are released by oxidases such as cytochrome oxidase. This is indeed fortunate: were the reverse the case, the description 'terminal oxidase' would take on quite a different meaning!

If ground-state O_2 were to accept two electrons in a single reaction step in order to fill its π^* antibonding orbitals, then both of these electrons must be

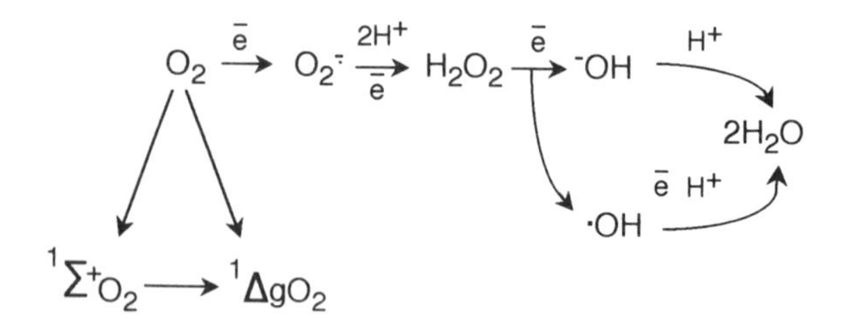

Figure 5.2. Pathways of oxygen reduction. The sequential reduction of molecular oxygen, ultimately to water, is shown. *Abbreviations:* $O_2^{-\cdot}$, superoxide free radical; H_2O_2, hydrogen peroxide; ^{-}OH, hydroxide ion; $\cdot OH$, hydroxyl free radical; *e*, electron; H^+, hydrogen ion. Singlet states of oxygen are also shown as $^1\Sigma^+O_2$ and $^1\Delta gO_2$.

of antiparallel spin -- that is, the reacting molecule must have the electronic configuration ↓X↓. This so-called *spin restriction* severely limits the number of species with which ground-state O_2 can react. Hence, even though O_2 is a free radical by virtue of its unpaired electrons, it is relatively unreactive. This spin restriction is overcome because ground-state oxygen can accept electrons one at a time. Indeed, many of the hundreds of O_2-utilising enzymes that exist within aerobic cells have transition metals at their active sites (sometimes in the form of haem) because these metals can accept and donate single electrons.

5.2.1 Singlet oxygen

Two forms of singlet oxygen exist. The first of these, $^1\Delta gO_2$ (or ↑↓O_2) has an energy of 22.4 kcal above ground-state O_2, whilst the energy of $^1\Sigma g^+O_2$ (or ↑O_2↓) is 37.5 kcal above ground state. The latter is so unstable that it rapidly decays to $^1\Delta gO_2$, which is thus the form of singlet oxygen commonly detected in biological systems. Singlet oxygen states may be formed by photosensitisation reactions when certain molecules, such as dyes, riboflavin, FAD, FMN, chlorophylls, bile pigments or porphyrins, are illuminated with light of the appropriate wavelength. They may also be formed by mixing hypochlorite with H_2O_2. As both these reactants are generated by neutrophils, such a reaction may lead to singlet-oxygen formation during the respiratory burst.

Singlet oxygen may react with biological molecules either by incorporating itself into the target molecule or else by transferring its energy to the target molecule, itself decaying to ground-state oxygen. Compounds with double-bond carbons (–C=C–) are susceptible to chemical reactions with singlet oxygen, a phenomenon that may be important in lipid peroxidation or in eicosanoid formation (see §6.3.2.2).

5.2.2 Superoxide formation

If ground-state oxygen is reduced by a single electron, then the *superoxide radical* ($\uparrow\downarrow O_2\downarrow$) is produced. Because this has an unpaired electron, it is a free radical, and it is denoted O_2^- or $O_2^{-\cdot}$, with the dot ($\cdot$) indicating the unpaired electron and the minus sign ($-$) the negative charge. The superoxide radical is generated by all aerobic cells because these cells possess the enzyme (or group of related enzymes) superoxide dismutase (SOD), which limits the concentration of this radical that can accumulate within cells. In aqueous solutions, particularly at acidic pH values, it can become protonated ($HO_2\cdot$); but because the pKa of $HO_2\cdot$ is 4.8, at physiological pH values, O_2^- will be almost completely in the unprotonated form. O_2^- can dismutate as shown below:

$$2O_2^- + 2H^+ \rightarrow H_2O_2 + O_2 \tag{5.1}$$

This is a dismutation reaction because one molecule of O_2^- is reduced whilst the other is oxidised:

$$O_2^- + 2H^+ + \text{electron} \rightarrow H_2O_2 \tag{5.2}$$

$$O_2^- - \text{electron} \rightarrow O_2 \tag{5.3}$$

This reaction occurs slowly at neutral pH ($5 \times 10^5\ M^{-1}\ s^{-1}$) but more rapidly at acidic pH. In aqueous solutions, O_2^- can act as a reducing agent by donating an electron. This forms the basis for an assay to detect O_2^- production because it is able to donate an electron to, and hence reduce, ferri-(oxidised) cytochrome c (§5.4.2.1). Reduced cytochrome c has an absorption maximum at 550 nm, and this increase in absorption can be monitored spectrophotometrically. In aqueous solutions, O_2^- can also act as a weak oxidising agent, and its reducing properties are greatly increased in organic solvents. Because of its negative charge, O_2^- cannot readily permeate biological membranes.

5.2.3 Hydrogen peroxide

When ground-state oxygen accepts two electrons, or when O_2^- accepts a single electron, the *peroxide ion* ($\uparrow\downarrow O_2\uparrow\downarrow$) is formed. Because this has no unpaired electrons, this is not by definition a free radical. The protonated form of this ion is H_2O_2, a weak oxidising agent that can oxidise -SH groups, thus inactivating some proteins. High concentrations of H_2O_2 are bactericidal. It may be utilised by peroxidases such as glutathione peroxidase (Fig. 5.3) or myeloperoxidase (§5.4.1), or else may be degraded by the activity of catalase:

$$2H_2O_2 \rightarrow 2H_2O + O_2 \tag{5.4}$$

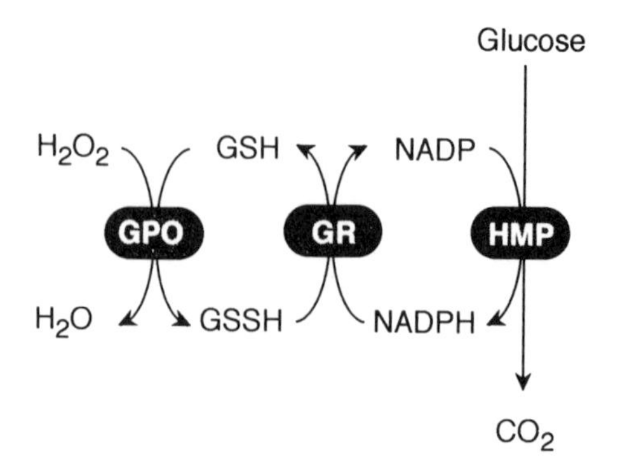

Figure 5.3. The glutathione cycle. *Abbreviations:* HMP, hexose monophosphate shunt; GSH, reduced glutathione; GSSH, oxidised glutathione; GR, glutathione reductase; GPO, glutathione peroxidase. In the system shown, H_2O_2 is converted into H_2O, but the system is also effective in breaking down organic peroxides.

5.2.4 Hydroxyl radical

It has been known for many years that homolytic fusion of the O–O bond in H_2O_2 will yield the *hydroxyl free radical* ·OH. This may be produced by exposure of H_2O_2 solutions to heat or ionising radiation and hence may be formed, for example, after accidental or therapeutic exposure to radiation (e.g. during cancer therapy):

$$H_2O_2 \rightarrow 2 \cdot OH \qquad (5.5)$$

It was also known from much earlier studies that the addition of transition-metal salts (e.g. iron or copper salts) to H_2O_2 solutions will also result in the formation of ·OH via Fenton's reaction:

$$H_2O_2 + Fe^{2+} \rightarrow Fe^{3+} + \cdot OH + {-OH} \qquad (5.6)$$

Other redox reactions can also occur in such mixtures, and some can lead to the formation of O_2^- and the regeneration of Fe^{2+}:

$$Fe^{3+} + H_2O_2 \rightarrow Fe^{2+} + O_2^- + H^+ \qquad (5.7)$$

Overall, the sum of these reactions is

$$2H_2O_2 + \text{Fe salts} \rightarrow O_2 + 2H_2O \qquad (5.8)$$

However, more recently Halliwell and Gutteridge (1984, 1985) have supported the idea that mixtures of O_2^- and H_2O_2, together with transition-metal salts (e.g. iron and copper salts) will also result in the formation of ·OH by the so-called *metal catalysed Haber–Weiss reaction:*

$$O_2^- + H_2O_2 \xrightarrow{\text{Fe/Cu salt}} \cdot OH + O_2 + {}^-OH \quad (5.9)$$

Because all aerobic cells generate O_2^-, which will dismutate (either spontaneously or enzymically) into H_2O_2, then, provided a suitable transition-metal salt is available for reaction (5.9), there is the possibility that $\cdot OH$ formation may occur in biological systems.

5.2.5 *Oxygen toxicity*

For many years, it was believed that the toxic effects of oxygen were directly due to O_2^- formation, and this belief led to the superoxide theory of oxygen toxicity. However, it is now appreciated that the superoxide radical is not very reactive in biological systems. It is now thought that most, if not all, of the damage associated with O_2 toxicity is due, not to O_2^- formation, but instead to the $\cdot OH$ generated from it via reaction (5.9), in the metal-catalysed Haber–Weiss reaction. Experiments measuring the toxic effects of O_2^- and H_2O_2 will inevitably contain transition-metal salts (as contaminants of biological buffers) that are capable of catalysing $\cdot OH$ formation. Indeed, the toxic effects of O_2^- and H_2O_2 are dramatically decreased if great care is taken to remove or chelate transition metals from experimental buffers. The hydroxyl free radical $\cdot OH$ is the most reactive species known in biological systems, reacting indiscriminately with the first molecule that it encounters. Thus, $\cdot OH$ will react at, or very close to, its site of formation, which is the site where the oxidant (such as H_2O_2) encounters a transition metal. Typical rate constants for the reaction of $\cdot OH$ with proteins are between 10^{-9} and 10^{-10} M^{-1} s^{-1}. It may react with biological molecules via a variety of mechanisms, including hydrogen abstraction, addition and electron transfer. Its formation in biological systems thus requires the availability of H_2O_2 (inevitably generated when O_2^- is formed) and transition metals in a form capable of catalysing reaction (5.9).

As we shall see later (§5.4), there is some evidence that many, if not all of these reactive oxygen species are generated by phagocytosing neutrophils during the respiratory burst. The production of some of these is established beyond doubt, whereas the production of others, such as $\cdot OH$, is controversial.

5.3 The respiratory-burst enzyme

The respiratory burst of phagocytes is catalysed by a membrane-bound NADPH oxidase that is responsible for the reaction:

$$NADPH + 2O_2 \rightarrow 2O_2^- + H^+ + NADP^+ \quad (5.10)$$

The enzyme is located on the plasma membrane (although some components are also located in the cytosol and on the membranes of specific granules, gelatinase-containing granules and secretory vesicles), and it is dormant in resting cells. Upon phagocytosis (or other forms of cell stimulation) the dormant oxidase becomes activated by a series of complex intracellular processes and then becomes capable of reducing O_2 to O_2^-, which is the primary product of the enzyme. The enzyme will utilise both NADPH and NADH in vitro, but the K_m values of 45 μM and 450 μM, respectively, indicate that NADPH is the substrate used in vivo. This NADPH is made available by the increased activity of the hexose monophosphate (HMP) shunt (which is activated concomitantly with the oxidase), which generates the reduced cofactor via the activities of glucose-6-phosphate dehydrogenase (Fig. 5.4) and 6-phosphogluconate dehydrogenase (Fig. 5.5). HMP activity may thus be measured as the release of $^{14}CO_2$ from glucose labelled at C1. Continued activity of the HMP shunt depends upon the supply of $NADP^+$, which is either generated via the activity of the oxidase or else is derived from the glutathione cycle (see Fig. 5.3).

The NADPH oxidase is in fact a multicomponent enzyme system that constitutes an electron transport chain from NADPH to O_2. The components of this oxidase complex are now almost completely defined, and experiments performed primarily with CGD neutrophils have helped to identify these major constituents.

5.3.1 Properties of the oxidase

The identification of the components of the NADPH oxidase has not been easy because the enzyme complex has been difficult to solubilise and is extremely unstable, particularly in the presence of salts. In spite of these methodological problems, much careful and patient experimentation has now begun to unravel the extraordinary complexity of the enzyme system that has an apparently simple task to perform, namely, to transfer a single electron to O_2. The enzyme complex requires a number of membrane-bound and cytosolic components, which come together during cell activation to assemble into an active electron transport chain. The membrane components may reside on the plasma membrane or on the membranes of subcellular organelles (i.e. specific granules, gelatinase-containing granules and secretory vesicles), and multiple intracellular signal transduction pathways are required to initiate the assembly of the oxidase in response to specific activating signals. Apart from these subcellular translocations, the oxidase components may also be modified by phosphorylations, and cofactors such as GTP and arachidonic acid are re-

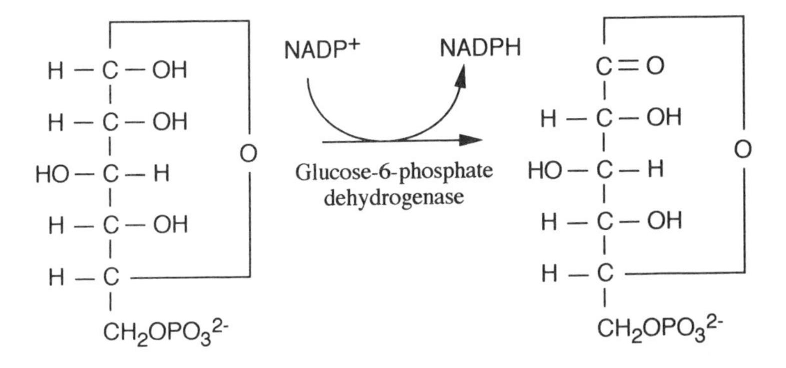

Figure 5.4. First site of NADPH generation in the pentose phosphate cycle.

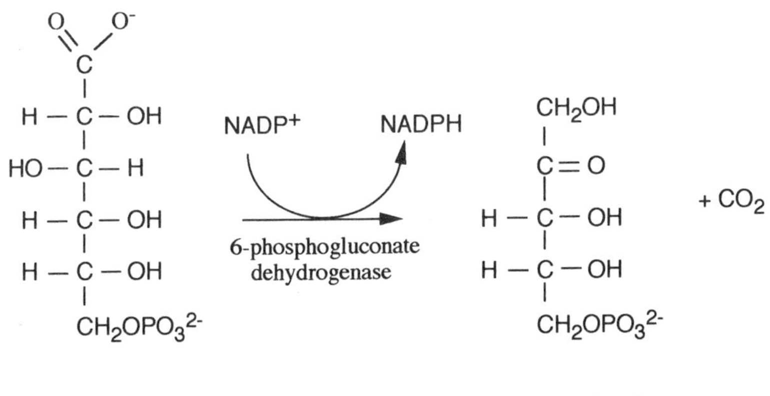

Figure 5.5. Second site of NADPH generation in the pentose phosphate cycle.

quired for activation. Active oxidase complexes probably have a short half-life and are auto-inactivated, presumably because such a 'self-destruct' mechanism avoids the need for a specific deactivation or disassembly process. This inactivation prevents overproduction of oxidants, which could result in damage to host tissues.

For many years, the literature was confused with various reports of 'purified NADPH oxidase preparations', which contained a variety of unidentified components. Here, again, the CGD neutrophil has proved to be invaluable: defective activity or amount of a putative oxidase component in the CGD neutrophil is very strong evidence that the component is a genuine constituent

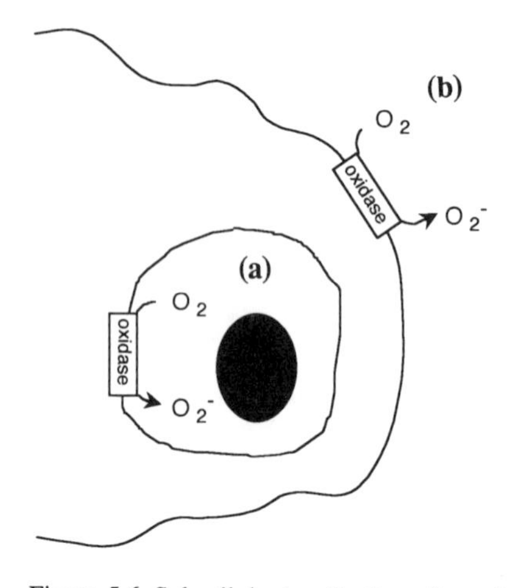

Figure 5.6. Subcellular localisation of reactive oxidant production in activated neutrophils: (a) the NADPH oxidase has been activated during phagocytosis, and so the O_2^- is generated within a phagocytic vesicle; (b) the NADPH oxidase molecules on the plasma membrane have been activated (e.g. in response to a soluble agonist), and so O_2^- is released.

of the oxidase. Rapid progress in identifying the oxidase components, and the mechanisms of its assembly and activation, have come from the experimental breakthrough that the oxidase can be activated in neutrophil extracts (details in §5.3.2.3). Recent developments and refinements of these cell-free systems have led to the reconstitution of a functional oxidase from purified components and cofactors.

Why does the oxidase need so many components, and why is the activation process so complicated? The products of the respiratory burst, especially the oxidising radicals, are extremely reactive but indiscriminate in the targets they attack. They are highly bactericidal but can, if released from activated neutrophils, attack host tissues and structures with equal potency; indeed, phagocyte-derived products are probably instrumental in the host-tissue damage seen in some inflammatory conditions, such as rheumatoid arthritis (see §8.9). Therefore, the NADPH oxidase is a two-edged sword: it must be sufficiently potent to kill a wide range of microbial pathogens before they have a chance to multiply, spread and overwhelm the body's defences, but damage to by-stander host cells and tissues must be minimalised. In part, this selective toxicity is achieved during bacterial killing because reactive oxidants are produced within phagolysosomes (Fig. 5.6); but additionally, the oxidase only becomes activated in response to specific signals in the neutrophil environment. Thus, the chances of inadvertent or non-specific activation of the oxidase occurring are kept to an absolute minimum.

5.3.2 Components of the oxidase

5.3.2.1 Cytochrome b

In the early 1960s in Japan, a b-type cytochrome was found in horse neutrophils and, because it bound CO, it was proposed to be functional during the respiratory burst. This work went largely unnoticed, but in 1978 Segal and Jones in the United Kingdom discovered that a b-type cytochrome became incorporated into phagolysosomes; furthermore, this cytochrome was absent in some patients with CGD. These workers correctly proposed that it was a key component of the NADPH oxidase. This cytochrome was a landmark discovery in phagocyte research for a number of reasons:

i. Its physicochemical properties could be established and its function identified.
ii. It predicted a molecular defect responsible for impaired neutrophil function in most patients with CGD. (This patient group was later described as X-linked CGD; see §8.2.)
iii. Because it appeared to be present in normal amounts in autosomal recessive CGD patients, other oxidase components were likely to exist.
iv. It could now be used as a molecular 'marker' to locate the oxidase within the cells and to assess the success of purification procedures: purified oxidase preparations must contain this cytochrome b.
v. It inspired a great many other research workers around the world to study this fascinating system.

Throughout the late 1970s and 1980s, the cytochrome b was found in neutrophils from all species so far examined. It has been purified to homogeneity and partially characterised, and the gene encoding it has been cloned. It has been found in all so-called professional phagocytes including neutrophils, monocytes, macrophages and eosinophils, and is probably also present in the protozoan phagocyte, *Acanthamoeba castellanii*, where it presumably also plays a key role in bacterial killing.

The cytochrome (by virtue of its ability to accept and donate electrons during its function in electron transport) can exist in either the oxidised or the reduced state. In reduced-minus-oxidised difference spectra, it has absorption maxima at 426, 530 and 558 nm, typical of many b-type cytochromes. The ease with which the cytochrome can accept and donate electrons is expressed by its redox (reduction–oxidation) potential, which is measured in millivolts. Unlike most mammalian b cytochromes, which have much higher midpoint potentials, that of the cytochrome of the NADPH oxidase is –245 mV. Because this is very close to the value of the O_2^-/O_2 couple, the cytochrome is capable of directly reducing O_2 to O_2^-. The cytochrome is therefore often referred to in the literature as *cytochrome b-245* or b_{-245} (to indicate its midpoint oxidation–reduction potential) or *cytochrome* b_{558} (to indicate its absorption maximum in the reduced state).

There has been a great deal of research aimed at determining whether or not the cytochrome is the terminal component of the NADPH oxidase electron transport chain that reacts with O_2 to form O_2^-. Some of the evidence in the literature supports this idea, but other evidence has cast doubt upon it. For a redox component to function as a terminal oxidase, it must satisfy several physicochemical parameters. For example, it must be shown to react with O_2 rapidly enough to support electron flow through the electron transport chain. Carbon monoxide (CO) is a respiratory poison and interacts in the dark with the haem group of many haemoproteins when they are in the reduced state. This prevents O_2-binding (CO is a competitive inhibitor), and the reduced haemoprotein cannot be reoxidised. Furthermore, the CO-ligand can be identified spectroscopically because it has a characteristic absorption spectrum. The cytochrome b of the NADPH oxidase binds CO (although the affinity of binding is low, 1.2 mM), and this binding is light-reversible, the CO rapidly recombining in the dark. When the cytochrome is chemically reduced, its half-time of oxidation by O_2 is 4.7 ms. NADPH reduces the cytochrome only very slowly under anaerobic conditions, but in the presence of O_2 the rate of reduction by this cofactor is very close to the kinetics of O_2^- formation. The cytochrome is absent from almost all cases of X-linked CGD neutrophils; hence, it can react quickly enough with O_2 to function as a terminal oxidase. Thus, current ideas favour its role as the terminal component of the electron transport chain of the NADPH oxidase.

Early reports describing the properties of the 'purified ' cytochrome b were confusing, but Segal and co-workers (Harper, Dunne & Segal, 1984) described a highly-purified cytochrome b preparation. The cytochrome consists of two subunits: a 76–92-kDa glycoprotein (the β-subunit) and a smaller 23 kDa (α) subunit. The β-subunit, because it is heavily glycosylated, runs as a broad, diffuse band on SDS-PAGE, thus making a precise assignment of its relative molecular mass quite difficult. Purified preparations of the cytochrome have been used to raise polyclonal and monoclonal antibodies, which are extremely useful experimental tools. These may be used to locate the cytochrome in sections of cells by immunocytochemical techniques, to quantify levels of the cytochrome by immunoblotting procedures (ELISA, Western blotting) and to screen cDNA-expression libraries (to identify transformed bacterial clones synthesising the cytochrome protein).

The gene for the β-subunit was identified by the elegant technique of reverse genetics (see §8.2). The identity of the 'CGD' gene as the β-subunit of the cytochrome was confirmed by comparison of the predicted amino acid sequence of the cloned gene with the actual amino acid sequence of the purified polypeptide. The sequences of both the α- and β-subunits possess little homology with other known b-cytochromes, but small regions of the α-subunit have some homology to a region of the mitochondrial cytochrome oxidase complex I and to bovine chromaffin granule cytochrome b. The β-subunit is only transcribed in myeloid cell lines – that is, the mRNA for the β-subunit can only be detected (by Northern blotting) in cells of the myeloid lineages – and these levels of mRNA in neutrophils can be up-regulated by exposure to γ-interferon. Curiously, the α-subunit, which probably contains the O_2-binding haem group, is transcribed in a number of non-phagocytic cell lines, even though these cells do not possess a spectrally-identifiable cytochrome b_{-245}: γ-interferon exposure has little effect on the levels of mRNA for this subunit in phagocytic cells.

Recently, it has been shown that the cytochrome b (and some other oxidase components) are present in human B lymphocyte cell lines that have been transformed with the Epstein–Barr virus (and hence can grow in culture). These cell lines also generate O_2^- when exposed to immunoglobulin aggregates, cytokines (e.g. TNF-α and IL-1β) or bacterial lipopolysaccharide. The cytochrome is also expressed on the cell surface of peripheral blood B lymphocytes. These observations suggest that the NADPH oxidase of blood cells may serve two distinct purposes: in phagocytes it is involved in the process of microbial killing during phagocytosis; in B lymphocytes it may play a role in regulating the function and/or differentiation of these cells during certain parts of their life cycle.

The cytochrome is found on the plasma membrane of unstimulated cells, but the majority of the total cellular pool (about 90%) is present intracellularly on the membranes of specific granules, gelatinase-containing granules or secretory vesicles. Experiments in which plasma membranes and granules were isolated both before and after cell stimulation have shown that during activation the cytochrome becomes translocated from the granules to either the plasma membrane or the membrane of the phagolysosome. For a long time it was believed that this membrane fusion was required in order to bring together components of the oxidase to assemble an active, multicomponent complex. Whilst this process may indeed occur, it is not the only way in which the oxidase is activated. In fact, the process of translocation is probably more important in the recruitment of new, latent cytochrome molecules from these intracellular stores to the site where they are required (i.e. on the plasma membrane or the phagosomal membrane). Thus, in unstimulated cells, the plasma membrane contains few latent cytochrome molecules, as the major pool of these are located intracellularly. This intracellular pool is translocated upon activation in order either to increase the number of active molecules or else to replenish the membrane with new cytochrome molecules after oxidase complexes become inactivated (Fig. 5.7). This limitation of the number of cytochrome molecules present on the plasma membrane probably restricts the levels of reactive oxidants that can be secreted upon activation. Thus, excessive oxidant secretion, which may lead to damage to surrounding cells and tissues, is usually avoided, and the possibility that non-specific factors may stimulate neutrophils to secrete large quantities of reactive oxidants is minimised. Indeed, experiments using anti–cytochrome b antibodies have shown that in blood neutrophils very few molecules of cytochrome b are normally present on the plasma membrane. Levels of expression of cytochrome b on the cell surface increase rapidly upon stimulation or after exposure to cytokines.

5.3.2.2 Flavins

The involvement of flavins in the respiratory burst was first proposed by Babior and Kipnes (1977) when they found that the activity of a detergent extract of the oxidase could be increased by the addition of flavin adenine dinucleotide (FAD). Furthermore, flavin analogues such as 5-carbodeaza-FAD or flavoprotein inhibitors such diphenylelene iodonium (DPI) inhibit oxidase activity. Oxidase-rich membrane preparations contain flavins, and a flavin semiquinone electron paramagnetic resonance (EPR) spectrum is observed in membranes from activated neutrophils. The subcellular distribution of FAD in the neutrophil is similar to that of the cytochrome b – namely, in the plasma membrane and membranes of specific or gelatinase-containing granules or secretory vesicles – and most oxidase preparations described contain both FAD and cytochrome b. FAD levels are also low in some patients with autosomal recessive CGD, being about half the amount present in normal neutrophils.

Flavoprotein dehydrogenases usually accept electrons from reduced pyridine nucleotides and donate them to a suitable electron acceptor. The oxidation–reduction midpoint potential of the FAD of the oxidase has been determined by ESR spectroscopy and shown to be -280 mV. The $NADP^+$/NADPH redox potential is –320 mV and that of the cytochrome b is –245 mV; hence, the flavin is thermodynamically capable of accepting electrons from NADPH and transferring them to cytochrome b. As two electrons are transferred from NADPH, although O_2 reduction requires only one electron, the scheme of electron transfer shown in Figure 5.8 has been proposed by Cross and Jones (1991).

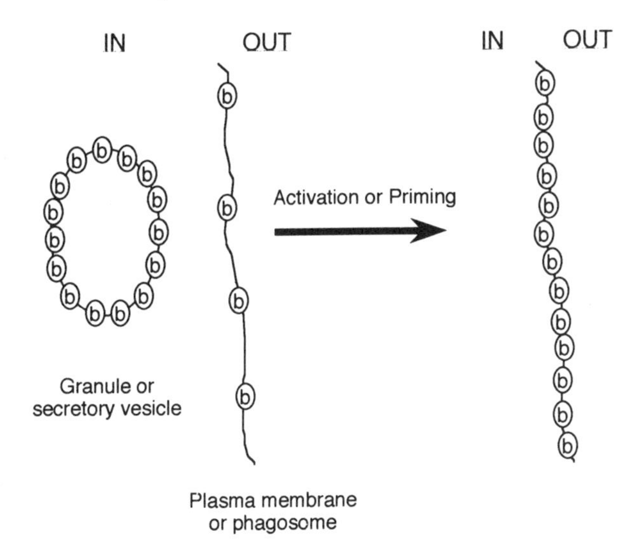

Figure 5.7. Translocation of cytochrome b to the plasma membrane. In non-stimulated cells, only a small proportion of the total cellular pool of cytochrome b is present on the plasma membrane. The major pool of this cytochrome is located on the membranes of specific granules, gelatinase-containing granules and secretory vesicles. During activation (e.g. by fMet-Leu-Phe or PMA) or priming (e.g. by cytokines), some of these subcellular pools of cytochrome b molecules are translocated to the plasma membrane, thereby increasing the level of surface cytochrome b.

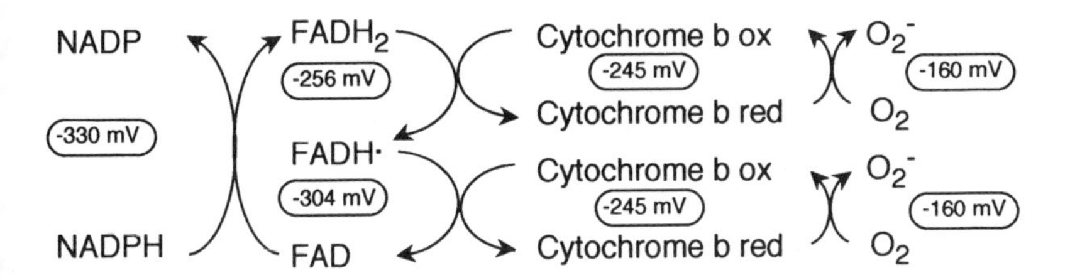

Figure 5.8. Proposed scheme of electron transport in neutrophils. In this scheme, electrons are transferred from NADPH → FAD→ cytochrome b → O_2. The values shown in the boxes are the redox midpoint potentials for each of the constituents of the pathway. NADPH donates two electrons upon oxidation, but single electrons are transferred from the flavin to the cytochrome b via the formation of FADH·. *Source:* Redrawn from Cross and Jones (1991).

However, it must be stressed that not everyone is in agreement with such a scheme. For example, flavoproteins are thermodynamically capable of transferring electrons directly to O_2 to form O_2^-, and not all functional oxidase preparations contain substantial amounts of FAD. Indeed, in one series of experiments the ratio of cytochrome b to FAD decreased from about 1 : 1 to 19 : 1 as the oxidase became progressively more purified. It may be, however, that in the intact oxidase molecules within membranes there is a large molar excess of FAD compared with cytochrome b.

Several groups have claimed to identify a specific flavoprotein. For example, it has been reported that polypeptides of 65–66 kDa, which copurify with oxidase activity, also contain flavin. A polypeptide of 45 kDa is specifically labelled with the flavoprotein- and oxidase-inhibitor DPI, and NADPH appears to compete for the DPI-binding site. This component has been purified, is present in neutrophils, macrophages and eosinophils and is induced during γ-interferon treatment of a monocytic cell line (U937) in parallel with the ability of these cells to generate O_2^-. Polyclonal antibodies raised to this purified polypeptide immunoprecipitate both the 45-kDa polypeptide and the small (α) subunit of the cytochrome b. More recently, however, Segal's group (1992) has proposed that the flavin-binding site of the oxidase is actually on the cytochrome b molecule itself. The cytochrome b_{-245} is thus a flavocytochrome, so specific flavoprotein molecules are probably not present in the oxidase.

5.3.2.3 Cytosolic components

Experiments in the mid-1980s, analysing the phosphorylation profiles of neutrophils from autosomal recessive forms of CGD, indicated that a component of around 45–47 kDa failed to become phosphorylated during activation with PMA. In normal neutrophils, this component is phosphorylated in the cytoplasm and then translocates to the plasma membrane. Curiously, in CGD neutrophils of patients that lacked the cytochrome b, the 47-kDa component was phosphorylated normally but failed to become incorporated into the plasma membrane. This strongly implied the following:

i. The 47-kDa component was an important component of the oxidase.
ii. It was normally a cytosolic component that translocated to the plasma membrane during activation.
iii. It probably associated with the cytochrome b after translocation.
iv. Phosphorylation of this component was a key event for either its translocation, its activation or its assembly with the cytochrome.
v. This component was defective in some forms of CGD.

At around the same time as these discoveries, several groups were trying to determine the ways in which the NADPH oxidase could be activated in vitro (i.e. activated in broken-cell suspensions). The ultimate aim of such studies was to determine the minimal components necessary for assembly and activation of the oxidase in in vivo experiments. Thus, once these minimal constituents were identified, oxidase activity could then be reconstituted in vitro from the individual component parts. The first breakthrough in these studies was the discovery that anionic detergents, such as arachidonic acid, *cis*-unsaturated fatty acids or SDS, could activate O_2^- production when added to broken-neutrophil homogenates.

Further refinement of this system, by fractionation of these homogenates, showed that the plasma membranes and cytosol (but not the granules) were required for activity, and that whilst Mg^{2+} was essential, neither Ca^{2+} nor ATP were required. This was a somewhat surprising finding because it was known that the specific and gelatinase-containing granules contain the bulk of the cellular pool of cytochrome b, and that both Ca^{2+} and ATP (required for protein-kinase-C-dependent phosphorylation reactions) were intimately involved in oxidase activation in intact neutrophils. Thus, for a while, the physiological significance of this system was questioned, but it was subsequently shown that this 'cell-free' NADPH oxidase activation system was defective in CGD patients. However, this defect could be corrected in experiments in which the cytosol and membranes from different types of CGD patients were mixed with each other or with fractions from control neutrophils: the defective activity could be restored, confirming the importance of this system. Thus, in autosomal recessive CGD neutrophils the *cytosol* was defective, whereas in X-linked, cytochrome-b-deficient CGD neutrophils the *membranes* were defective. Substitution of the defective cytosol or membrane fractions with fractions from normal controls could restore activity.

Attempts to purify the cytosolic components by Mono Q anion exchange chromatography yielded three fractions (termed NCF-1, -2 and -3) that were active in oxidant production. Independently, a polyclonal antiserum was raised against the cytosolic components that eluted from a GTP-affinity column (see also §8.2.4). Fortunately, this antiserum (which was raised against a mixture of many proteins) only recognised two major proteins in immunoblots, and these had relative molecular masses of 47 kDa and 66 kDa. NCF-1 was then shown to contain the 47-kDa component (now termed p47-*phox*), whilst NCF-2 contained the 66-kDa component (now termed p66-*phox*). The antiserum (B1 antiserum) has since been used to isolate cDNA clones for these proteins from an HL-60 cDNA library, and these clones have been sequenced. The p47-*phox* clone has several predicted phosphorylation sites, and amino acid residues 156–221 show some homology to regions of v-*src*, phosphatidylinositol-specific phospholipase C and α-fodrin. A 33-amino-acid stretch from 233 to 265 shows 50% homology to GTPase-activating protein. cDNA for p66-*phox* has been isolated and sequenced. This clone encodes a 526-amino-acid protein that contains sequences similar to a motif found in the non-catalytic region of *src*-related protein tyrosine kinases. Thus, both of these cytosolic factors contain two SH3 domains (i.e. regions of *src* homology), and these domains are present in many types of proteins involved in cell signalling. These domains are probably required for protein–protein interactions and are involved in interactions between proteins and small GTP-binding proteins (§6.2.2).

Most of the patients with autosomal recessive CGD lack p47-*phox* whilst the remainder lack p66-*phox* (details are given in §8.2). Recombinant proteins derived from these full-length cDNA clones can restore activity when added to extracts of autosomal recessive CGD neutrophils, in the cell-free oxidase activation system.

The active components of this cell-free system were thus becoming defined: cytochrome b_{-245}, p47-*phox,* p66-*phox,* fatty acids (such as arachidonic acid or SDS) and NADPH. However, even when recombinant proteins were used, trace amounts of cytosol were always needed for activity, and the 'active' constituent of NCF-3 was, for a long time, undefined. It was also noted that during purification of the cytosolic components, the greater the purity, the greater the requirement of the reconstitution system for exogenously-added GTP. Thus, crude cytosol contains sufficient GTP for activity, but this is lost from the active fractions during purification. This observation implied a GTP-binding protein in oxidase activation. Other, independent evidence for the involvement of GTP-binding proteins came from the discoveries that a 23-kDa ras-related protein (rap1A) copurifies with cytochrome b_{-245} and that a ras-related protein binds to an anti–cytochrome b affinity column.

The third cytosolic component, recently purified from guinea-pig macrophages, has been shown to contain two proteins, $p21^{rac1}$ and GDP-dissociation inhibition factor (GDI). In the presence of GTP, only $p21^{rac1}$ is needed for activity. Thus, the cell-free oxidase system has now been reconstituted using purified components. A mixture of recombinant p66-*phox,* p47-*phox,* $p21^{rac1}$ plus purified cytochrome b can, in the presence of NADPH, FAD, SDS and Mg^{2+}, reconstitute O_2^- generation. Thus, the system confirms that the β-subunit of the cytochrome b can bind both FAD and NADPH. The protein $p21^{rac2}$ has been proposed to be the third cytosolic factor present in the cytoplasm of human neutrophils, and it is 92% homologous to $p21^{rac1}$ at the amino acid level. Whilst the molecular components of the NADPH oxidase have now been fully defined, precisely how these are arranged into an active electron transport chain is not yet established. These experiments also indicate that intracellular Ca^{2+} and protein-kinase-C-dependent phosphorylation reactions are not required for oxidase activation per se. Instead, they are more likely to be required for the translocation events (e.g. movement of cytosolic factors and granules through the cytoskeletal network) and for the generation of lipid mediators, such as arachidonic acid (see Chapter 6), that are required for oxidase activation.

5.3.2.4 Other components

Over the years, there have been numerous reports of 'oxidase preparations' that contain polypeptide components, additional to those described above. As yet no molecular probes are available for these, and so their true association with the oxidase is unconfirmed. There are many reports in the literature describing the role of ubiquinone as an electron transfer component of the oxidase, but its involvement is controversial. Quinones (ubiquinone-10) have reportedly been detected in some neutrophil membrane preparations, but other reports have shown that neither plasma membranes, specific granules nor most oxidase preparations contain appreciable amounts of quinone, although some is found in either tertiary granules or mitochondria. Still other reports suggest that ubiquinone, flavoprotein and cytochrome b are present in active oxidase preparations. Thus, the role of ubiquinone and other quinones in oxidase activity is in doubt, but the available evidence weighs against their involvement. Indeed, the refinement of the cell-free activation system described above obviates the requirement for any other redox carriers for oxidase function.

5.4 Reactive oxidant production during the respiratory burst

Many soluble and particulate agonists, such as fMet-Leu-Phe, PAF, LTB_4, PMA, IgG-immune complexes, opsonised zymosan and opsonised bacteria, can stimulate reactive oxidant production in neutrophils. In many cases, the production of reactive oxygen metabolites is quite low, unless the cells have been pretreated with agents such as cytochalasin B or cytokines such as GM-CSF. Furthermore, there are considerable variations in the detected rates of reactive oxidant production. One reason for this variation is that unstimulated neutrophils in the blood are only poorly responsive to receptor-mediated activation because the plasma membrane possesses few cytochrome b molecules and few receptors that are coupled to activation of the oxidase. During treatment with cytochalasin B (which shortens actin filaments and hence does not restrain movement of granules) or cytokines such as GM-CSF (which stimulates the translocation of specific or gelatinase-containing granules or secretory vesicles to the plasma membrane), there is an up-regulation of some receptors and cytochrome b molecules (and perhaps some enzymes involved in signal transduction) onto the cell surface. Thus, the past history of the neutrophil, in terms of its in vivo or in vitro exposure to agents that can cause such an up-regulation, will affect its ability to generate reactive oxidants in response to specific agonists.

Another consideration is the method used to detect the respiratory burst. Several such methods are available (described in §5.4.2), each with its own advantages and disadvantages. There is no ideal method for the detection of the respiratory burst that has perfect sensitivity and selectivity. In order to detect all the oxidants generated in response to all classes of pathophysiological stimuli, two important factors should be considered. Firstly, which of the range of reactive oxidant species generated by activated neutrophils is to be measured? Secondly, are intra- or extracellular oxidants to be measured? Many probes used to detect these oxidants are large molecules that do not gain access into intracellular vesicles. These are thus inefficient for detecting oxidants produced inside the neutrophil (e.g. within phagolysosomes during phagocytosis).

5.4.1 Myeloperoxidase

5.4.1.1 Properties

Myeloperoxidase greatly augments the bactercidal activity of H_2O_2. In the presence of halide ions such as I^-, Br^- or Cl^-, *hypohalous acids,* such as HOI, HOBr and HOCl, may be formed. Of these hypohalous acids, HOI is the most potent antibacterial compound, but the I^- concentrations available in vivo for this reaction are extremely low. Similarly, Br^- levels are low (but higher than I^-), whilst Cl^- levels are abundant in physiological conditions. Hence, it is believed that the major halide ion oxidised by the myeloperoxidase–H_2O_2 system in vivo is Cl^-, and so the major product is HOCl. The H_2O_2 for these reactions is usually provided during the respiratory burst, but some micro-organisms can produce and secrete H_2O_2 and hence may contribute to their own destruction. For example, organisms that lack catalase (which normally breaks down H_2O_2 into H_2O and O_2), or those that generate H_2O_2 during electron transport via flavoprotein oxidases rather than cytochromes, can secrete sufficient quantities of peroxide to allow the myeloperoxidase–H_2O_2 system to operate. This may be an important reason why organisms of this type are killed by patients with chronic granulomatous disease, even though their neutrophils are incapable of generating reactive oxidants during the respiratory burst.

The reaction of myeloperoxidase with H_2O_2 is complex and depends upon the concentration of H_2O_2 and the presence of other factor(s) within the microenvironment (Fig. 5.9). The Fe in native myeloperoxidase is in the Fe III (oxidised) state (MPO^{3+}, ferric myeloperoxidase), and this reacts with low (equimolar) concentrations of H_2O_2 to form *compound I,* a short-lived inter-

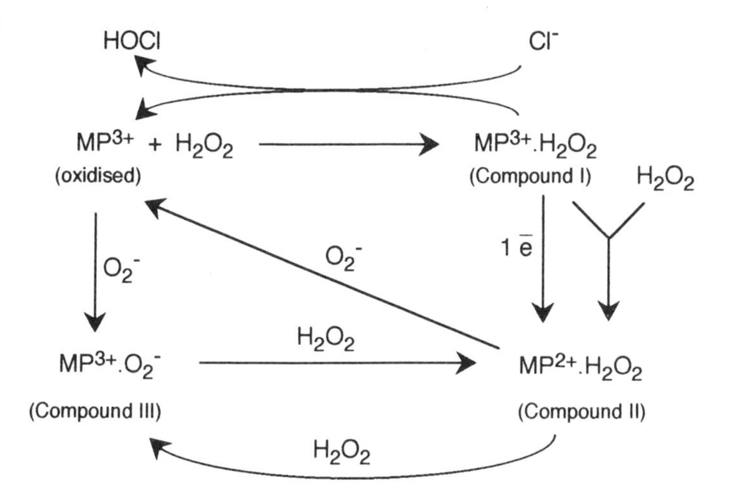

Figure 5.9. Reactions of myeloperoxidase with H_2O_2 and O_2^-. See text for details.

mediate. Compound I is very reactive and capable of catalysing the oxidation of chloride to hypochlorous acid, with the concomitant re-formation of native myeloperoxidase. Formation of compound I is very rapid (2.3×10^7 M^{-1} s^{-1}), and the reaction is pH dependent.

Compound I may alternatively be converted into compound II by reducing agents, or else via its further reaction with H_2O_2. *Compound II* is inactive in the formation of HOCl, but it is functional in other myeloperoxidase-dependent processes (e.g. oxidations, peroxidations). In compound II the Fe is probably in the ferryl (IV) state. Compound II may then be converted into compound III by its reaction with H_2O_2, which may also be generated via the reaction of ferrous myeloperoxidase with oxygen or by a reaction of ferric myeloperoxidase with O_2^- . *Compound III* is unstable and is converted back to the native enzyme; it may also react with electron donors and electron acceptors. Compound III may react with O_2^- to form compound I (which is active in HOCl formation), thus exhibiting SOD activity. Furthermore, the reaction of myeloperoxidase with H_2O_2 can result in the formation of O_2^-. These changes in redox state of myeloperoxidase as it reacts with H_2O_2 (or O_2^-) can be followed spectrophotometrically.

The transient production of compounds II and III has been reported during stimulation of neutrophils by fMet-Leu-Phe and PMA, respectively. Ferric myeloperoxidase and compound III show catalase activity, even in the presence of Cl^-, when H_2O_2 concentrations are in excess of 200 μM. Thus, under these conditions, O_2 formation will occur at the expense of HOCl formation. Thus, depending upon the concentrations of H_2O_2, halide ions, O_2^- and reductants in its microenvironment, myeloperoxidase can exhibit catalase, peroxidase, oxidase and superoxide dismutase activities, as well as produce large quantities of HOCl.

It has been reported that, depending upon the stimulus used, 40–70% of the O_2 consumed during the respiratory burst appears in HOCl via the reaction:

$$H_2O_2 + Cl^- \rightarrow HOCl + H_2O \quad (5.11)$$

Additional products of the myeloperoxidase system depend upon the local environment. For example, HOCl can react with nitrogen-containing compounds such as biological amines to form chloramines:

$$HOCl + R\text{-}NH_2 \rightarrow R\text{-}NHCl + H_2O \quad (5.12)$$

Whilst chloramines are less reactive than HOCl, they are longer-lived and so can diffuse away from their site of production. Those formed from lipophilic amines are especially toxic because they can permeate membranes. Chloramines are toxic for a number of reasons: they can oxidise sulphydryl or sulphur-ether groups, they are unstable and can be hydrolysed to release chlorine in the form of HOCl or NH_2Cl, they can react with iodide to form iodine and they can covalently bind proteins.

It has been proposed that other forms of active oxygen can be generated via the myeloperoxidase–H_2O_2 system. The extremely-reactive ·OH *may* be generated from a reaction between HOCl and O_2^- as follows:

$$HOCl + O_2^- \rightarrow Cl^- + O_2 + \cdot OH \quad (5.13)$$

and singlet oxygen ($^1O_2$) may be generated:

$$OCl^- + H_2O_2 \rightarrow Cl^- + H_2O + {}^1O_2 \quad (5.14)$$

However, the formation of either $^1O_2$ or ·OH by the myeloperoxidase system is debatable, and it has even been proposed that myeloperoxidase can actually lower ·OH formation by decreasing the concentration of H_2O_2 available for the Haber–Weiss reaction (§5.4.2.5).

5.4.1.2 Function of myeloperoxidase

The myeloperoxidase–H_2O_2 system is extremely potent in the killing of a wide range of target cells and in the inactivation of biological molecules, such as the following:

i. bacteria (e.g. *Mycoplasma, Chlamydia, Mycobacteria, Legionella pneumophila, Staphylococcus aureus*);

ii. fungi (e.g. *Aspergillus, Candida, Rhizopus*);
iii. protozoa (e.g. *Leishmania donovani, Trypanosoma*);
iv. multicellular organisms (e.g. *Schistosoma mansonii*);
v. mammalian cells (e.g. tumours, neutrophils, platelets, lymphocytes, erythrocytes and NK cells);
vi. humoral substances (e.g. chemotaxins [fMet-Leu-Phe and C5a], α_1-proteinase inhibitor, toxins [diphtheria and *Clostridia* toxins], SRS, prostaglandins, leukotrienes);
vii. neutrophil granule enzymes (e.g. some lysosomal enzymes, vitamin-B_{12}-binding protein)

Furthermore, the myeloperoxidase systems can stimulate secretion of serotonin from platelets, histamine release from mast cells and the activation of latent collagenase and latent gelatinase of neutrophils.

Additionally, the myeloperoxidase system even regulates the duration of the respiratory burst because neutrophils from patients with myeloperoxidase deficiency (see §8.3) generate more reactive oxidants than control cells. Also, when myeloperoxidase is inhibited with a specific antibody or a specific inhibitor such as salicylhydroxamic acid, the duration of the respiratory burst, but not the maximal rate of oxidant production, is extended. This indicates that a product of the myeloperoxidase system inhibits the NADPH oxidase and so self-regulates reactive oxidant production during inflammation.

There is much debate as to the mechanisms by which myeloperoxidase destroys its targets. However, its abilities to generate HOCl and chloramines and to catalyse directly the peroxidation and halogenation of the target may all be involved. Biochemical processes that may be targets for myeloperoxidase-dependent bacterial destruction include:

i. electron transport systems,
ii. depletion of adenine nucleotide pools,
iii. oxidation of iron-sulphur centres and haemoproteins,
iv. oxidation of thiol (-SH) groups to disulphides and sulphur oxides,
v. oxidation of thio-ethers to sulphoxides,
vi. oxidation of amino groups to chloramines and dichloramines,
vii. halogenation of pyridine nucleotides and
viii. lipid peroxidation.

Overall, bacteria attacked by the myeloperoxidase system undergo a loss of selective permeability prior to death.

In order to function efficiently, myeloperoxidase must be translocated from its intracellular location, the azurophilic granule, to the site of NADPH oxidase activity, which is either on the plasma membrane or within phagosomes. Thus, there must be subcellular movement of azurophilic granules and discharge of their contents to enable this system to operate. During phagocytosis of micro-organisms, the granules migrate to the phagosome, the membranes fuse and the contents of the granules, including the myeloperoxidase, are discharged into the vacuole (see Fig. 1.6). Thus, the myeloperoxidase–H_2O_2 system is reconstituted within this vacuole and so can attack the target. As discussed above (§5.4.1.1), myeloperoxidase can also utilise H_2O_2 that may be secreted from certain types of microbes, and so this mechanism may be important in neutrophils whose NADPH oxidase activity is impaired.

Alternatively, myeloperoxidase can be released from activated neutrophils, and if NADPH oxidase molecules on the plasma membrane are activated (e.g. by soluble factors) or if the pathogenic target is too large to be engulfed within a phagosome, then an extracellular myeloperoxidase–H_2O_2 system can operate. This process may thus be important in the destruction of large targets (e.g. multicellular organisms of mammalian cells) or when neutrophils become activated by soluble factors during inflammation. Such extracellular activation of the myeloperoxidase system may thus be instrumental in the inactivation of humoral substances that may be beneficial (e.g. toxins) or potentially harmful (e.g. α_1-proteinase inhibitor, which occurs via oxidation of a critical methionine residue on the molecule).

5.4.2 Assays used to measure reactive oxidant production

5.4.2.1 Cytochrome c reduction assay

The basis of this assay was first used to measure the activity of superoxide dismutase (SOD) using a xanthine/xanthine oxidase O_2^--generating system. O_2^- generated via this enzyme will reduce ferri (oxidised)-cytochrome c, but SOD (which has a much higher affinity for O_2^- than cytochrome c) will prevent this reduction. Babior, Kipnes and Curnutte (1973) modified this technique to provide a specific assay to measure O_2^- production by activated neutrophils. Thus, O_2^- reduces cytochrome c (measured by an absorbance increase at 550 nm), but this reduction will be blocked by the addition of exogenous SOD (Fig. 5.10).

SOD-inhibitable reduction of cytochrome c is thus now widely used to measure O_2^- production by activated neutrophils. It is an extremely-selective method, being specific for O_2^-, and is easily measured spectrophotometrically. Because SOD can convert two molecules of O_2^- into one molecule of H_2O_2 and one molecule of O_2 (reaction 5.1), catalase (which degrades H_2O_2

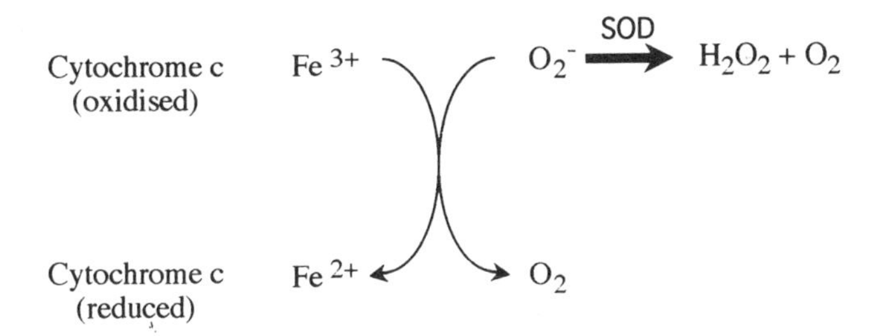

Figure 5.10. Cytochrome c reduction by O_2^-. Production of O_2^- from activated neutrophils may be assayed using cytochrome c. Oxidised (Fe^{3+}) cytochrome c can be reduced by O_2^- to form Fe^{2+}–cytochrome c, which absorbs at 550 nm; thus, in a mixture of activated neutrophils and cytochrome, absorption increases at 550 nm are due to O_2^- production. Superoxide dismutase (SOD) has a higher affinity for O_2^- than does cytochrome c; thus, the addition of SOD to activated neutrophil suspensions will prevent the reduction of cytochrome c. SOD-inhibitable cytochrome c reduction is therefore a direct measure of the rate of O_2^- formation.

into H_2O and O_2; see reaction 5.4) is often added to assays to prevent H_2O_2 from reoxidising the reduced cytochrome c. Such H_2O_2-dependent oxidation of reduced cytochrome c will give inaccurate, falsely-low apparent rates of O_2^- formation.

Reduction of cytochrome c by neutrophils stimulated with agonists such as fMet-Leu-Phe can be detected within seconds of addition (Fig. 5.11b). Maximal rates of cytochrome c reduction, and hence O_2^- production, occur within 1 min of stimulation with this agonist and then decline to zero levels within a few minutes. This decline is not due to depletion of substrates (e.g. O_2 or cytochrome c) nor to the reoxidation of the probe by H_2O_2, but appears genuinely to be due to diminished rates of O_2^- secretion. On the other hand, this assay is not very useful in measuring intracellular oxidant production that may occur (e.g. during phagocytosis of large particles), because cytochrome c is a large, basic protein that does not permeate neutrophil membranes. Furthermore, experiments have shown that insignificant amounts of this probe are co-internalised into phagocytic vesicles during particle uptake.

Hence, this assay is an extremely useful and selective assay to measure O_2^- secretion. Because of this selectivity and because it measures the initial product of O_2 reduction, it is often used as the method of choice to detect NADPH oxidase activity. It is suitable for semi-automation because assays can be performed in 96-well microtitre plates (using ELISA plate readers with a suitable filter), or cytochrome c reduction can be detected using simple spectrophotometers. The assay, however, is not suitable for measuring O_2^- that may be generated intracellularly within activated neutrophils.

5.4.2.2 Oxygen uptake

Dioxygen uptake can be measured using conventional O_2 electrodes, such as Clark-type electrodes consisting of platinum cathodes and silver anodes. These electrodes (bathed in saturated KCl solution) are separated from the cell suspension by an O_2-permeable membrane such as Teflon®. Dioxygen, which diffuses from the suspension through the membrane, is reduced at the cathode to generate a current. Thus, if a voltage of 0.5–0.8 V is applied across the electrodes, then the current generated at the cathode is proportional to the O_2 tension in the solution. Because O_2 diffuses freely into neutrophils, this assay can efficiently measure O_2 reduction that occurs at the plasma membrane or within intracellular sites. At low O_2 tensions, however, intracellular O_2 reduction may be limited because O_2 gradients can occur across cells, even those as small as neutrophils.

Because electrode measurements of O_2 uptake can detect intra- and extracellular oxidase activity, this assay can be used to measure the respiratory burst elicited by soluble and particulate stimuli. What is somewhat surprising is that, during stimulation of neutrophils with agonists such as fMet-Leu-Phe, the activated O_2 uptake profile is biphasic (Fig. 5.11c). A rapid burst of O_2 uptake (which coincides with measurements of cytochrome c reduction) is followed by a more sustained activity of lower magnitude.

Thus, whilst this assay is selective (i.e. it is specific for O_2 reduction) and can measure oxidase activity at intra- and extracellular sites, it does have some disadvantages. It is not suitable for automation, it is relatively insensitive, requiring large numbers of cells (although some newer microelectrodes offer improvements on this) and it measures all the O_2 consumed by the activated cell. Because neutrophils (unlike monocytes) do not have significant numbers of mitochondria, no O_2 is required for genuine respiration. However, other O_2-utilising processes, such as eicosanoid formation (see §6.3.3.2), will contribute to the total O_2 consumed by activated neutrophils. Moreover, O_2 can be both utilised and regenerated during the formation of reactive oxygen metabolites: O_2 is consumed during reaction (5.10), but may be regenerated by reactions (5.1), (5.4), (5.8) and (5.9).

5.4.2.3 Detection of hydrogen peroxide formation

Hydrogen peroxide (H_2O_2) is generated via the dismutation (reaction 5.1) of the O_2^- that is generated via the NADPH oxidase, although there is some debate as to whether the oxidase can generate H_2O_2 directly via a two-electron reduction of O_2. Methods commonly used to detect H_2O_2 formation are based on fluorescence assays. One such method is to use a fluorescent indicator

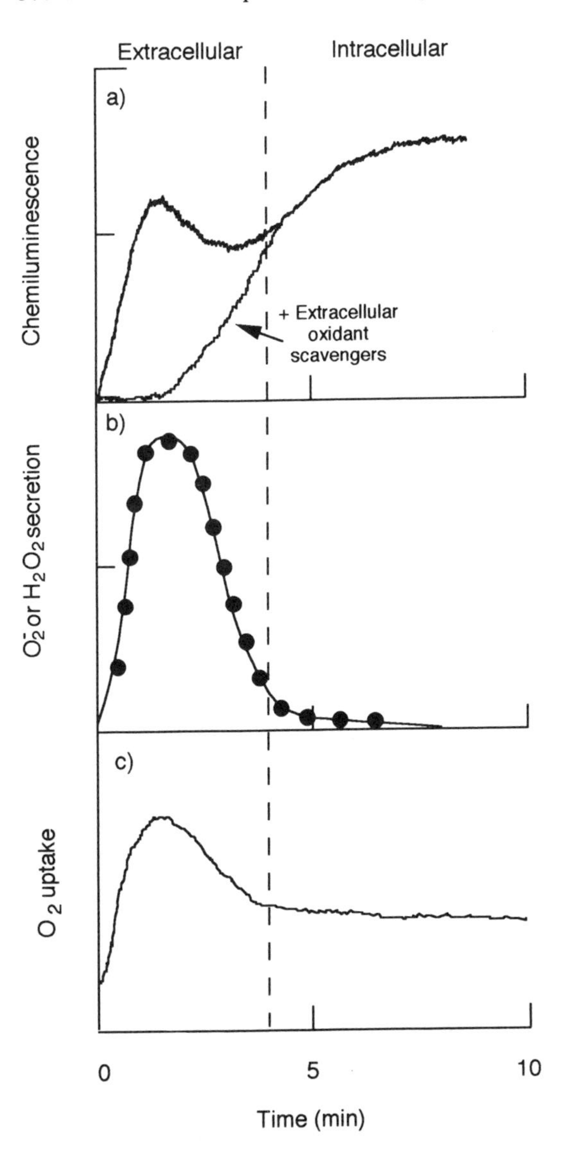

Figure 5.11. Reactive oxidant production by human neutrophils. In (a), neutrophils (5×10^5/ml) were suspended in buffer containing 10 μM luminol in the presence and absence of a mixture of the extracellular oxidants scavengers SOD (1 μg/ml), catalase (2 μg/ml) and methionine (0.25 mg/ml, to scavenge HOCl). In (b), neutrophils (1×10^6/ml) were suspended in buffer containing 75 μM cytochrome c (to measure O_2^- production) or 4 μM scopoletin plus 5 μg/ml horseradish peroxidase (to measure H_2O_2 production). In (c), neutrophils (2×10^6/ml) were placed in the chamber of a Clark-type O_2 electrode. All measurements were made at 37 °C; cell suspensions were stimulated by the addition of 1 μM fMet-Leu-Phe.

such as scopoletin, which loses its fluorescence when oxidised by a peroxidase and H_2O_2. Thus, an assay system comprising scopoletin, horseradish peroxidase and neutrophils has high fluorescence which then decreases during cell stimulation as the H_2O_2 is generated (see Fig. 5.11b). Such assays have good specificity for H_2O_2 but, again, a drawback is that the indicators (scopoletin and horseradish peroxidase) are large molecules and do not permeate neutrophils. Thus, only H_2O_2, which is either generated at the plasma membrane or secreted, may be detected by such assays. During H_2O_2 generation, the fluorescence of the solution decreases, and these decreases must be calibrated by the addition of exogenous H_2O_2 to blank solutions that do not contain activated neutrophils. Such assays are suitable for semi-automated measurements of large numbers of neutrophil samples because assays may be performed in microtitre plate formats using fluorescence plate readers.

Another method for measuring H_2O_2 production utilises 2´,7´-dichlorofluorescin. This compound has low fluorescence but may be oxidised by a peroxidase and H_2O_2 to produce 2´,7´-dichlorofluorescein, which has high fluorescence. This assay may be used to measure fluorescence of individual neutrophils, for example, by fluorescence 'viewing' on a microscope slide using a fluorescence microscope or, more commonly, using *fluorescence-activated cell sorting* (FACS). In this latter technique the scatter and fluorescence properties of individual cells can be measured as they pass through a highly-focused laser beam. Thus, fluorescence distributions of cell populations are obtained that indicate (i) the average fluorescence per cell of the population and (ii) the fluorescence intensity of individual cells. This approach has been used to measure the heterogeneity of the ability of neutrophil populations to generate reactive oxidants in response to stimulation.

Neutrophil preparations are incubated with dichlorofluorescin diacetate, which can diffuse into the cells. Esterase activity within the neutrophil then generates 2´,7´-dichlorofluorescin, which becomes trapped within the neutrophil. Upon activation, the H_2O_2 generated then oxidises the probe to dichlorofluorescein, and a fluorescence signal is generated. However, during stimulation, very little fluorescence is detected on the plasma membrane, presumably because the oxidants and/or the probe diffuse away from the cell. The fluorescence that is detected thus comes from within the cell, and fluorescence microscopy indicates that even soluble stimuli such as fMet-Leu-Phe generate H_2O_2 intracellularly. The fluorescence detected comes from within granules. Precisely why soluble agonists should generate oxidants intracellularly is unknown, but further studies (described in §5.4.2.4) indicate that such intracellular oxidants are not generated on the plasma membrane and then diffused back into the cell; rather, they are generated at intracellular sites.

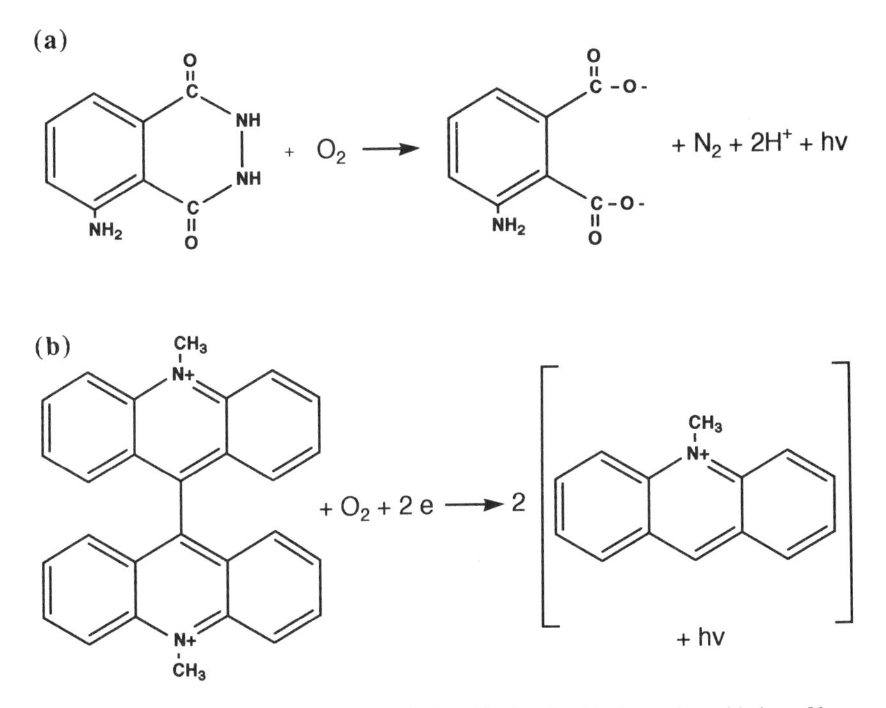

Figure 5.12. Generation of light from luminol and lucigenin: (a) shows the oxidation of luminol to generate light, whereas (b) shows light production from the oxidation of lucigenin; hv represents a photon.

5.4.2.4 Chemiluminescence

In the early 1970s Robert Allen (reviewed in Allen 1986) discovered that activated neutrophils could generate light or chemiluminescence. This light generation is detected by sensitive photomultiplier tubes, either in scintillation counters (operating in the 'out of coincidence' mode) or within specialised luminometers. The molecular species responsible for 'native' (unamplified) chemiluminescence is not known with certainty, but sources of such light emission include singlet oxygen (which releases some of its energy as a photon as it decays to the ground state), reactions of haem with H_2O_2, and carbonyl compounds in excited states. Furthermore, this light emission can be considerably enhanced by the use of chemiluminescence probes such as luminol and lucigenin. These become oxidised by the products of the respiratory burst and then emit photons as they decay to lower-energy states (Fig. 5.12).

When neutrophils are stimulated with fMet-Leu-Phe, they generate a biphasic luminol chemiluminescence response. The first phase of chemiluminescence coincides with the measurements of activated O_2^- secretion (see Fig. 5.11a). The second phase is associated not with O_2^-/H_2O_2 secretion but with the sustained period of O_2 uptake. The addition of extracellular oxidant scavengers, such as SOD (to destroy O_2^-), catalase (to break down H_2O_2) and methionine (to react with HOCl), block the first phase of chemiluminescence (indicating its extracellular nature) but have no effect on the second (indicating its intracellular location). Experiments by Claus Dahlgren and colleagues (Dahlgren, Aniansson & Magnusson, 1985; Dahlgren 1987) have suggested that luminol freely permeates neutrophils (and so can measure intra- and extracellular oxidants) whereas lucigenin, a larger molecule, does not permeate the cell and can only measure secreted products of the respiratory burst.

The use of chemiluminescence techniques therefore raises two questions. Firstly, what is the nature of the intracellular chemiluminescence? Secondly, what reactive oxidant species are detected by this technique?

The intracellular oxidants do not arise from oxidants produced at the plasma membrane diffusing back into the cell: scavenging the extracellular oxidants has no effect on the production of intracellular oxidants. Subcellular fractionation studies and direct visualisation of dichlorofluorescein-loaded cells indicate that the intracellular oxidants appear to be generated within cytoplasmic granules. The nature of these granules is not defined, but they may represent a compartment derived from granule–granule or granule–plasma membrane fusion. Further evidence for the involvement of granules comes from experiments with *cytoplasts,* neutrophils treated with cytochalasin B and then centrifuged through stepped gradients. Cytoplasts retain many functional properties of intact neutrophils but contain neither granules nor other cytoplasmic organelles. Cytoplasts secrete near-normal levels of reactive oxidants but do not generate these compounds intracellularly. There is emerging evidence to show that the activation of oxidant secretion (i.e. oxidase molecules activated on the plasma membrane) and intracellular products are regulated by distinct signal transduction processes. Initiation of oxidant secretion is prevented when increases in intracellular Ca^{2+} are blocked, but is unaffected when the activities of protein kinase C or phospholipase D (PLD) are inhibited. Intracellular oxidant production is, however, dependent on the activities of both protein kinase C and PLD.

The molecular species responsible for luminol and lucigenin chemiluminescence are not fully defined. Clearly, luminol chemiluminescence requires the combined activities of both the NADPH oxidase *and* myeloperoxidase: inhibitors of either enzyme (e.g. diphenylene iodonium for the oxidase and salicylhydroxamic acid for myeloperoxidase) completely prevent luminol chemiluminescence. There are two major ways in which products of myeloperoxidase may generate luminol chemiluminescence. The first of these is a reaction of luminol with H_2O_2 and HOCl; the second is an oxidation of luminol, catalysed by myeloperoxidase and H_2O_2 (or possibly O_2^-). Thus, the assay requires activation of the oxidase and degranulation of myeloperoxidase, and so does not directly measure the activity of the oxidase. However, the inclusion of saturating amounts of peroxidase (e.g. horseradish peroxidase or purified myeloperoxidase) into chemiluminescence assays can overcome the need for degranulation so that the products of the oxidase (O_2^- and H_2O_2) can be measured more directly. The addition of extracellular oxidant scavengers can distinguish between oxidants generated on the plasma membrane and those formed within the cell.

The molecular species responsible for lucigenin chemiluminescence are not fully defined. Because this probe does not permeate the cell, only secreted oxidants are detected. Several reports have suggested that this probe measures O_2^- secretion or else H_2O_2 secretion; on the other hand, some evidence indicates that this cannot be so. Certainly, the oxidant(s) detected by lucigenin is dependent on O_2^- formation, but the kinetics of activated O_2^- and H_2O_2 formation by neutrophils do not parallel those of lucigenin chemiluminescence. Thus, this probe is at least an indirect measure of O_2^- secretion.

In summary, chemiluminescence is a sensitive, non-invasive technique that can measure reactive oxidant production by small numbers of neutrophils; indeed, neutrophil-derived chemiluminescence can be detected in as little as 5 μl of unfractionated human blood. The assay is suitable for automation using either multichannel luminometers or luminescence microtitre plate readers. Many researchers, however, have questioned the usefulness of this technique because of the uncertainty of the nature of the oxidant(s) that are detected. Nevertheless, in view of the recent developments made towards the identification of the oxidants measured and the assay's ability to detect intracellular oxidant production, it is has an important place in the phagocyte research laboratory.

5.4.2.5 Electron paramagnetic resonance (EPR) spectroscopy

This technique can be used to measure the production of free radicals because the unpaired electron in a free radical has magnetic resonance. However, because the radicals are unstable, owing to their high chemical reactivity, the technique of 'spin-trapping' is used. In this technique, the generated radicals react with a suitable probe, and the EPR spectra arising from the reaction of the probe with different radical species can then be identified.

Because neutrophils generate substantial quantities of O_2^- and H_2O_2, and because most biological systems contain Fe or Cu salts in a form capable of catalysing the Haber–Weiss reaction (reaction 5.9), it has been proposed that they also generate ·OH. Several approaches to examine this possibility have been explored, including the use of so-called specific ·OH scavengers. The problem with this approach is that ·OH is so reactive that specific scavenging is impossible: any ·OH generated will react with the first molecule it encounters, so it can never diffuse away from its site of production. The use of EPR spectroscopy, coupled with spin trapping, has been used to explore the possibility of ·OH generation by activated neutrophils.

There is some evidence, from EPR spectroscopy and analysis of spin-trapped adducts, to suggest that ·OH may indeed be formed by activated neutrophils. However, caution must be exercised in interpreting such data because O_2^--generated adducts may decay to form adducts that resemble those generated directly from ·OH. For example, 5,5-dimethyl-1-pyrroline-1-oxide (DMPO) can react with O_2^- to form DMPO-OOH, and with ·OH to form DMPO-OH; formation of the latter adduct in phagocytosing neutrophils is taken as evidence for ·OH formation. However, two facts must be considered:

i. DMPO-OH formation must be catalase sensitive, as ·OH formation requires H_2O_2, as in reaction (5.9).
ii. DMPO-OH may be formed from the decomposition of DMPO-OOH.

It has recently been shown that catalase-sensitive (DMPO-OOH-independent) DMPO-OH formation in neutrophils can be detected, and that ·OH formation may be regulated by the extent of specific and azurophilic degranulation: lactoferrin (from specific granules) can decrease ·OH formation by chelating iron, whereas myeloperoxidase (from azurophilic granules) can decrease ·OH formation by utilising H_2O_2. However, it must be stressed that ·OH formation in these experiments is only detected when neutrophil suspensions are supplemented with fairly high concentrations of iron in the form of iron-diethyletriaminepenta-acetic acid. In the absence of this iron supplement, ·OH formation is not detected.

Thus, it appears from these experiments that neutrophils do not actively generate substantial levels of ·OH.This highly-reactive free radical may, however, be generated in small amounts during phagocytosis, for example, when O_2^- and/or H_2O_2 encounter a suitable transition metal catalyst present on the surface of or within the phagocytosed bacterium. The potential pathways for the generation of reactive oxidants by activated neutrophils are depicted in Figure 5.13.

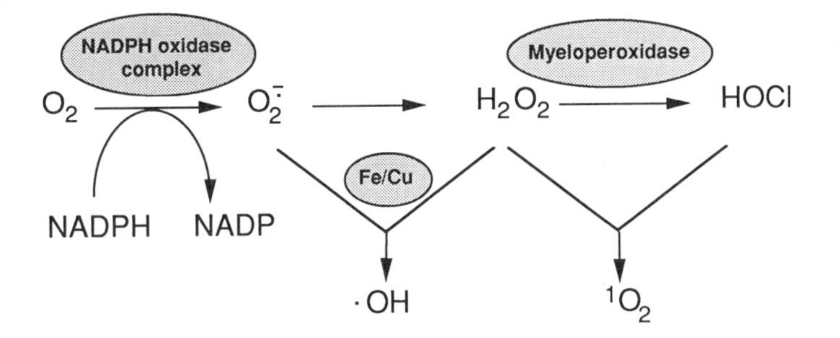

Figure 5.13. Pathways of reactive oxidant generation by neutrophils. See text for details.

5.5 Reactive nitrogen intermediates

The ability of mammalian cells to synthesise and secrete nitric oxide (NO) has only recently been discovered. The first observation of this phenomenon came from the discovery that endothelium-derived relaxing factor (EDRF), a compound that plays an important role in the regulation of blood vessel tone, blood coagulation and platelet aggregation, was in fact NO. Nitric oxide relaxes vascular smooth muscle and inhibits both platelet adhesion and platelet aggregation. It is thought to mediate its effects by freely permeating the plasma membrane of target cells and forming a complex with the haem group of soluble guanylate cyclase, which activates it. The target cell therefore generates increased levels of cGMP, which affects the activity of phospholipase C, thus perturbing the regulation of intracellular Ca^{2+} levels.

The enzyme responsible for NO production is an NO synthase that catalyses the oxidation of a guanido nitrogen of L-arginine (Fig. 5.14). The enzyme requires NADPH, molecular O_2 and another reductant, which in vitro can be tetrahydrobiopterin (BH_4). In vitro assays indicate that FAD is also required for activity. Indeed, NO synthase activity can be inhibited by the flavoprotein inhibitor diphenylene iodonium. More recently, purified preparations of NO synthase have shown that the enzyme contains both FAD and FMN. The products of NO synthase activity are citrulline and NO, and this latter product can then form NO_2^- and NO_3^-. Assays for NO synthase activity thus include the direct measurement of NO (using a specialised chemiluminescence technique; see Schmidt, Seifert & Bohme, 1989) and the chemical determination of NO_2^- and NO_3^-, the NO_2^- being measured with the Greiss reagent. Pre-incubating cells with L-arginine analogues (e.g. *N*-monomethyl-L-arginine, NMMA) or depleting the cells of exogenous L-arginine decreases the rate of production of NO and NO-derived products.

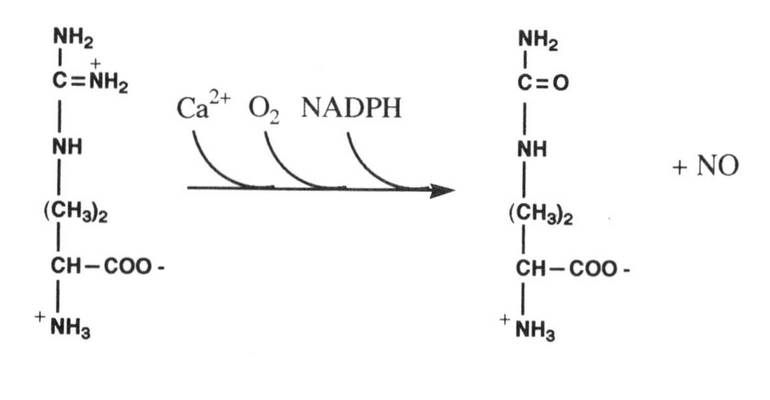

Figure 5.14. NO formation from arginine. The enzyme NO synthase catalyses the formation of NO from the amino acid arginine.

It appears that there are two major forms of NO synthase in mammalian cells. One of these is constitutively expressed (e.g. in endothelial cells and brain) and requires Ca^{2+} and a Ca^{2+}-binding protein, such as calmodulin. This enzyme is responsible for the secretion of fairly low levels of NO and functions as a regulator of cellular function via its ability to activate guanylate cyclase. The second enzyme is inducible and is expressed in cells such as macrophages after exposure to γ-interferon and LPS for several hours. This enzyme generates much higher, cytotoxic levels of NO, which is used by these cells to kill tumour cells and certain types of micro-organisms. The constitutive and inducible enzymes are antigenically distinct, although they catalyse the same reaction.

Neutrophils have been shown to generate NO via an NO synthase. Detection of this molecule is hampered, however, because O_2^- generated during the respiratory burst can degrade NO. Indeed, the detection of NO in neutrophil suspensions is enhanced by the addition of SOD. Neutrophil-derived NO has been implicated in the induction of vascular smooth-muscle contraction and in the inhibition both of platelet aggregation and of the release of ATP from platelets. It has also been shown to be important in the destruction of *Staphylococcus aureus* under conditions where the production of reactive oxidants is decreased. Production is stimulated by agents such as fMet-Leu-Phe, PAF and LTB_4 (under conditions where O_2^- production is scavenged by exogenous SOD). The NO synthase from rat neutrophils is a 150-kDa, monomeric protein that is dependent upon Ca^{2+} (but not calmodulin), NADPH, FAD and BH_4 for its activity.

5.6 Bibliography

Abo, A., & Pick, E. (1991). Purification and characterisation of a third cytosolic component of the superoxide-generating NADPH oxidase of macrophages. *J. Biol. Chem.* **266**, 23577–85.

Abo, A., Boyhan, A., West, I., Thrasher, A. J., & Segal, A. W. (1992). Reconstitution of neutrophil NADPH oxidase activity in the cell-free system by four components: p67-*phox*, p47-*phox*, $p21^{rac1}$, and cytochrome b_{-245}. *J. Biol. Chem.* **267**, 16767–70.

Abo, A., Pick, E., Hall, A., Totty, N., Teahan, C. G., & Segal, A. W. (1991). Activation of the NADPH oxidase involves the small GTP-binding protein $p21^{rac1}$. *Nature* **353**, 668–70.

Akard, L. P., English, D., & Gabig, T. G. (1988). Rapid deactivation of NADPH oxidase in neutrophils: continuous replacement by newly activated enzyme sustains the respiratory burst. *Blood* **72**, 322–7.

Allen, R. C. (1986). Phagocytic leukocyte oxygenation activities and chemiluminescence: A kinetic approach to analysis. *Methods Enzymol.* **133 B**, 449–93.

Aviram, I., & Sharabani, M. (1989). Kinetics of cell-free activation of neutrophil NADPH oxidase. *Biochem. J.* **261**, 477–82.

Babior, B. M. (1978). Oxygen-dependent microbial killing by phagocytes. *New Eng. J. Med.* **298**, 659–66.

Babior, B. M. (1984a). Oxidants from phagocytes: Agents of defense and destruction. *Blood* **64**, 959–66.

Babior, B. M. (1984b). The respiratory burst of phagocytes. *J. Clin. Invest.* **73**, 599–610.

Babior, B. M., & Kipnes, R. S. (1977). Superoxide-forming enzyme from human neutrophils: Evidence for a flavin requirement. *Blood*, **50**, 517–24.

Babior, B. M., Kipnes, R. S., & Curnutte, J. T. (1973). Biological defense mechanisms: The production by leukocytes of superoxide, a potential bacteriocidal agent. *J. Clin. Invest.* **52**, 741–4.

Baggiolini, M., & Wymann, M. P. (1990). Turning on the respiratory burst. *TIBS* **15**, 69–72.

Baldridge, C. W., & Gerard, R. W. (1933). The extra respiration of phagocytosis. *Am. J. Physiol.* **103**, 235–6.

Barber, D. L. (1991). Mechanisms of receptor-mediated regulation of Na–H exchange. *Cell. Signalling* **3**, 387–97.

Bellavite, P., Corso, F., Dusi, S., Grzeskowiak, M., Della-Bianca, V., & Rossi, F. (1988). Activation of NADPH-dependent superoxide production in plasma membrane extracts of pig neutrophils by phosphatidic acid. *J. Biol. Chem.* **263**, 8210–14.

Bellavite, P., Cross, A. R., Serra, M. C., Davoli, A., Jones, O. T. G., & Rossi, F. (1983). The cytochrome b and flavin content and properties of the $O_2^{\cdot-}$-forming NADPH oxidase solubilized from activated neutrophils. *Biochim. Biophys. Acta* **746**, 40–7.

Bjerrum, O. W., & Borregaard, N. (1989). Dual granule localization of the dormant NADPH oxidase and cytochrome b_{559} in human neutrophils. *Eur. J. Haematol.* **43**, 67–77.

Borregaard, N., & Tauber, A. I. (1984). Subcellular localization of the human neutrophil NADPH oxidase. *J. Biol. Chem.* **259**, 47–52.

Briheim, G., Stendahl, O., & Dahlgren, C. (1984). Intra- and extracellular events in luminol-dependent chemiluminescence of polymorphonuclear leukocytes. *Infect. Immun.* **45**, 1–5.

Britigan, B. E., Cohen, M. S., & Rosen, G. M. (1987). Detection of the production of oxygen-centered free radicals by human neutrophils using spin trapping techniques: A critical perspective. *J. Leuk. Biol.* **41**, 349–62.

Britigan, B. E., Hassett, D. J., Rosen, G. M., Hamill, D. R., & Cohen, M. S. (1989). Neutrophil degranulation inhibits potential hydroxyl-radical formation. *Biochem. J.* **264**, 447–55.

Bromberg, Y., & Pick, E. (1984). Unsaturated fatty acids stimulate NADPH-dependent superoxide production by cell-free system derived from macrophages. *Cell. Immunol.* **88**, 213–21.

Bromberg, Y., & Pick, E. (1985). Activation of NADPH-dependent superoxide production in a cell-free system by sodium dodecyl sulfate. *J. Biol. Chem.* **60**, 13539–45.

Clark, R. A., Leidal, K. G., Pearson, D. W., & Nauseef, W. M. (1987). NADPH oxidase of human neutrophils. *J. Biol. Chem.* **262**, 4065–74.

Cohen, M. S., Britigan, B. E., Hassett, D. J., & Rosen, G. M. (1988). Phagocytes, O_2 reduction, and hydroxyl radical. *Rev. Infect. Dis.* **10**, 1088–96.

Cross, A. R., Harper, A. M., & Segal, A. W. (1981). Oxidation–reduction properties of the cytochrome b found in the plasma-membrane fraction of human neutrophils. *Biochem. J.* **194**, 599–606.

Cross, A. R., Higson, F. K., Jones, O. T. G., Harper, A. M., & Segal, A. W. (1982). The enzymic reduction and kinetics of oxidation of cytochrome b_{-245} of neutrophils. *Biochem. J.* **204**, 479–85.

Cross, A. R., & Jones, O. T. G. (1991). Enzymic mechanisms of superoxide production. *Biochim. Biophys, Acta* **1057**, 281–98.

Cross, A. R., Jones, O. T. G., Garcia, R., & Segal, A. W. (1982). The association of FAD with the cytochrome b_{-245} of human neutrophils. *Biochem. J.* **208**, 759–63.

Cuperus, R. A., Muijsers, A. O., & Wever, R. (1986). The superoxide dismutase activity of myeloperoxidase; formation of Compound III. *Biochim. Biophys. Acta* **871**, 78–84.

Curnutte, J. T. (1985). Activation of human neutrophil nicotinamide adenine dinucleotide phosphate, reduced (triphosphopyridine nucleotide, reduced) oxidase by arachidonic acid in a cell-free system. *J. Clin. Invest.* **75**, 1740–3.

Dahlgren, C. (1987). Difference in extracellular radical release after chemotactic factor and calcium ionophore activation of the oxygen radical-generating system in human neutrophils. *Biochim. Biophys. Acta* **930**, 33–8.

Dahlgren, C., Aniansson, H., & Magnusson, K.-E. (1985). Pattern of formylmethionyl-leucyl-phenylalanine-induced luminol- and lucigenin-dependent chemiluminescence in human neutrophils. *Infect. Immun.* **47**, 326–8.

Davies, B., & Edwards, S. W. (1989). Inhibition of myeloperoxidase by salicylhydroxamic acid. *Biochem. J.* **258**, 801–6.

Dinauer, M. C., Orkin, S. H., Brown, R., Jesaitis, A. J., & Parkos, C. A. (1987). The glycoprotein encoded by the X-linked chronic granulomatous disease locus is a component of the neutrophil cytochrome b complex. *Nature* **327**, 717–20.

Dinauer, M. C., Pierce, E. A., Bruns, G. A. P., Curnutte, J. T., & Orkin, S. H. (1990). Human neutrophil cytochrome b light chain (p22-*phox*). Gene structure, chromosomal location, and mutations in cytochrome-negative autosomal recessive chronic granulomatous disease. *J. Clin. Invest.* **86**, 1729–37.

Edwards, S. W. (1987). Luminol- and lucigenin-dependent chemiluminescence of neutrophils: Role of degranulation. *J. Clin. Lab. Immunol.* **22**, 35–9.

Edwards, S. W. (1991). Regulation of neutrophil oxidant production. In *Calcium, Oxygen Radicals and Tissue Damage* (Duncan, J., ed.), Cambridge University Press, U.K.

Eklund, E. A., Marshall, M., Gibbs, J. B., Crean, C. D., & Gabig, T. G. (1991). Resolution of a low molecular weight G protein in neutrophil cytosol required for NADPH oxidase activation and reconstitution by recombinant Krev-1 protein. *J. Biol. Chem.* **266**, 13964–70.

Fridovich, I. (1978). The biology of oxygen radicals: The superoxide radical is an agent of oxygen toxicity; superoxide dismutases provide an important defence. *Science* **201**, 875–80.

Garcia, R. C., & Segal, A. W. (1984). Changes in the subcellular distribution of the cytochrome b_{-245} on stimulation of human neutrophils. *Biochem. J.* **219**, 233–42.

Garcia, R. C., & Segal, A. W. (1988). Phosphorylation of the subunits of cytochrome b_{-245} upon triggering of the respiratory burst of human neutrophils and macrophages. *Biochem. J.* **252**, 901–4.

Grinstein, S., Swallow, C. J., & Rotstein, O. D. (1991). Regulation of cytoplasmic pH in phagocytic cell function and dysfunction. *Clin. Biochem.* **24**, 241–7.

Gutteridge, J. M. C., Rowley, D. A., & Halliwell, B. (1982). Superoxide-dependent formation of hydroxyl radicals and lipid peroxidation in the presence of iron salts. *Biochem. J.* **206**, 605–9.

Halliwell, B., & Gutteridge, J. M. C. (1984). Oxygen toxicity, oxygen radicals, transition metals and disease. *Biochem. J.* **219**, 1–9.

Halliwell, B., & Gutteridge, J. M. C. (1985). *Free Radicals in Biology and Medicine.* Clarendon Press, Oxford.

Harper, A. M., Dunne, M. J., & Segal, A. W. (1984). Purification of cytochrome b_{-245} from human neutrophils. *Biochem. J.* **219**, 519–27.

Henderson, L. M., Chappell, J. B., & Jones, O. T. G. (1988). Internal pH changes associated with the activity of NADPH oxidase of human neutrophils: Further evidence for the presence of an H^+ conducting channel. *Biochem. J.* **251**, 563–7.

Heyneman, R. A., & Vercauteren, R. E. (1984). Activation of a NADPH oxidase from horse polymorphonuclear leukocytes in a cell-free system. *J. Leuk. Biol.* **36**, 751–9.

Heyworth, P. G., & Segal, A. W. (1986). Further evidence for the involvement of a phosphoprotein in the respiratory burst oxidase of human neutrophils. *Biochem. J.* **239**, 723–31.

Higson, F. K., Durbin, L., Pavlotsky, N., & Tauber, A. I. (1985). Studies of cytochrome b_{-245} translocation in the PMA stimulation of the human neutrophil NADPH-oxidase. *J. Immunol.* **135**, 519–24.

Jesaitis, A. J., Buescher, E. S., Harrison, D., Quinn, M. T., Parkos, C. A., Livesey, S., & Linner, J. (1990). Ultrastructural localization of cytochrome b in the membranes of resting and phagocytosing human granulocytes. *J. Clin. Invest.* **85**, 821–35.

Klebanoff, S. J. (1968). Myeloperoxidase–halide–hydrogen peroxide antibacterial system. *J. Bacteriol.* **95**, 2131–8.

Knaus, U. G., Heyworth, P. G., Kinsella, T., Curnutte, J. T., & Bokoch, G. M. (1992). Purification and characterisation of Rac2. A cytosolic GTP-binding protein that regulates human neutrophil NADPH oxidase. *J. Biol. Chem.* **267**, 23575–82.

Kobayashi, S., Imajoh-Ohmi, S., Nakamura, M., & Kanegasaki, S. (1990). Occurrence of cytochrome b_{558} in B-cell lineage of human lymphocytes. *Blood* **75**, 458–61.

Kolb, H., & Kolb-Bachofen, V. (1992). Nitric oxide: A pathogenetic factor in autoimmunity. *Immunol. Today* **13**, 157–60.

Leto, T. L., Lomax, K. J., Volpp, B. D., Nunoi, H., Sechler, J. M. G., Nauseef, W. M., Clark, R. A., Gallin, J. I., & Malech, H. L. (1990). Cloning of a 67-kD neutrophil oxidase factor with similarity to a noncatalytic region of p60c-*src*. *Science* **248**, 727–30.

Lomax, K. J., Leto, T. L., Nunoi, H., Gallin, J. I., & Malech, H. L. (1989). Recombinant 47-kilodalton cytosol factor restores NADPH oxidase in chronic granulomatous disease. *Science* **245**, 409–12.

Malawista, S. E., Montgomery, R. R., & Van Blaricom, G. (1992). Evidence for reactive nitrogen intermediates in killing of Staphylococci by human neutrophil cytoplasts: A new microbicidal pathway for polymorphonuclear leukocytes. *J. Clin. Invest.* **90**, 631–6.

Maly, F. E., Cross, A. R., Jones, O. T. G., Wolf-Vorbeck, G., Walker, C., Dahinden, C. A., & de Weck, A. L. (1988). The superoxide generating system of B cell lines. *J. Immunol.* **140**, 2334–9.

Merenyi, G., Lind, J., & Eriksen, T. E. (1990). Luminol chemiluminescence: Chemistry, excitation, emitter. *J. Biolum. Chemilum.* **5**, 53–6.

Nakanishi, A., Imajoh-Ohmi, S., Fujinawa, T., Kikuchi, H., & Kanegasaki, S. (1992). Direct evidence for interaction between COOH-terminal regions of cytochrome b_{558} subunits and cytosolic 47-kDa protein during activation of an O_2^--generating system in neutrophils. *J. Biol. Chem.* **267**, 19072–4.

Odajima, T., & Yamazaki, I. (1972). Myeloperoxidase of the leukocyte of normal blood: IV. Some physicochemical properties. *Biochim. Biophys. Acta* **284**, 360–7.

Odell, E. W., & Segal, A. W. (1988). The bactericidal effects of the respiratory burst and the myeloperoxidase system isolated in neutrophil cytoplasts. *Biochim. Biophys. Acta* **971**, 266–74.

Ohtsuka, T., Okamura, N., & Ishibashi, S. (1986). Involvement of protein kinase C in the phosphorylation of 46 kDa proteins which are phosphorylated in parallel with activation of NADPH oxidase in intact guinea-pig polymorphonuclear leukocytes. *Biochim. Biophys. Acta* 332–7.

Okamura, N., Babior, B. M., Mayo, L. A., Peveri, P., Smith, R. M., & Curnutte, J. T. (1990). The p67-*phox* cytosolic peptide of the respiratory burst oxidase from human neutrophils. *J. Clin. Invest.* **85**, 1583–7.

Parkos, C. A., Dinauer, M. C., Walker, L. E., Allen, R. A., Jesaitis, A. J., & Orkin, S. H. (1988). Primary structure and unique expression of the 22-kilodalton light chain of human neutrophil cytochrome b. *Proc. Natl. Acad. Sci. USA* **85**, 3319–23.

Patel, A. K., Hallett, M. B., & Campbell, A. K. (1987). Threshold responses in production of reactive oxygen metabolites in individual neutrophils detected by flow cytometry and microfluorimetry. *Biochem. J.* **248**, 173–80.

Pou, S., Cohen, M. S., Britigan, B. E., & Rosen, G. M. (1989). Spin-trapping and human neutrophils. *J. Biol. Chem.* **264**, 12299–302.

Root, R. K., Metcalf, J., Oshino, N., & Chance, B. (1975). H_2O_2 release from human granulocytes during phagocytosis. *J. Clin. Invest.* **55**, 945–55.

Rossi, F. (1986). The O_2^--forming NADPH oxidase of the phagocytes: Nature, mechanisms of activation and function. *Biochim. Biophys. Acta* **853**, 65–89.

Sbarra, A. J., & Karnovsky, M. L. (1959). The biochemical basis of phagocytosis: I. Metabolic changes during the digestion of particles by polymorphonuclear leukocytes. *J. Biol. Chem.* **234**, 1355–62.

Schmidt, H. H. H. W., Seifert, R., & Bohme, E. (1989). Formation and release of nitric oxide from human neutrophils and HL-60 cells induced by a chemotactic peptide, platelet activating factor and leukotriene B_4. *FEBS Lett.* **244**, 357–60.

Segal, A. W. (1989). The electron transport chain of the microbicidal oxidase of phagocytic cells and its involvement in the molecular pathology of chronic granulomatous disease. *Biochem. Soc. Trans.* **17**, 427–34.

Segal, A. W., & Abo, A. (1993). The biochemical basis of the NADPH oxidase of phagocytes. *TIBS* **18**, 43–7.

Segal, A. W., & Jones, O. T. G. (1978). Novel cytochrome b system in phagocytic vacuoles of human granulocytes. *Nature* **276**, 515–17.

Segal, A. W., West, I., Wientjes, F., Nugent, J. H. A., Chavan, A. J., Haley, B., Garcia, R. C., Rosen, H., & Scrace, G. (1992). Cytochrome b_{-245} is a flavocytochrome containing FAD and the NADPH-binding site of the microbicidal oxidase of phagocytes. *Biochem. J.* **284**, 781–8.

Stuehr, D. J., Kwon, N. S., Cross, S. S., Thiel, B. A., Levi, R., & Nathan, C. F. (1989). Synthesis of nitrogen oxides from L-arginine by macrophage cytosol: Requirement for inducible and constitutive components. *Biochem. Biophys. Res. Commun.* **161**, 420–6.

Teahan, C. G., Totty, N., Casimir, C. M., & Segal, A. W. (1990). Purification of the 47 kDa phosphoprotein associated with the NADPH oxidase of human neutrophils. *Biochem. J.* **267**, 485–9.

Umei, T., Babior, B. M., Curnutte, J. T., & Smith, R. M. (1991). Identification of the NADPH-binding subunit of the respiratory burst oxidase. *J. Biol. Chem.* **266**, 6019–22.

Volpp, B. D., Nauseef, W. M., Donelson, J. E., Moser, D. R., & Clark, R. A. (1989). Cloning of the cDNA and functional expression of the 47-kilodalton cytosolic component of human respiratory burst oxidase. *Proc. Natl. Acad. Sci. USA* **86**, 7195–9.

Watson, F., Robinson, J. J., & Edwards, S. W. (1991). Protein kinase C dependent and independent activation of the NADPH oxidase of human neutrophils. *J. Biol. Chem.* **266**, 7432–9.

Winterbourn, C. C., Garcia, R. C., & Segal, A. W. (1985). Production of the superoxide adduct of myeloperoxidase (compound III) by stimulated human neutrophils and its reactivity with hydrogen peroxide and chloride. *Biochem. J.* **228**, 583–92.

Yui, Y., Hattori, R., Kosuga, K., Eizawa, H., Hiki, K., Ohkawa, S., Ohnishi, K., Terao, S., & Kawai, C. (1991). Calmodulin-independent nitric oxide synthase from rat polymorphonuclear neutrophils. *J. Biol. Chem.* **266**, 3369–71.

6

Neutrophil activation: The production of intracellular signalling molecules

6.1 Introduction

Whilst the neutrophil has one primary function – namley, to recognise and destroy microbial pathogens – many separate but interrelated processes are required to control this function. For example, the neutrophil must be able to detect a variety of pro- and anti-inflammatory signals in its microenvironment and then respond accordingly. The responses to these signals include up-regulation of receptor number and function, adherence, diapedesis, chemotaxis, particle binding, phagocytosis, degranulation and activation of the NADPH oxidase. There may also be, where appropriate, the generation of further pro-inflammatory mediators, such as eicosanoids and secondary cytokines. Furthermore, these responses must be terminated when the noxious agent or pathogen has been eliminated. Such deactivation mechanisms of neutrophil function are almost as important as the activation mechanisms, because prolonged or inappropriate neutrophil activity can lead to inflammatory damage via the secretion or overproduction of cytotoxic products.

All known physiological and pathological agents activate neutrophils by first binding to plasma membrane receptors, details of which are given in Chapter 3. The cells must respond either to soluble agonists, which generally signal events such as changes in receptor expression, adherence, aggregation and chemotaxis, or else to particulate stimuli (e.g. opsonised bacteria and immune complexes), which generally stimulate phagocytosis. However, some soluble agonists (e.g. fMet-Leu-Phe, C5a and LTB_4) stimulate adherence and chemotaxis (which are early events in the inflammatory response of neutrophils) at low concentrations, but at higher concentrations can activate degranulation and reactive oxidant production (which are later functional responses). How are such separate functional events activated in response to different concentrations of the same agonist? In part, this is due to the fact that many receptors exist on the plasma membrane in high- and low-affinity states. High-affinity receptors will thus be occupied by low concentrations of agonists and may activate chemotaxis and adherence; low-affinity receptors will become occupied only at higher agonist concentrations and may activate degranulation and the NADPH oxidase. Clearly, though, these two forms of the same receptor must be coupled to different signal transduction processes, which activate different end responses.

Many neutrophil studies in vitro utilise the synthetic peptide fMet-Leu-Phe (whose synthesis is based upon the structure of small peptides released during bacterial protein biosynthesis) and phorbol esters (e.g. PMA), which are structural analogues of diacylglycerol (DAG), the activator of the enzyme protein kinase C. Thus, fMet-Leu-Phe is used as a means of assessing receptor-mediated cell activation, whilst phorbol esters are generally used to evaluate the role of protein kinase C in neutrophil function. Increasingly, more studies are being performed with more physiologically-important agonists such as LTB_4, C5a and IL-8. With these latter two agents this is now possible because of the availability of recombinant proteins. Fewer studies, however, are performed with particulate stimuli such as opsonised zymosan, opsonised bacteria and immune complexes. In general terms, the soluble agonists activate neutrophils very rapidly (within seconds or minutes after addition), and events such as oxidase activation occur on the plasma membrane, so that reactive oxidants and granule enzymes are released into the extracellular medium, thus making them relatively easy to measure. In contrast, particulate stimuli tend to activate neutrophils more slowly (responses might be detected up to 60 min after addition of stimulus), activate oxidase molecules within phagolysosomes and cause the degranulation of enzymes into these phagocytic vesicles. Such intracellular processes are much more difficult to study, but in a sense are more important if the hope is to understand the mechanisms regulating phagocytosis and pathogen killing.

This chapter will therefore discuss the molecular mechanisms that have been implicated in neutrophil activation following the binding of agonists to specific receptors on the plasma membrane.

6.2 G-proteins

Almost all receptor-mediated neutrophil functions are mediated via GTP-binding proteins (*G-proteins*), which provide the link between occupancy of plasma membrane receptors and the activation of intracellular enzymes, such as phospholipases and protein kinases. There are two groups of G-proteins: those that are *heterotrimeric* and those with *low molecular weight.*

Both of these types play important roles in the activation of various neutrophil functions.

6.2.1 Heterotrimeric G-proteins

These proteins were discovered in 1971, when it was observed that GTP was needed for the activation of adenylate cyclase by β-adrenergic agonists. It had been known for some time that activation of some cells resulted in the stimulation of adenylate cyclase, whereas activation of cells with other types of agonists could inhibit the activity of this enzyme. For example, binding of epinephrine (adrenaline) to β-adrenergic receptors activates adenylate cyclase, but binding of the same hormone to α_2-adrenergic receptors can inhibit this enzyme. It was then shown that the link between occupancy of the receptor and activation of adenylate cyclase was provided by these G-proteins, and later that several types of G-proteins exist. Those that activate (stimulate) adenylate cyclase were termed G_s whilst those that inhibit the enzyme were termed G_i (Fig. 6.1). In addition to regulating the activity of adenylate cyclase, these G-proteins also regulate the activities of several phospholipases.

These types of G-proteins are trimeric structures comprising a large α-subunit (relative molecular masses of 39–52 kDa), a β-subunit (35–36 kDa) and a smaller (<10 kDa) γ-subunit (Fig. 6.2). The β- and γ-subunits are tightly associated with each other and require denaturing conditions to separate them. The α-subunit is more easily dissociated from the other subunits and becomes freed from the complex during receptor occupancy; it may then interact with its target enzyme (either adenylate cyclase or a phospholipase) either to activate or inhibit it. The $\beta\gamma$ dimer is hydrophobic because the γ-subunit is isoprenylated, and thus this dimer functions to anchor $G\alpha$ into the membrane. The guanine nucleotide-binding site is on the α-subunit, which is classified as either 's' or 'i' type (i.e. $G\alpha_s$ or $G\alpha_i$) depending upon whether it stimulates or inhibits adenylate cyclase, respectively (Fig. 6.1). Other α-subunit classifications have now been described. For example, at least three different types of $G\alpha_i$ have been identified ($G\alpha_{i-1}$, $G\alpha_{i-2}$ and $G\alpha_{i-3}$) in different cell types, and Go is found in brain cells and adrenal chromaffin cells; *transducin,* a G-protein (termed $G\alpha_t$) found exclusively within the photosensitive rod cells of the retina, functions to regulate phosphodiesterase activity. The β- and γ-subunits are more homologous in different heterotrimeric G-proteins in different types of cells, although small variations in their structures may exist. All $G\alpha$-subunits bind GTP and possess GTPase activity; hence, activation of $G\alpha$-subunits occurs via cycles of GTP binding and GTP hydrolysis.

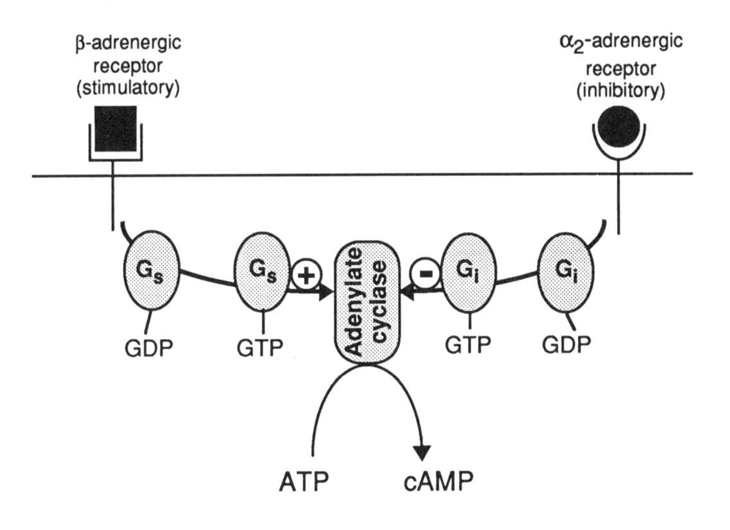

Figure 6.1. Regulation of adenylate cyclase activity by G-proteins. Occupancy of receptors such as the β-adrenergic receptor result in the activation (+) of adenylate cyclase via coupling through stimulatory G-proteins (G_s). Alternatively, occupancy of receptors such as the α_2-adrenergic receptor inhibit (–) adenylate cyclase via coupling through inhibitory G-proteins (G_i).

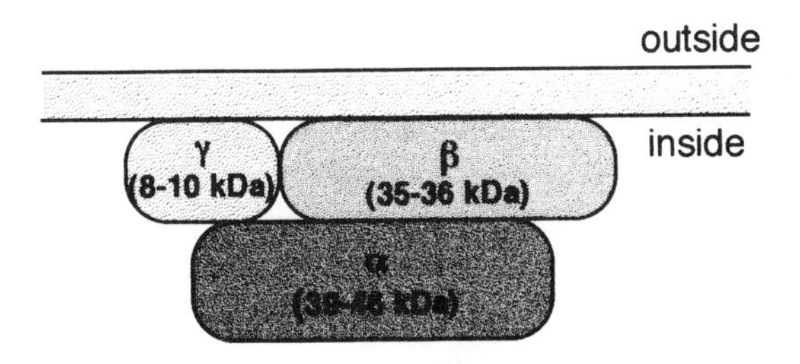

Figure 6.2. Structure of heterotrimeric G-proteins. The generalised structure of the α-, β-, and γ-subunits of the heterotrimeric G-proteins and their organisation in the plasma membrane are shown.

The sequences of events that occur during activation of adenylate cyclase after receptor occupancy are shown in Figure 6.3. This scheme thus shows activation of a $G\alpha_s$-type protein (i.e. a process that leads to the activation of adenylate cyclase), whereas similar processes will occur with a $G\alpha_i$ protein, except that the interaction with adenylate cyclase will result in its inactivation. In the same way, activation of phospholipases by mobile $G\alpha$-type subunits will occur via similar mechanisms. In the unstimulated state, $G\alpha_s$ (or $G\alpha_i$) is bound to GDP. Binding of the receptor with its agonist induces a conformational change in the receptor that activates its G-protein. This stimulates the dissociation of GDP from the α-subunit, which is then replaced with GTP. This binding of Gα with GTP promotes its dissociation from the β- and γ-subunits (which remain in the membrane), and Gα moves through the membrane until it encounters a molecule of its target enzyme (e.g. adenylate cyclase or a phospholipase), which then becomes activated (or inhibited). The activated enzyme then acts upon its substrate to produce the signalling molecules (e.g. cAMP from adenylate cyclase or inositol trisphosphate [IP_3], diacylglycerol (DAG) from the action of phospholipase C on PIP_2; see §6.3.1). Because the Gα-subunit has inherent GTPase activity (which may be augmented by the action of a cytosolic factor), the bound GTP becomes hydrolysed to GDP. The GDP-Gα subunit dissociates from its target enzyme (thereby inactivating it in the case of Gα_s, or activating it in the case of Gα_i) and recombines with the β- and γ-subunits. This G-protein complex can be reactivated in the presence of continued receptor occupancy by further exchange of bound GDP with GTP. This will result in the activation of more target enzymes and hence prolong the activation state of the cell. In the absence of continued receptor occupancy, GDP is not exchanged with GTP, the G-proteins are no longer active and the generation of intracellular signalling molecules (e.g. cAMP, IP_3 and DAG) is terminated. The key steps in these processes are thus GTP–GDP exchange and GTP hydrolysis.

Experimentally, it is possible to manipulate the activity of Gα_s, thereby producing a cell in a continuously-activated or -deactivated state. GTP-binding can be substituted by binding with compounds such as GTPγS, a non-hydrolysable analogue of GTP that actively exchanges with GDP on Gα but cannot be hydrolysed by GTPase activity to form GDPγS; hence, Gα becomes permanently stimulated and it continuously interacts with its target enzyme. However, GTPγS does not readily permeate intact cells such as neutrophils, so experiments with this compound must be performed either with cell extracts or with cells made permeable to small molecules (e.g. by treatment with saponin or after electroporation). In addition, other experimental techniques can be employed to investigate the role of G-proteins in cell activation. For example, several bacterial toxins (e.g. *Vibrio cholera* toxin and *Bordatella pertussis* toxin) interact with G-proteins and affect their function. Cholera toxin cleaves NAD^+ and transfers the ADP-ribose group to a site on Gα_s (in a process termed *ADP-ribosylation*), thus converting G_s into an irreversible activator. On the other hand, pertussis toxin performs a similar ADP-ribosylation of Gα_i that leads to the formation of an irreversible inactivator. Also, the fluoraluminate complex (AlF_4^-) activates G-proteins in the absence of either receptor occupancy or GTP-binding.

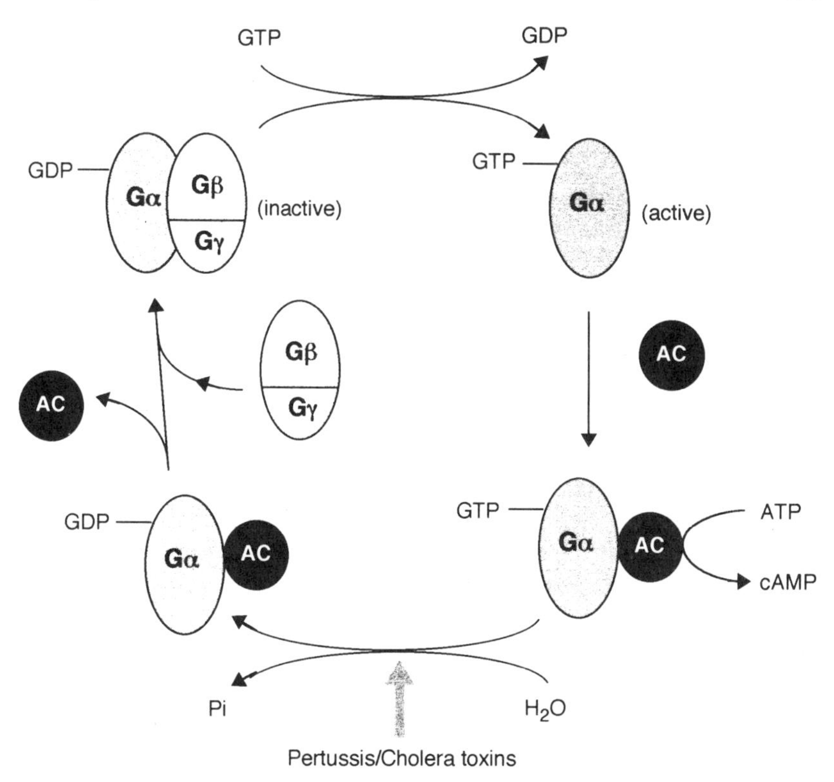

Figure 6.3. Mechanism of action of heterotrimeric G-proteins. Upon receptor occupancy, the $G\alpha$-subunit binds GTP in exchange for GDP, and then moves in the membrane until it encounters its target enzyme, shown here as adenylate cyclase (alternatively, a phospholipase). The activated target enzyme then becomes functional. Inherent GTPase activity within the α-subunit then hydrolyses bound GTP to GDP, and the α-subunit dissociates from its target enzyme (which becomes inactive) and rebinds the β- and γ-subunits. Upon continued receptor occupancy, further catalytic cycles of GTP exchange and target enzyme activation may occur. The scheme shown is for a stimulatory G-protein ($G\alpha_s$), but similar sequences of events occur with inhibitory G-proteins ($G\alpha_i$) except that the interaction of the α-subunit with adenylate cyclase will result in its inhibition. The sites of action of pertussis and cholera toxins are shown.

All of these experimental approaches have been adopted in neutrophil studies to show that activation of several receptor-mediated functions occurs via the participation of heterotrimeric G-proteins. In many cases, the conventional $G\alpha_i/G\alpha_s$ nomenclature is used to describe these G-proteins, even though the subunits may not be linked to either inhibition or activation of adenylate cyclase. The nomenclature used is based on structural and functional similarities to other $G\alpha$-subunits in other cell types, and also on their sensitivities to cholera and pertussis toxins. Several of these G-proteins are linked to phospholipase activation. Those that have been described in neutrophils include:

$G\alpha_s$ (52 + 45 kDa), a substrate for cholera toxin;

$G\alpha_{i\text{-}2}$ (40 kDa), a substrate for cholera and pertussis toxin;

$G\alpha_{i\text{-}3}$ (41 kDa), a substrate for pertussis toxin.

The G-protein that has been termed Gp, and that is linked to phospholipase C activation, may in fact be $G\alpha_{i\text{-}2}$ or $G\alpha_{i\text{-}3}$. Ga is designated as the G-protein responsible for activation of phospholipase A_2, which results in arachidonic acid release. Some experimental evidence indicates that, at least in HL-60 cells, different agonists can preferentially activate different phospholipases, and some of these are responsible for the activation of secretion. In neutrophils, the two pertussis-toxin-sensitive $G\alpha$-proteins ($G\alpha_{i\text{-}2}$ and $G\alpha_{i\text{-}3}$) have been identified by peptide mapping of proteolytic digests of the proteins, by peptide sequencing and by immunoblotting. Complementary-DNA clones for the mRNA of these two molecules have also been isolated from an HL-60 cDNA library. $G\alpha_{i\text{-}2}$ is five to ten times more abundant than $G\alpha_{i\text{-}3}$, the former component comprising 3% of the total plasma membrane proteins. It is possible that these two different $G\alpha$-subunits are coupled to different phospholipases (e.g. phospholipases C and D). Pertussis toxin inhibits the secretion of O_2^- after stimulation of neutrophils by fMet-Leu-Phe, but pertussis-toxin-insensitive G-proteins are also present in neutrophils. These may be members of the Gq family and may be involved in the activation of phospholipase $C_{\beta 1}$ (see §6.3.1).

The receptors for fMet-Leu-Phe, C5a and PAF have all been cloned (see Chapter 3) and possess the predicted seven membrane-spanning domains present in other G-protein-linked receptors of the rhodopsin superfamily (see Fig. 3.2). The pertussis-toxin sensitivity of the G-proteins associated with these receptors arises from the ADP-ribosylation of a cysteine residue that is four amino acids from the COOH-terminus of the molecule. Some other pertussis-toxin-insensitive G-proteins that exist lack this critical cysteine residue.

6.2.2 Low-molecular-weight G-proteins

In addition to the large, heterotrimeric G-proteins that have been shown to play a role in signal transduction, a superfamily of low-molecular-weight (LMW) GTP-binding proteins, which are structurally related to ras, has been shown also to play important roles in the regulation of cell function. These proteins were initially identified by molecular cloning techniques and shown to have sequence homology to the ras proteins, which function in both normal and pathological cell growth and differentiation. Alternatively, such proteins can be identified following gel electrophoresis of cell extracts, electrophoretic transfer to nitrocellulose and reaction with radiolabelled GTP. As with the heterotrimeric G-proteins, the low-molecular-weight G-proteins mediate their effects via cycles of GTP binding, GTP hydrolysis and guanine nucleotide exchange (i.e. GDP replaced by GTP). They thus all bind GTP and possess GTPase activity. There is some evidence to show that bound GDP can be converted into GTP via the activity of nucleoside diphosphate kinase. This enzyme may thus reactivate the G-protein without the need for GDP–GTP exchange. Some 30 or so distinct ras-related proteins have now been identified in mammalian cells.

The first members of this superfamily discovered were the 21-kDa proteins encoded by the Harvey and Kirsten rat sarcoma viruses, $p21^{V\text{-}Ha\text{-}ras}$ and $p21^{V\text{-}Ki\text{-}ras}$, respectively. The protein $p21^{N\text{-}ras}$ was later discovered as an oncogene product in neuroblastoma. Normal $p21^{C\text{-}ras}$ binds GTP and catalyses its hydrolysis to GDP: the GTP-bound form is 'active' whilst the GDP-bound form is 'inactive'. The oncogenes encode forms of p21 that are activated by point mutations, either at or near the sites where the proteins bind the phosphates of the nucleotides, or else at sites of their interaction with the guanine base. These mutations have the effect of decreasing the GTPase activity or decreasing the affinity of the G-protein for the nucleotide, hence increasing its exchange rate for bound GDP with cytosolic GTP. Thus, the net effect of these mutations is to increase the amount of GTP bound to p21 and, hence, to increase the amount of 'active' protein. Putative roles for such LMW GTP-binding proteins in neutrophils include degranulation, phagolysosome fusion, directed locomotion and granule movement (which is required for the up-regulation of receptors and cytochrome b molecules onto the plasma membrane). These processes all involve intracellular traffic, sorting and subsequent fusion of membranes. Several lines of evidence show that guanine nucleotides can stimulate events such as degranulation in permeabilised neutrophils by mechanisms that are independent of the activities of heterotrimeric G-proteins.

The ras superfamily of proteins in mammalian cells is divided into three subgroups, depending upon the degree of homology between the proteins (Table 6.1):

i. The *ras proteins* are 85% homologous to each other. Three other subfamilies within this group of ras proteins can be distinguished: the rap, R-ras and ral proteins, which are about 50% homologous to Ha-ras,

Ki-ras and N-ras. Expression of rap1A (also known as Krev-1) causes the reversion of cells that have been transformed by Ki-ras, and so the function of p21 ^{rap1A} may counteract the action of ras proteins.

ii. The *rab proteins* may be present in the Golgi, and may be involved in protein trafficking between subcellular compartments. A guanine nucleotide dissociation inhibitor is thought to regulate their function. The protein type rab3 is involved in neurotransmitter granule exocytosis and, to date, has only been identified in nervous tissue. Mutations in yeast have identified the SEC4 protein and the YPT1 protein as being important for vesicular transport of proteins from the endoplasmic reticulum to the plasma membrane via the Golgi. The protein SEC4 (23.5 kDa) is located on the cytoplasmic face of the plasma membrane and secretory granules, and is thus involved in the late events of secretion. On the other hand, YPT1 protein (23 kDa) is present in the Golgi and is responsible for the early events in secretion. Mammalian rab proteins share 38–75% homology at the amino acid level with these yeast proteins, and the putative effector-binding site of p21ras (amino acid residues 35–40) is highly conserved in the rab proteins, YPT1p and SEC4p. Other small GTP-binding proteins do not share such homology with these yeast proteins. In mammalian cells, rab1A, rab2, rab4, rab5 and rab6 are ubiquitous.

iii. The *rho proteins* (e.g. rhaA, rac1) may be involved in processes such as actin microfilament assembly. The C3 exoenzyme of *Clostridium botulinum* can ADP-ribosylate rhaA, rhoC and some rac proteins, and has been shown capable of collapsing actin filaments. Also identified in the regulation of rho-protein function are GTPase-activating proteins and nucleotide exchange proteins. The proteins rac1 and rac2 have been identified in leukemic myeloblasts.

Proteins of the rap type have been identified in brain, platelets and neutrophils, and at least four different types have been described: rap1 and rap2 are about 70% homologous at the amino acid level whilst rap1A and rap1B differ from each other in only 9 out of 184 amino acids (six of these differences being in the 13 COOH-terminal residues). There is 90% homology between rap2 and rap2b. In its effector (32–42-amino-acid) domain, rap1A protein is identical to ras protein; thus rap proteins may antagonise the effects of ras proteins by competing for ras-GAP in transformed cells. The full repertoire and functions of rap proteins in normal cells such as the neutrophil are, however, largely unknown. Protein types rap1 and rap2 are present in neutrophils but rap1A, which is phosphorylated at serine residues by protein kinase A (PKA, cAMP-dependent protein kinase) is the major form

Table 6.1. *Some small GTP-binding proteins of mammalian cells*

ras	*rho*	*rab*
Ha-ras	rhoA	rab1A
Ki-ras	rhoB	rab1B
N-ras	rhoC	rab2
rap1A (Krev-1, smg p21)	rac1	rab3A (smg p25A)
rap1B	rac2	rab3B
rap2	G25K	rab4
R-ras		rab5
ralA		rab6
ralB		rab7
		BRL-ras

present. Phosphorylation of rap1A has no effect on its guanine nucleotide binding kinetics, its rate of GTP hydrolysis or its ability to be stimulated by rap-GAP, but phosphorylation of rap1B affects its subcellular location and ability to bind rap guanine nucleotide-dissociation stimulator (GDS). Thus, rap1A may function to mediate the inhibitory effects resulting from the elevation of cAMP levels that is observed during neutrophil activation by some agonists.

Several proteins are involved in the ras signal-transduction pathway. Regulation of GTP binding and hydrolysis involves GAP protein (GTPase-activating protein), proteins that inhibit GDP-dissociation and proteins that stimulate guanine nucleotide exchange. GAP is an abundant 120-kDa protein that stimulates the hydrolysis of GTP bound to 'active' normal, but not to mutant, p21. Hence, the activity of GAP will down-regulate p21 function in normal, but not malignant, cells. The protein that stimulates guanine nucleotide exchange does so by stimulating the binding of GTP to p21, which in its absence occurs at extremely low rates (10^{-2} min^{-1}). The activity of this protein, which may be stimulated during receptor signalling, will thus 'activate' p21 (Fig. 6.4).

Several lines of evidence implicate rap proteins in the signal-transduction events regulating neutrophil activation. A 23-kDa rap protein (shown to be rap1A) copurifies with the cytochrome b_{-245} of the NADPH oxidase, and a ras-related protein binds to an anti–cytochrome b affinity column. Further evidence for the role of this protein in regulating oxidase function comes from the fact that the GTPγS form of rap1A binds to the cytochrome b more tightly than the GDP form; also, phosphorylation of rap1A by protein kinase A prevents its binding to the cytochrome. There is also limited sequence homology between *ras*GAP and both p47-*phox* and p66-*phox*, indicating that,

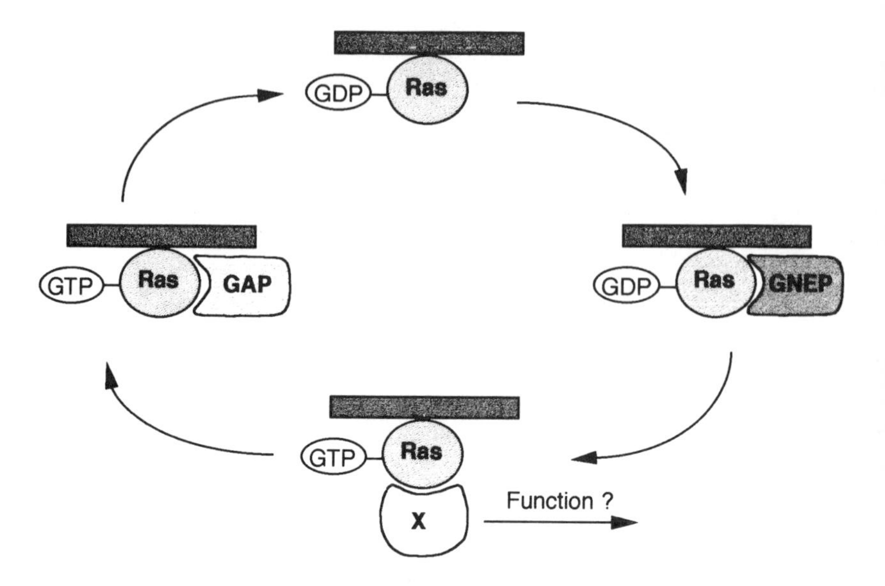

Figure 6.4. Mode of action of low-molecular-weight G-proteins. The raslike proteins normally bind GDP, but this may be exchanged for GTP via a process that may be assisted by guanine nucleotide exchange protein (GNEP). The GTP-bound ras protein may then interact with and activate its target protein (X). The activity of GTPase activating protein (GAP) may then assist to hydrolyse GTP to GDP, to inhibit ras activity.

in addition to the cytochrome, rap1A may also interact with the cytosolic oxidase factors. A further role for small GTP-binding proteins in NADPH oxidase activation has come from experiments involving the purification of the cytosolic components required for activation of the oxidase in a cell-free system (§5.3.2.3). In addition to the cytosolic factors p47-*phox* and p66-*phox,* a complex of two proteins existing as a heterodimer of 46 kDa is required. This complex has been shown to comprise the small GTP-binding protein p21^{rac1} and the GDP-dissociation inhibitor *rho*GDI. (Further details of oxidase activation are given in §5.3.2.) The addition of recombinant p21^{rac1} to the other purified or recombinant oxidase components can activate the oxidase in vitro.

Guanine nucleotides also stimulate degranulation, and so it has been suggested that small GTP-binding proteins may also play a role in this process. Both azurophilic and specific granules possess GTP-binding proteins that differ in their relative molecular masses and thus may be independently regulated. Indeed, the degranulation of specific and azurophilic granules can be activated somewhat independently of each other. Using specific monoclonal antibodies it has been shown that rap2 proteins are present in specific granules, whereas rap1 proteins (rap1A) are present in specific granules and in

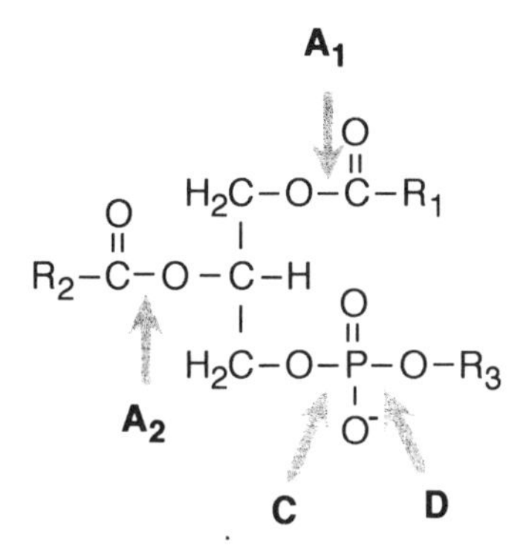

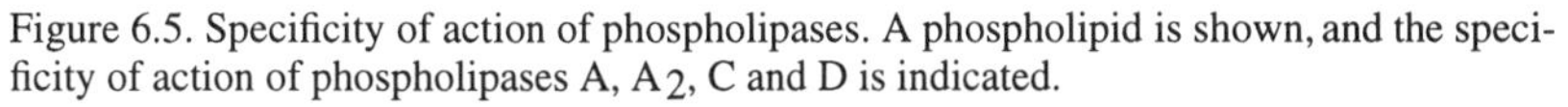
Figure 6.5. Specificity of action of phospholipases. A phospholipid is shown, and the specificity of action of phospholipases A, A_2, C and D is indicated.

the plasma membrane. Independent studies have shown that rap1A colocalises with the cytochrome b in the plasma membrane and also in specific granules. Upon stimulation, both proteins cotranslocate from the specific granules to the plasma membrane or phagolysosome. Whilst neutrophils possess rab1A, rab2, rab4 and rab6, none of these has been shown to be localised on the membranes of azurophilic or specific granules. The proteins rab1A and rab6 are probably associated with the Golgi, rab2 is probably on a structure associated with an intermediate compartment of the endoplasmic reticulum and Golgi, whilst rab5 is on the cytoplasmic face of the plasma membrane and early endosomes.

6.3 Phospholipase activation

Several enzymes involved in the activation of neutrophil function have been shown to be regulated by G-proteins. These include the phospholipases, which cleave membrane phospholipids as shown in Figure 6.5.

6.3.1 Phospholipase C

6.3.1.1 Properties

Phosphoinositase C (i.e. phosphoinositide-specific phospholipase C [PLC]) enzymes are found in the vast majority of mammalian cells. Molecular cloning of these enzymes, analysis of their predicted amino acid sequences and immunological cross-reactivity indicate that at least three major forms of the enzyme exist: PLC-β, -δ and -γ. Each of these enzyme types is encoded by a distinct gene. More recent experiments using the polymerase chain reaction and molecular cloning have revealed even greater enzyme di-

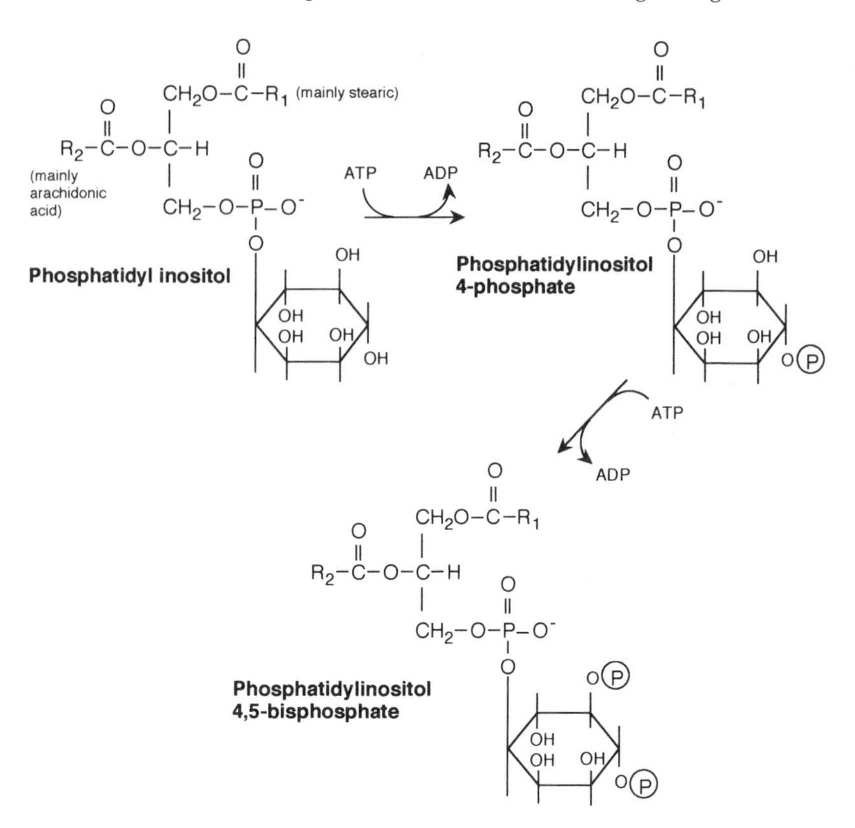

Figure 6.6. Structure and formation of phosphatidylinositol (PtdIns), phosphatidylinositol 4-phosphate (PIP) and phosphatidylinositol 4,5-bisphosphate (PIP_2). See text for details.

versity. Thus, PLC-β_1 and -β_2, PLC-δ_{1-3} and PLCγ_{1-2} have now been identified. More recent studies indicate that PLC-ε and PLC-α families may exist in some cells, although more detailed structural and sequence analyses are required to confirm this. All of these enzymes are single peptide molecules of relative molecular masses 150–154 kDa (PLC-β), 145–148 kDa (PLC-γ) and 85–88 kDa (PLC-δ). All have similar catalytic properties (which is dependent upon Ca^{2+}) and can hydrolyse phosphatidylinositol (PtdIns), phosphatidylinositol 4-phosphate (PIP) and phosphatidylinositol 4,5-bisphosphate (PIP_2; Fig. 6.6). At concentrations of Ca^{2+} of 0.1–1 μM, the enzyme preferentially hydrolyses PIP_2, whereas at higher concentrations the enzyme will begin to hydrolyse PtdIns and PIP. Thus, the preferred substrates in cells with physiological intracellular Ca^{2+} concentrations are PIP and PIP_2. Hydrolysis of PIP_2 by PLC yields two potential second-messenger molecules (Fig. 6.7), inositol 1,4,5-trisphosphate (Ins 1,4,5 P_3) and diacyl-

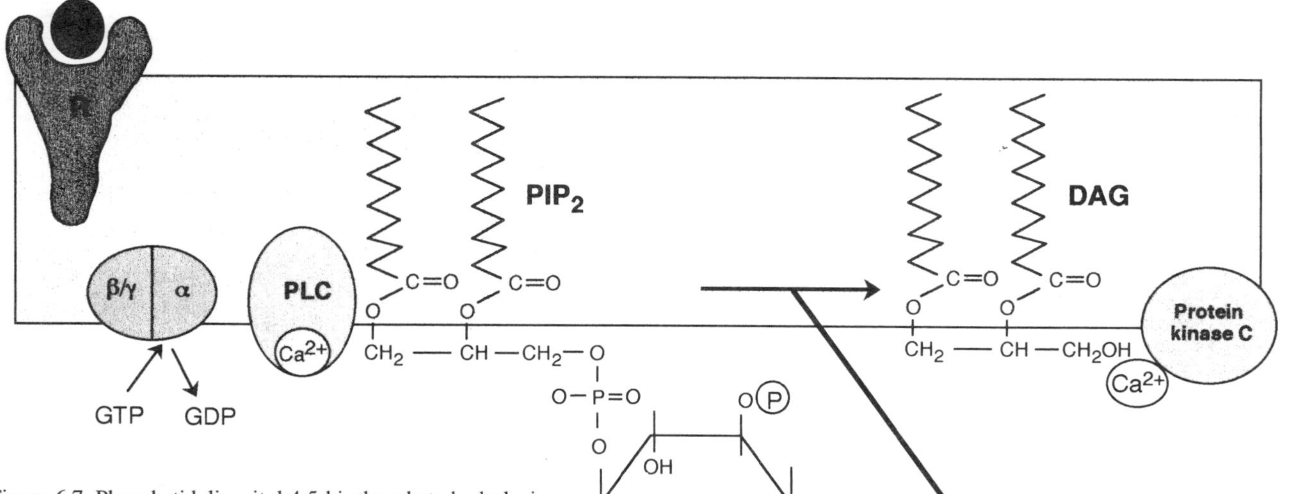

Figure 6.7. Phosphatidylinositol 4,5-bisphosphate hydrolysis by phospholipase C. Occupancy of receptors (R) results in exchange of bound GDP for GTP on the α-subunit of a heterotrimeric G-protein. The α-subunit then dissociates from the β- and γ-subunits and activates phospholipase (PLC). This enzyme is calcium dependent and, upon activation, can hydrolyse phosphatidylinositol 4,5-bisphosphate (PIP_2). The products of this hydrolysis are inositol 1,4,5-trisphosphate (Ins 1,4,5-P_3), which is released into the cytoplasm, and diacylglycerol (DAG), which remains in the membrane. The DAG is an activator of protein kinase C, which moves from the cytoplasm to the membrane, where it forms a quaternary complex with DAG and Ca^{2+}.

glycerol (DAG), whose properties are described in sections 6.3.1.2–5. Hydrolysis of PIP generates DAG and inositol 1,4-bisphosphate.

The three major forms of PLC catalyse the hydrolysis of phosphoinositides, and the products may be cyclic and non-cyclic inositol phosphates (Fig. 6.8). The rank order of generation of cyclic inositol phosphates (e.g. 1,2-cyclic 4,5-trisphosphate) by these enzymes is PLC-β > PLC-δ > PLC-γ. Thus, the possession of different types of PLC enzymes by different cells may be necessary to generate different quantities of cyclic and non-cyclic inositol phosphates. Furthermore, these different forms of PLC are regulated by different control mechanisms. For example, PLC-γ is preferentially phosphorylated by cAMP-dependent protein kinase (PKA), which down-regulates its function; thus, elevations in cAMP can inhibit PLC activity. Also, some enzyme forms in different cells are either activated or inhibited via G-protein coupling (e.g. PLC-β_1) whilst others are activated via phosphorylation on tyrosine residues (e.g. PLC-γ_1). In these latter examples, the receptor coupled to PLC activation may have intrinsic protein tyrosine kinase activity (e.g. receptors for insulin, platelet-derived growth factor [PDGF], fibroblast growth factor [FGF]), or else the PLC may be phosphorylated via non-receptor protein tyrosine kinase activity (e.g. after occupancy of some Fc receptors).

Neutrophil membranes contain inositol lipids, which comprise about 5–6% of the total membrane lipids. About 80% of these inositol lipids possess stearic acid (C18 : 0) at C1 and arachidonic acid (C20 : 4) at C2 positions. Phosphatidylinositol accounts for most of these lipids (90%), with smaller amounts of PIP (6%) and PIP_2 (4%), which are synthesised sequentially by the action of 4- and 5-specific kinases, respectively (see Fig. 6.6). Neutrophil membranes also possess a phosphatidylinositol-specific phospholipase C which cleaves phosphatidylinositol 4,5-bisphosphate (PIP_2) into Ins 1,4,5 P_3 and DAG (Fig. 6.7). Both PLC-β (β_2) and PLC-γ (γ_2) families appear to be present in neutrophils. The coupling of receptor occupancy to PLC activation in neutrophils can be through a heterotrimeric G-protein, the mobile subunit of which has been termed $G\alpha_p$. Evidence for this G-protein link comes from the following facts:

i. GTPγS can stimulate phosphoinositide breakdown in isolated plasma membranes.
ii. The fluoraluminate complex (AlF_4^-) stimulates the formation of inositol phosphates.
iii. Some receptor-mediated activation can be inhibited by pertussis toxin.

$G\alpha_p$-coupling lowers the apparent affinity of PLC for Ca^{2+} to about 0.1 μM – that is, the level found in the cytosol of resting cells.

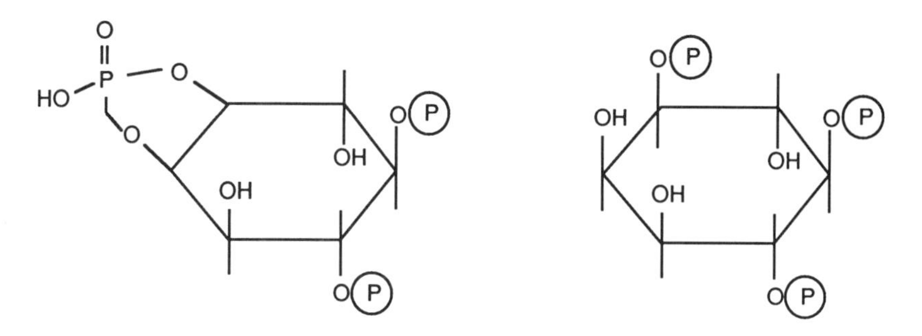

1,2 -cyclic inositol 4,5-trisphosphate Inositol 1,4,5-trisphosphate

Figure 6.8. Cyclic and non-cyclic inositol phosphates. Hydrolysis of phosphatidyl 4,5-bisphosphate (PIP_2) by phospholipase C can generate cyclic and non-cyclic inositol phosphates.

Whilst PLC can be directly activated by Ca^{2+} in vitro, the concentrations required (up to 1 mM) are usually far higher than can be achieved in vivo. Therefore, receptor occupancy and $G\alpha_p$ activity are required to stimulate PLC in the absence of a rise in cytoplasmic Ca^{2+} above basal (resting) levels. However, once PLC is activated, one of its products, Ins 1,4,5 P_3, can increase intracellular Ca^{2+} concentrations to about 1 μM. These elevated Ca^{2+} levels can synergise with $G\alpha_p$ to activate PLC further. However, another product of PLC is DAG, which in turn can activate the enzyme protein kinase C (see §6.3.1.6). This enzyme, on the other hand, can inhibit $G\alpha_p$-stimulated PLC activity, thereby limiting the hydrolysis of inositol phosphates and so decreasing the rate of formation of second-messenger molecules via this route. The majority of PLC activity is located within the cytosol of most cells. The membrane-bound forms of the enzyme are easily removed, indicating that they are not integral membrane proteins. It is likely that the cytosolic enzymes are recruited and attached to the plasma membrane during cell activation.

6.3.1.2 Inositol phosphates and cell activation

The formation of inositol phosphates is a good measure of the activity of PLC. Within seconds of addition of fMet-Leu-Phe to neutrophils, Ins 1,4,5-P_3 formation is detected, and levels of this compound increase for about 20–30 s, falling to their original levels within 2 min of stimulation. It is likely that this decrease in Ins 1,4,5-P_3 occurs because it is phosphorylated to inositol 1,3,4,5-tetrakisphosphate (Ins 1,3,4,5-P_4) and then subsequently dephosphorylated to inositol 1,3,4-trisphosphate (Ins 1,3,4-P_3). This latter molecule can be converted into inositol 1,4-bisphosphate, then into inositol

1-phosphate and ultimately back into PIP_2 (Fig. 6.9). Hence, the Ins 1,4,5-P_3 that is released into the cytoplasm can be metabolised by two separate routes.

More recently, the importance of a group of highly polar inositol lipids, present in neutrophils and many other cell types, has been recognised. Activation of neutrophils by fMet-Leu-Phe results in the transient accumulation of phosphatidylinositol 3-phosphate (PtdIns 3-P), phosphatidylinositol 3,4-bisphosphate (PtdIns 3,4-P_2) and phosphatidylinositol 3,4,5-trisphosphate (PtdIns 3,4,5-P_3). Apparently, the enzyme phosphatidylinositol 3-hydroxy (3-OH) kinase plays a key role in the formation of these novel lipids. This enzyme can catalyse the formation of these lipids from phosphatidylinositol, phosphatidylinositol 4-phosphate (PtdIns 4-P) and phosphatidylinositol 4,5 bisphosphate (PtdIns 4,5-P_2) in vitro (Fig. 6.10). Alternatively, it is possible that PtdIns 3,4-P_2 and PtdIns 3-P are derived from the sequential dephosphorylation of PtdIns 3,4,5-P_3.

The biological functions of these lipids are not fully appreciated, but they are likely to represent a new class of signalling molecules. The concentration of PtdIns 3,4,5-P_3 transiently rises from 5 μM in resting neutrophils to 200 μM in fMet-Leu-Phe-stimulated cells. This lipid is not a substrate for PLC, and so the decrease in its intracellular concentration is not due to its metabolism by this enzyme; nor is it broken down into Ins 1,4,5-P_3 and DAG. In other cell types, the activity of the PtdIns 3-OH kinase is associated with the receptors for several growth factors (e.g. insulin, PDGF, EGF, CSF-1) and oncogene products (e.g. $pp60^{v\text{-}src}$, $p68^{v\text{-}ros}$), which have been shown to possess intrinsic protein tyrosine kinase activity. In these situations, the PtdIns 3-OH kinase is implicated in mitogenic stimulation or oncogenic transformation. It is thought that the lipids themselves, rather than products derived from them, are the signalling or effector molecules per se. Indeed, there is some evidence to show that PtdIns 3,4-P_2 levels correlate with actin polymerisation, and that this lipid may anchor actin filaments to the plasma membrane.

6.3.1.3 Inositol phosphates and regulation of intracellular Ca^{2+}

The idea that stimulated inositide metabolism was involved in increases in cytoplasmic Ca^{2+} was first proposed by Bob Michell (1975). To date, over 20 different inositol phosphates can be isolated from stimulated cells, but so far only one of these molecules can be ascribed a definite function: Ins 1,4,5-P_3, which is released into the cytoplasm following PLC hydrolysis of PIP_2, liberates Ca^{2+} from intracellular stores. A role for Ins 1,3,4,5-P_4 in opening a Ca^{2+} 'gate', thus allowing the influx of Ca^{2+} from the external

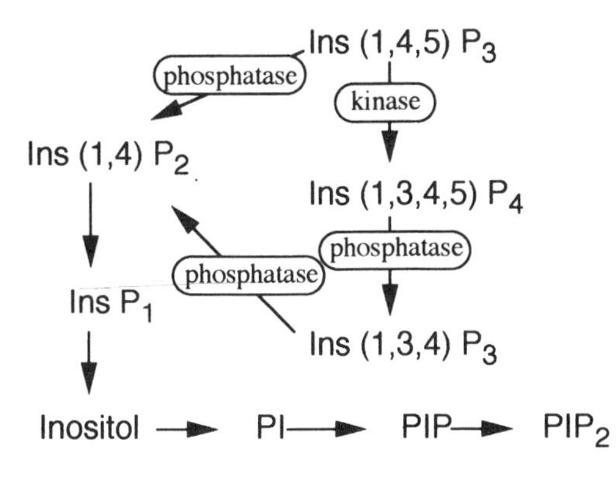

Figure 6.9. Pathways of inositol phosphate metabolism. Ins 1,4,5-P_3, generated via the hydrolysis of phosphatidyl 4,5-bisphosphate by phospholipase C, can be metabolised by a kinase (to generate Ins 1,3,4,5-P_4) or via a phosphatase (to yield Ins 1,4-P_2). These products can be metabolised further to produce inositol, which itself may be sequentially phosphorylated to regenerate phosphatidylinositol 4,5-bisphosphate.

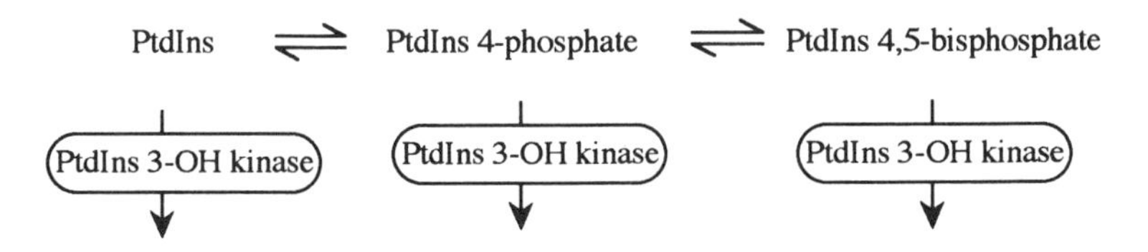

Figure 6.10. Phosphatidylinositol 3-hydroxy kinase activity. The enzyme phosphatidylinositol 3-hydroxy (PtdIns 3-OH) kinase may phosphorylate inositol lipids as shown, to generate a series of novel polar lipids that may function in cell activation. See text for details.

environment, has also been postulated, but this has not been reproduced in many studies. It may be that Ins 1,3,4,5-P_4 acts together with Ins 1,4,5-P_3 in regulating Ca^{2+} influx. The precise location of these intracellular stores has not been defined, although in neutrophils the endoplasmic reticulum and organelles called *calciosomes* have been proposed (see later in this section). The roles that Ins 1,4,5-P_3 and Ca^{2+} play in the activation of neutrophil function have been the subject of intense investigations.

Increases in the levels of Ins 1,4,5-P_3 can be measured in several ways (e.g. by radioimmunoassay), whilst intracellular levels of Ca^{2+} can be monitored by the inclusion into the neutrophil cytoplasm of suitable Ca^{2+} indicators. Some elegant work using the Ca^{2+}-sensitive photoprotein *obelin* (obtained from the luminous jellyfish *Obelia*) has shown that increases in intracellular Ca^{2+} can be detected within seconds of activation of neutrophils with soluble (but not particulate) stimulants, and that these increases precede activation of the oxidase. Such experiments led to one of the first proposals that the oxidase may be activated by multiple signalling systems. More recent research has utilised fluorescent Ca^{2+} indicators (such as Quin-2, Fura-2, Indo-2 and Fluo-3), which have advantages over photoproteins because they can more easily be incorporated into the neutrophil cytoplasm (Fig. 6.11). For example, the esterified form of the indicator is membrane permeable; once inside the cell, cytoplasmic esterase activity cleaves the molecule into a polar (and, hence, membrane-impermeable) product, so that the indicator becomes trapped intracellularly. Experiments using such intracellular indicators have shown that intracellular concentrations of Ca^{2+} are increased rapidly (within seconds of stimulation), and that such increases in the level of this cation are achieved via either the mobilisation of intracellular Ca^{2+} stores or else by opening the Ca^{2+} 'gate' on the plasma membrane to allow Ca^{2+} influx.

Basal levels of Ca^{2+} in resting neutrophils are around 100 nM. Upon stimulation with soluble agonists such as fMet-Leu-Phe, PAF and LTB_4, intracellular levels rise to about 1 μM within 20–30 s and then return to basal levels. Experiments in which neutrophils are suspended in media devoid of Ca^{2+} (and containing EGTA to chelate trace contaminants of this cation), show that although the initial rise in intracellular Ca^{2+} is unaffected, the elevation in concentration is not sustained when extracellular Ca^{2+} is absent. Thus, the *initial* rise in intracellular Ca^{2+} is due to mobilisation of intracellular stores; Ca^{2+} influx is then activated in order to sustain these levels (Fig. 6.11). Addition of Ins 1,4,5-P_3 to permeabilised neutrophils results in a rapid (within 2 s) rise in intracellular Ca^{2+}, which then declines to basal levels, possibly as a result of the released Ca^{2+} returning to the intracellular stores. These intracellular vesicles must therefore have independent Ca^{2+} efflux and influx.

In neutrophils, the intracellular Ca^{2+} stores have been termed *calciosomes.* These Ins 1,4,5-P_3-sensitive organelles may be structurally related to the endoplasmic reticulum (Fig. 6.12). Entry of Ca^{2+} into these stores is thought to be regulated by a Ca^{2+}-ATPase; release is activated upon cell stimulation via occupancy of the Ins 1,4,5-P_3-sensitive channel, Ins 1,4,5-P_3 being generated via the activity of PLC on PIP_2. However, it is also possible that there may be Ins 1,4,5-P_3-*in*sensitive stores (similar to those that have been described in other cell types), which may be responsible for Ca^{2+}-induced Ca^{2+} release. Also present in these stores is the Ca^{2+}-binding protein *calreticulin* (analogous to that found in muscle cells), each molecule

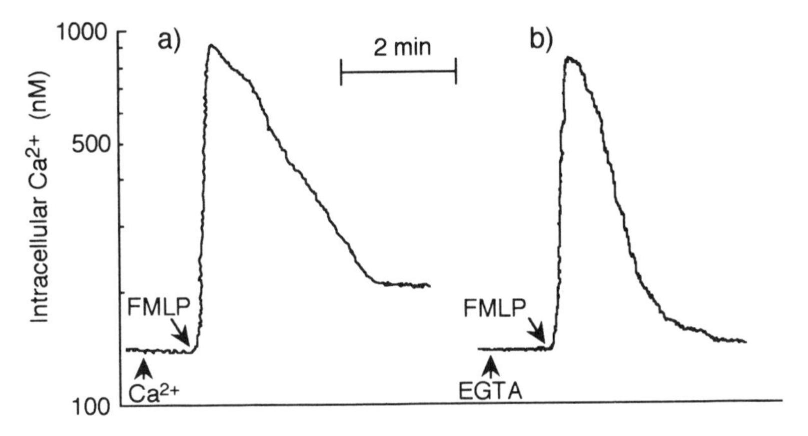

Figure 6.11. Intracellular Ca^{2+} levels during neutrophil activation with fMet-Leu-Phe. Neutrophil suspensions were loaded with Fluo-3 AM for 15 min. This molecule is membrane permeable but cleaved by intracellular esterase activity to yield the polar molecule Fluo-3, which is thus trapped within the cell. The neutrophils were then suspended in buffer that was devoid of Ca^{2+}, and treated as shown. In (a), 1 mM Ca^{2+} and 1 μM fMet-Leu-Phe were added to the suspension, as indicated by the arrows. In (b), 1 mM EGTA and 1 μM fMet-Leu-Phe were added as shown. Thus, in (a), the change in intracellular Ca^{2+} is due to mobilisation of intracellular Ca^{2+} stores and the influx of extracellular Ca^{2+}, whereas in (b), the Ca^{2+} rise is due solely to release of Ca^{2+} from intracellular stores.

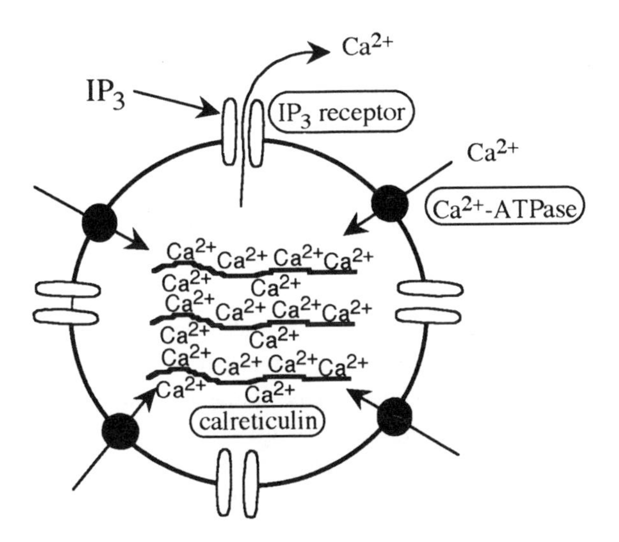

Figure 6.12. Calciosomes, the intracellular Ca^{2+} store in neutrophils. Calciosomes are believed to possess three important components: (i) the Ins 1,4,5-P_3 (IP_3) receptor, occupancy of which releases Ca^{2+} from the calciosome, (ii) a Ca^{2+}-dependent ATPase, responsible for loading the calciosome with cytoplasmic Ca^{2+} and (iii) calreticulin, a Ca^{2+}-binding protein.

of which has about 20 Ca^{2+}-binding sites. It is thought that calreticulin possesses both high- and low-affinity binding sites.

Calciosomes possess an Ins 1,4,5-P_3 receptor, a Ca^{2+}-ATPase and calreticulin, the Ca^{2+}-binding protein. Inside the calciosome, Ca^{2+} concentrations may be as high as 1 mM; Ca^{2+} efflux from these stores is regulated by occupancy of the Ins 1,4,5-P_3 receptor, whereas influx into these stores is regulated by the activity of the Ca^{2+}-ATPase. The Ins 1,4,5-P_3 receptor has many structural similarities to the ryanodine receptor in that it has four conserved cysteine residues and is a tetramer of four subunits. Stimulation of the head of this receptor thus opens the channel to release Ca^{2+} from these stores into the cytoplasm. Once these stores are emptied, Ca^{2+} influx into the cell is stimulated by the opening of the Ca^{2+} gate on the plasma membrane. This Ca^{2+} influx is thus necessary for maintaining the cytoplasmic Ca^{2+} at an elevated level, and also for replenishing the stores.

There is much debate as to the functional role of Ins 1,3,4,5-P_4. During cell activation, Ins 1,4,5-P_3 is generated before Ins 1,3,4-P_3 formation can be detected. If the intracellular Ca^{2+} store is depleted, then Ins 1,4,5-P_3 formation upon cell activation still occurs, but the rise in intracellular Ca^{2+} and the formation of Ins 1,3,4-P_3 are inhibited. Thus, it is thought that Ins 1,4,5-P_3 is converted initially into Ins 1,3,4,5-P_4 and then into Ins 1,3,4-P_3 via a Ca^{2+}-dependent process (Fig. 6.9). Ins 1,3,4,5-P_4 may regulate intracellular Ca^{2+} levels by activating reuptake of Ca^{2+} into intracellular stores, by controlling Ca^{2+} transfer between the Ins 1,4,5-P_3-sensitive and -insensitive pools, by modulation of Ca^{2+} influx across the plasma membrane and by potentiating the Ins 1,4,5-P_3-induced Ca^{2+} mobilisation. The plasma membrane of neutrophils contains Ca^{2+} channels that regulate Ca^{2+} influx. The rate of filling of the intracellular pools may thus have a role in regulating Ca^{2+} influx, and so Ins 1,4,5-P_3 may indirectly regulate the rate of Ca^{2+} entry into the cell.

Most early experiments measuring changes in intracellular Ca^{2+} levels during cell activation were performed using cells in suspension. However, the application of recent technological developments that allow intracellular Ca^{2+} measurements to be made in single neutrophils (using intracellular fluorescent dyes and measurements with fluorescence microscopes) has led to the discovery of important differences in the responses of suspended and adhered cells. In order to measure Ca^{2+} changes in single cells with this technology, the neutrophils must be adhered to a glass surface so that the fluorescence changes can be 'viewed'. Once neutrophils adhere to glass surfaces, a single subcellular Ca^{2+} gradient is generated; however, when they adhere to fibronectin- or albumin-coated surfaces, spontaneous oscillations in intracellular Ca^{2+} levels occur, with an amplitude of about 80 nM and a duration of 30 s. It is thought that Ca^{2+} oscillations are required for attachment and detachment from protein-coated surfaces when their adherence receptors (e.g. Mac-1) become occupied. Furthermore, different patterns of Ca^{2+} changes are observed at different concentrations of stimuli. For example, activation of adhered cells by low concentrations (10^{-10} M) of fMet-Leu-Phe results in sustained Ca^{2+} oscillations, whereas activation by higher concentrations (10^{-6} M) results in more continuous rises. It is thus possible that Ca^{2+} oscillations and sustained elevations may regulate different cellular functions. Control of secretion may be regulated by the amplitude and frequency of these oscillations. The molecular processes that regulate Ca^{2+} oscillations in neutrophils have not been fully delineated.

6.3.1.4 Role of Ca^{2+} in neutrophil function

Several lines of independent evidence point to a role for Ca^{2+} in the activation of the NADPH oxidase by some agonists:

i. Increases in intracellular Ca^{2+} levels often precede activation of oxidant production.
ii. In neutrophils in which Ca^{2+} levels are buffered using intracellular Ca^{2+} chelators (e.g. BAPTA) or high levels of the Ca^{2+} indicator Quin-2, reactive oxidant secretion is prevented.
iii. Inhibitors of calmodulin (e.g. trifluoperazine) prevent oxidase activation (although some of these inhibitors may not be absolutely specific for calmodulin).

However, intracellular Ca^{2+} increases in themselves are insufficient to activate the oxidase because levels of this cation can be artificially increased in the absence of oxidase activity (e.g. by the addition of Ca^{2+} ionophores such as ionomycin). Furthermore, activation of the oxidase by agonists such as PMA occurs in the absence of intracellular Ca^{2+} increases. Thus, if DAG is generated at sufficiently high concentrations to activate protein kinase C, then increases in intracellular Ca^{2+} are not required for oxidase activation. Furthermore, some agonists (e.g. LTB_4 and PAF) generate substantial increases in intracellular Ca^{2+} but are poor activators of the respiratory burst. Also, there is a marked discrepancy between the concentrations of some agonists (e.g. fMet-Leu-Phe) required to activate either Ca^{2+} increases or oxidase activity. For example, maximal elevations in Ca^{2+} can occur at concentrations of fMet-Leu-Phe that are too low to activate the respiratory burst. It therefore appears that intracellular Ca^{2+} increases are not required to activate the respiratory burst per se, but are involved in the activation of an enzyme(s) or process required for oxidase activation; thus, alternative, Ca^{2+}-independent pathways of activation also exist. At present, the role of Ca^{2+} in oxidase activation is not clearly defined, but it is likely to be required for the activation of enzymes such as phospholipase A_2, which itself generates molecules that are directly involved in oxidase activation.

Phagocytosis of C3bi-coated yeast particles by adherent neutrophils does not result in elevations in intracellular Ca^{2+}; in contrast, the phagocytosis of immunoglobulin-coated particles does activate changes in Ca^{2+}. Spreading can occur in the absence of extracellular Ca^{2+}, but adherence-associated actin polymerisation requires the presence of extracellular Ca^{2+}. The role of intracellular Ca^{2+} in degranulation is more clearly established. Changes in intracellular Ca^{2+} levels can be achieved by addition of different concentrations of Ca^{2+} ionophore, by varying the concentration of Ca^{2+} in the extracellular medium or by manipulation of intracellular Ca^{2+} levels using intracellular buffers. These approaches have shown that degranulation does not occur at intracellular concentrations <200 nM, whereas degranulation of specific and gelatinase-containing granules occurs at 610–650 nM Ca^{2+} and 2.5 μM Ca^{2+} is required for maximal release of azurophilic granules. However, these experiments were performed in populations of neutrophils in suspension, and it may be that these values are different for individual or adhered cells. Localised Ca^{2+} changes within the cell may also be involved in direction movement.

Experiments by Daniel Lew and his colleagues (Lew et al. 1984, 1985; Kraus et al. 1990) have shown that when neutrophils phagocytose particles, a rapid increase in cytosolic Ca^{2+} occurs across most of the cell, shortly declining to basal levels; Ca^{2+} levels then increase again, but this increase is now localised around the phagosome. In cells depleted of intracellular Ca^{2+}, phagocytosis can occur normally, but phagolysosomal fusion (i.e. fusion of phagocytic vesicles with granules) does not occur. Experiments using fluorescent antibodies to the Ca^{2+}-ATPase of calciosomes and to F-actin show that both F-actin and the calciosomes are spread throughout the resting cell. During phagocytosis, both the calciosomes and F-actin are enriched around the phagosome, forming a ring of calciosomes and actin bundles; the specific granules are also enriched around the phagosome. In Ca^{2+}-depleted cells, the granules accumulate, but fusion does not occur; in Ca^{2+}-containing cells, a localised release of Ca^{2+} from these calciosomes to the phagosome occurs, which is thought to solubilise F-actin and hence allow granule–phagosome fusion so that the granule contents are discharged. Hence, phagocytosis and granule movement are Ca^{2+}-independent, but fusion is critically dependent upon localised increases in the concentration of this cation.

6.3.1.5 Diacylglycerol

There are several pieces of evidence showing the involvement of diacylglycerol (DAG) in neutrophil activation, in particular in the activation of the NADPH oxidase:

i. The action of phospholipase C on PIP_2 yields, in addition to Ins 1,4,5-P_3, detectable amounts of DAG (see Fig. 6.7).
ii. Phorbol esters such as PMA and other structural analogues of DAG can activate the respiratory burst.
iii. Protein kinase C, which is activated by DAG (in the presence of Ca^{2+}), phosphorylates oxidase components.

DAG is the substrate for protein kinase C, which is normally present in the cytoplasm in an inactive form. The generation of DAG (which remains in the membrane after its synthesis from PIP_2 hydrolysis) allows protein kinase C to translocate to the plasma membrane, where it interacts with DAG and Ca^{2+} to form a quaternary complex (see Fig. 6.7). Thus, upon cell activation, protein kinase C levels in the cytosol decrease, and this is paralleled by an increase in its activity in the plasma membrane. Alternatively, DAG can be converted to phosphatidic acid by the action of DAG kinase at the expense of ATP (Fig. 6.13), and the phosphatidic acid so formed is an important signalling molecule in its own right (see §§6.3.3.1–2). In vitro, protein kinase C is activated by Ca^{2+} in the concentration range 1–10 μM. However, DAG binding lowers the affinity of protein kinase C for Ca^{2+} so that the kinase is active at physiological (i.e. non-elevated, or submicromolar) Ca^{2+} levels. Active protein kinase C phosphorylates target proteins at key serine or threonine residues, and the phosphorylation status of such proteins can regulate their activities: some enzymes are activated in the phosphorylated state; others become inactive when phosphorylated. A family of protein kinase C isoenzymes exists (α, βI, βII, γ, δ, ζ, ε, η), all of which are single-chain polypeptides (~80 kDa). They comprise two functional domains: the *COOH-terminal domain* (which contains the binding sites for the protein phosphate acceptor and the nucleotide phosphate donor) and a regulatory *NH_2-terminal domain* (which contains the binding sites for Ca^{2+}, phospholipid and DAG).

In a resting neutrophil, protein kinase C is normally inactive because (i) the intracellular Ca^{2+} levels are below the threshold for activation and (ii) DAG is largely unavailable. However, it has recently been shown that activated protein kinase C can be proteolytically cleaved from the membrane and, once released, no longer requires either phospholipid or Ca^{2+},

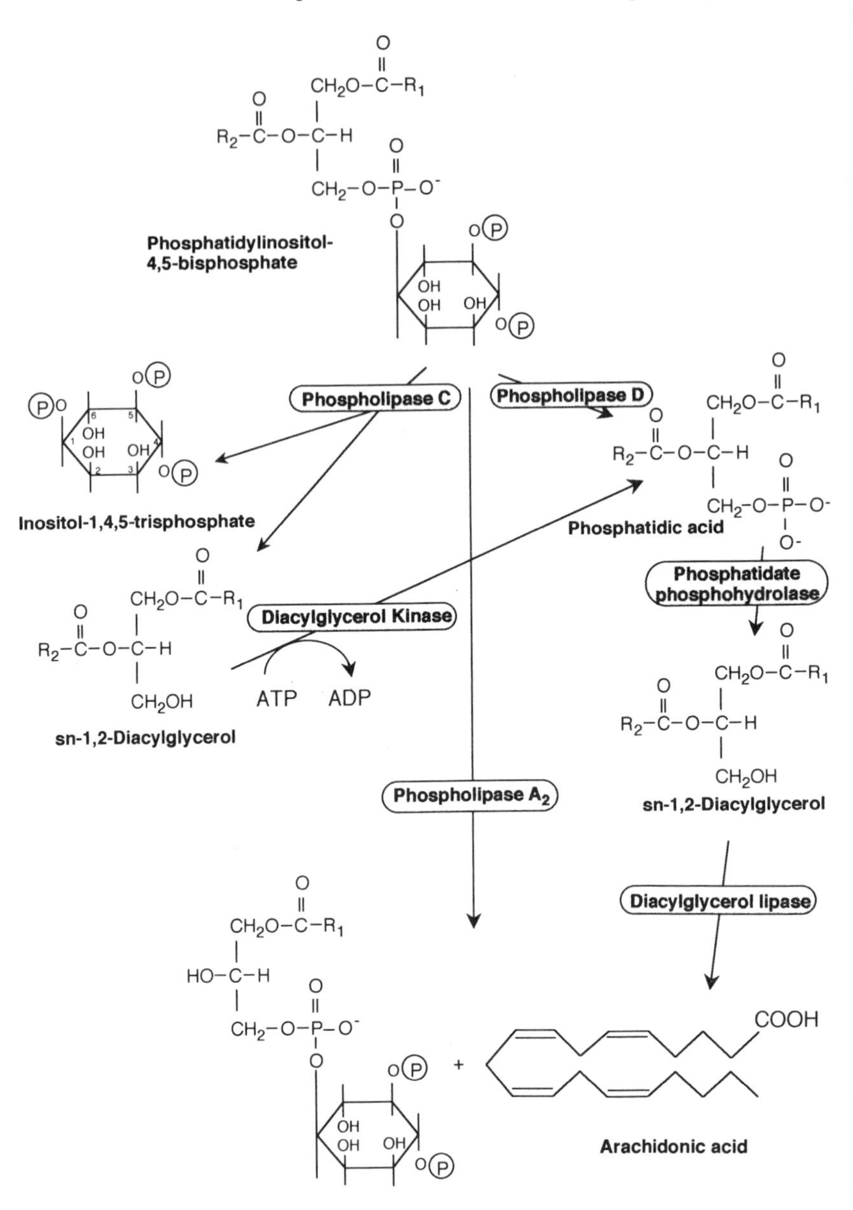

Figure 6.13. Major routes of phosphatidyl 4,5-bisphosphate metabolism in neutrophils. See text for details.

but still retains its capacity to phosphorylate. One such protease capable of this action is *calpain,* a Ca^{2+}-activated protease. Hence, elevated Ca^{2+} levels may serve to activate protein kinase C irreversibly, thus enabling it to act on both membrane and cytosolic targets.

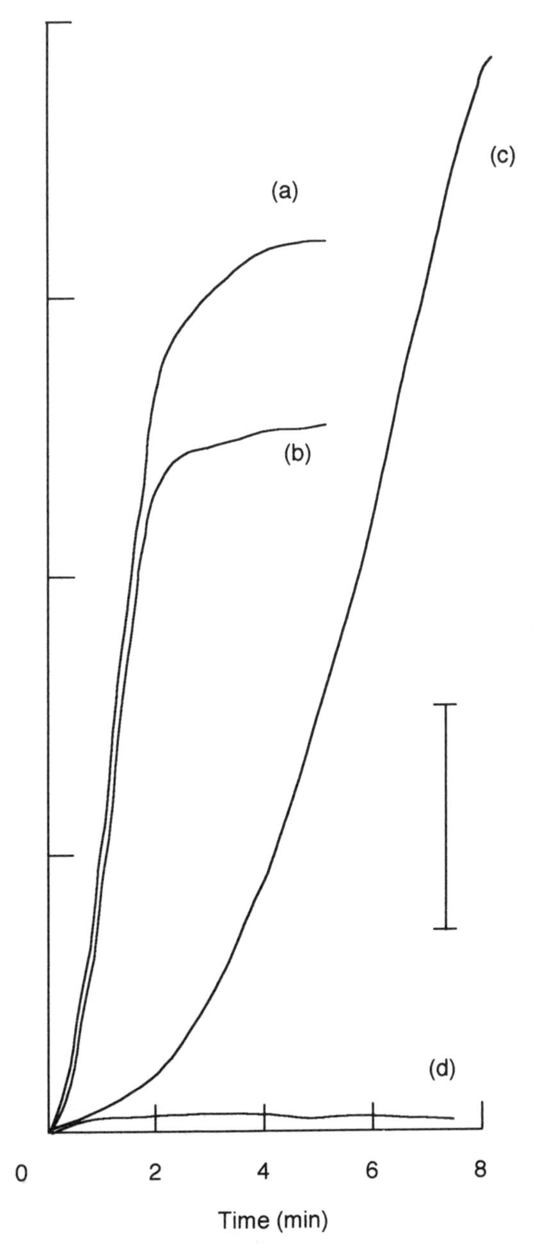

Figure 6.14. Kinetics of O_2^- secretion by activated neutrophils. Neutrophil suspensions (5×10^5/ml) were suspended in RPMI 1640 medium containing 75 μM cytochrome c. In (b) and (d), suspensions contained 100 nM staurosporine. At time zero, suspensions in (a) and (b) were stimulated by the addition of 1 μM fMet-Leu-Phe, whilst suspensions in (c) and (d) were stimulated by the addition of 0.1 μg/ml PMA. Reference cuvettes were identical to the sample cuvettes, but additionally contained 30 μg/ml SOD. The bar marker represents a ΔA of 0.03 in (a) and (b), or 0.08 in (c) and (d). Similar results were obtained using the more specific protein kinase C inhibitor, bisindolylmaleimide.

It is known that protein kinase C can phosphorylate a number of key oxidase components, such as the two cytochrome b subunits and the 47-kDa cytoplasmic factor. This process is prevented by protein kinase C inhibitors such as staurosporine (although it is now recognised that this inhibitor is not specific for protein kinase C), which also inhibits the respiratory burst activated by agonists such as PMA. However, when cells are stimulated by fMet-Leu-Phe, translocation of p47-*phox* to the plasma membrane can occur even if protein kinase C activity is blocked – that is, phosphorylation is not essential for the translocation of this component in response to stimulation by this agonist. Similarly, the kinetics of phosphorylation of the cytochrome subunits do not follow the kinetics of oxidase activation, and protein kinase C inhibitors have no effect on oxidase activity elicited by some agonists – for example, on the initiation of the respiratory burst elicited by agonists such as fMet-Leu-Phe (Fig. 6.14). Furthermore, the kinetics of DAG accumulation do not always follow those of oxidase activity. Hence, whilst protein kinase C is undoubtedly involved in oxidase activation by some agonists, oxidase function is not totally dependent upon the activity of this kinase.

DAG formation in activated neutrophils is biphasic. A small, transient increase in levels is observed within seconds of activation with agonists such as fMet-Leu-Phe (which closely coincides with inositol phosphate formation); this is followed by a larger and more sustained increase. Experiments in which neutrophils are loaded with radiolabelled phospholipids and the fatty acid constitution of the liberated DAG is analysed indicate that this molecule is produced by two separate routes. The phosphoinositide lipids (which are hydrolysed by PLC) predominately have stearate and arachidonate side chains, and these fatty acids are largely found in the DAG initially generated during activation. Phosphatidylcholine (which is hydrolysed by phospholipase D [PLD]; see §6.3.3.7) generally contains more saturated fatty acids in the side chains, and these are primarily found in the DAG produced later in the response. Thus, the sequential activities of PLC and PLD are responsible for the kinetic generation of DAG, but the bulk of the DAG that is generated during neutrophil activation is believed to come from the activity of PLD. It is possible, but not yet proven, that the different types of DAG generated from these separate phospholipid pools may activate different isoforms of protein kinase C. Furthermore, it has been shown that alkyl derivatives of DAG can stimulate the activity of protein kinase C, whilst ester forms of DAG may inhibit this kinase.

Apart from direct assays measuring the rates of formation of DAG and measurements of the activities of protein kinase C in membrane and cytosolic fractions, much use has been made of so-called specific inhibitors of protein kinase C. There is currently much interest in the development of such inhibitors, not only for their usefulness as analytical tools in in vitro studies, but for their potential use as anti-inflammatory agents. The most effective inhibitors to date are those that compete for the ATP-binding site on the COOH-terminal domain of this enzyme. One commonly-used inhibitor is *H-7* (1-(5 isoquinolinyl sulphonyl)-2-methylpiperazine), a member of the isoquinoline sulphonamide family of inhibitors. Whilst this has been useful for the analysis of protein kinase C activity in cell extracts, it is not selective for protein kinase C, and cellular concentrations of ATP are usually high enough to prevent effective competition of H-7 for the ATP-binding site on the enzyme. *Staurosporine,* a microbial product of the indolocarbazole family, has also been used extensively as a 'specific' protein kinase C inhibitor (Fig. 6.15a). This is an extremely potent inhibitor of this kinase (I_{50} = 9 nM), but its specificity has recently been questioned: it can also inhibit protein kinase A (i.e. cAMP-stimulated protein kinase [PKA]) and tyrosine kinases. However, the structure of staurosporine has been used as the basis for the chemical synthesis of new protein kinase C inhibitors, with the aim of developing derivatives that retain the potency of the parent compound but have improved selectivity in their action. This family of bis-indolylmaleimides was derived from the removal of the 12a–12b bond in staurosporine aglycone, the introduction of a second carbonyl moiety and subsequent modifications of the molecule. Several of these derivatives, notably Ro 31-8220 and Ro 31-8425 (Fig. 6.15b,c) appear to have the desired features of both potency (I_{50} values of ~8–10 nM) and selectivity for protein kinase C.

6.3.2 Phospholipase A_2

6.3.2.1 Properties

As in most cells, one of the major functions of phospholipase A_2 (PLA_2) in neutrophils is for the release of arachidonic acid from membrane phospholipids (see Fig. 6.13). Several mechanisms have been implicated in PLA_2 activation, such as increases in intracellular Ca^{2+} levels (brought about by the action of Ins 1,4,5-P_3 on intracellular stores or via increased Ca^{2+} influx) and the activity of protein kinase C (activated by the generation of DAG). These second-messenger systems may thus activate PLA_2 either independently or synergistically. Agonists that activate PLA_2 and subsequently lead to arachidonic acid release include fMet-Leu-Phe, C5a and LTB_4.

In the neutrophil, PLA_2 is located in the cytosol and in the secretory granules of resting cells. The cytosolic enzyme is likely to be involved in the generation of arachidonic acid from neutrophil phospholipids during cell activation, whilst the granule form (which may be secreted) probably plays an extracellular function or else is involved in microbial attack during phagocytosis. The cytosolic enzyme, with a relative molecular mass of 90–110 kDa, becomes associated with the plasma membrane when the intracellular Ca^{2+} levels exceed 300 nM (i.e. during cell activation that results in increases in intracellular Ca^{2+}). This translocation arises because the enzyme possesses a 45-amino-acid domain with a Ca^{2+}-dependent, phospholipid-binding motif. This motif is also present in protein kinase C, GAP and PLC-γ_1. Secretory PLA_2 is a small protein (14 kDa) and belongs to the group II PLA_2 family.

In addition to the importance of Ca^{2+}, PLA_2 activity is also regulated by *lipocortin* (also termed *lipomodulin*), which is a 40-kDa protein. The inhibitory effect of lipocortin is regulated by its phosphorylation status, acting as an inhibitor of the enzyme when in the dephosphorylated state. Upon cell activation (e.g. by fMet-Leu-Phe), the lipocortin becomes phosphorylated, and PLA_2 activity (usually detected as the release of arachidonic acid) increases. Protein kinase C can cause this phosphorylation, and so activation of this kinase may lead to the relief of PLA_2 inhibition via phosphorylation of lipocortin. Thus, elevations in the levels of intracellular Ca^{2+} and production of DAG (required for protein kinase C activation) may coordinately activate PLA_2.

Whilst PLA_2 possesses a Ca^{2+}-binding domain, in vitro experiments have shown that activation of this enzyme only occurs at Ca^{2+} concentrations over the range 10–100 μM. It is unlikely that such concentrations of this cation are ever achieved in vivo, even in localised areas of the cell (e.g. at or near the plasma membrane). Thus, apart from its regulation by protein kinase C activity, PLA_2 activity may also be regulated by receptor-linked G-proteins; such interactions may therefore decrease the concentration of Ca^{2+} required for activation. Indeed, several lines of evidence now exist to substantiate G-protein involvement in the activation of PLA_2. For example, agonist-stimulated arachidonic acid release is sensitive to pertussis toxin (indicating the involvement of a $G\alpha_i$-type protein), whilst GTP analogues (e.g. GTPγS) and fluoride can activate arachidonic acid release in permeabilised cells. However, care must be exerted in the interpretation of such experiments. For example, agonists that activate arachidonic acid release almost always activate PLC as well. Thus, these agonists activate increases both in intracellular Ca^{2+} concentrations and in DAG via G-protein-linked receptors. Because both Ca^{2+} and DAG play roles in PLA_2 activation,

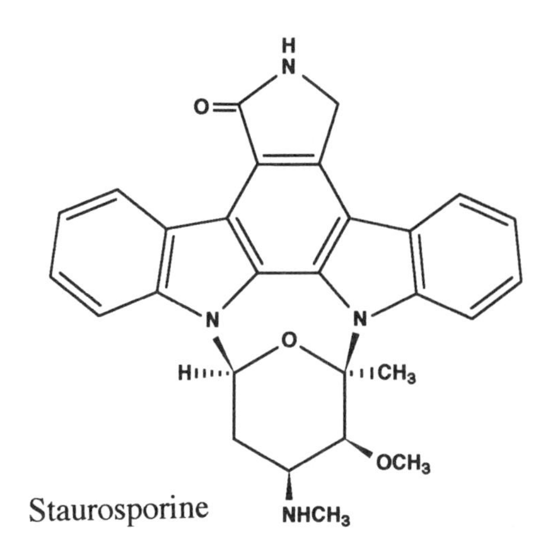

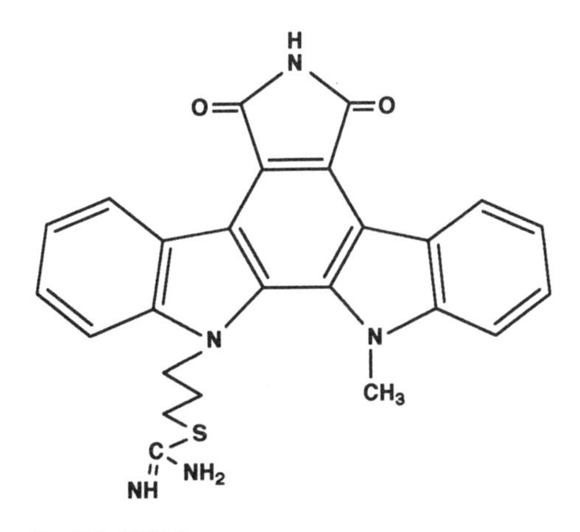

Ro31-8220

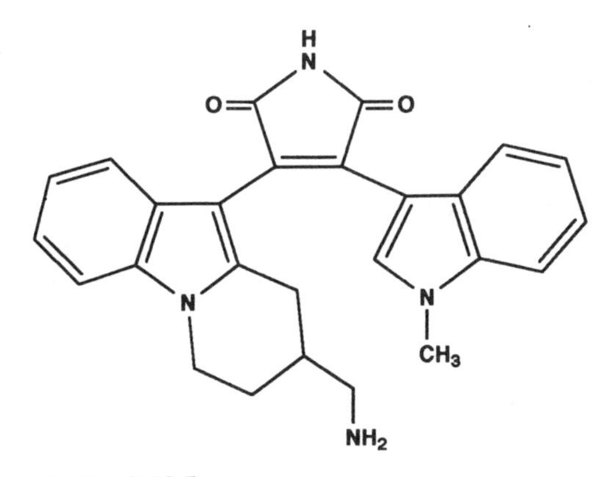

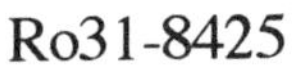

Figure 6.15. Structures of some protein kinase C inhibitors. Staurosporine is a potent inhibitor of protein kinase C, but is not selective. Some derivatives of protein kinase C, such as Ro31-8220 and Ro31-8425, retain a potency similar to that of the parent molecule in protein kinase C inhibition, but are much more specific.

pertussis-toxin sensitivity may represent inhibition of PLC activity rather than of PLA_2 activity. It is also possible that PLA_2 activation may be regulated by protein tyrosine kinase activity.

In the neutrophil, several phospholipids can serve as endogenous sources of arachidonic acid, such as phosphatidylethanolamine (60%), phosphatidylcholine (18%) and phosphoinositides (18%). Arachidonic acid release during activation comes primarily from phosphatidylinositol, particularly during the initial stages of cell activation (Fig. 6.16), and from phosphatidylcholine. Cleavage of phosphatidylcholine (1-*O*-alkyl-2-arachidonyl-*sn*-glycero-3-phosphocholine) by PLA_2 will thus yield arachidonic acid and lyso-PAF (§3.3). Phospholipase A_2 activity on all of these phospholipids can yield arachidonic acid, but this molecule may also be formed from inositol phosphates via an alternative route: phospholipase C activity will yield Ins 1,4,5-P_3 and DAG, and DAG lipase activity will generate arachidonic acid (see Fig. 6.13). Thus, the source of cellular arachidonic acid may depend upon the purpose for which it is required.

6.3.2.2 *Role of arachidonic acid in neutrophil function*

Several important roles of arachidonic acid in neutrophil function have been identified. A major one is as a substrate for either cyclooxygenase or lipoxygenase, in the generation of prostaglandins and leukotrienes, respectively. In neutrophils, the predominant eicosanoids generated during activation are the leukotrienes, and the types of leukotrienes generated are largely dependent upon both the animal source of neutrophils (e.g. rabbit or human) and the agonist used for cell stimulation. In human neutrophils, the primary leukotriene generated is leukotriene B_4 (LTB_4), which is produced in response to the calcium ionophore A23187, chemoattractants, IgG-containing immune complexes and serum-coated zymosan. Cytochalasin B augments the ability of chemoattractants to stimulate LTB_4 production.

There are many types of lipoxygenases in different cell types, and these enzymes are designated by the position on the arachidonic acid molecule at which they catalyse a dioxygenation (e.g. 5′-lipoxygenase, 12′-lipoxygenase). Neutrophils possess a 5′-lipoxygenase (80 kDa), which is present in the cytosol and may also be associated with specific granules. In vitro experiments indicate that its activity requires Ca^{2+}, ATP, hydroperoxide, two cytosolic factors and a membrane-bound factor. The enzyme catalyses the dioxygenation of arachidonic acid at position C5 (Fig. 6.17) to form 5-HPETE (5 hydroperoxy-6,8,11,14 eicosatetraenoic acid). This is then converted (possibly via a peroxidase) into 5-HETE (5 hydroxy-6,8,11,14 eicosatetraenoic acid) or leukotriene A_4 (5,6-epoxy-7,9,11,14-eicosatetraenoic

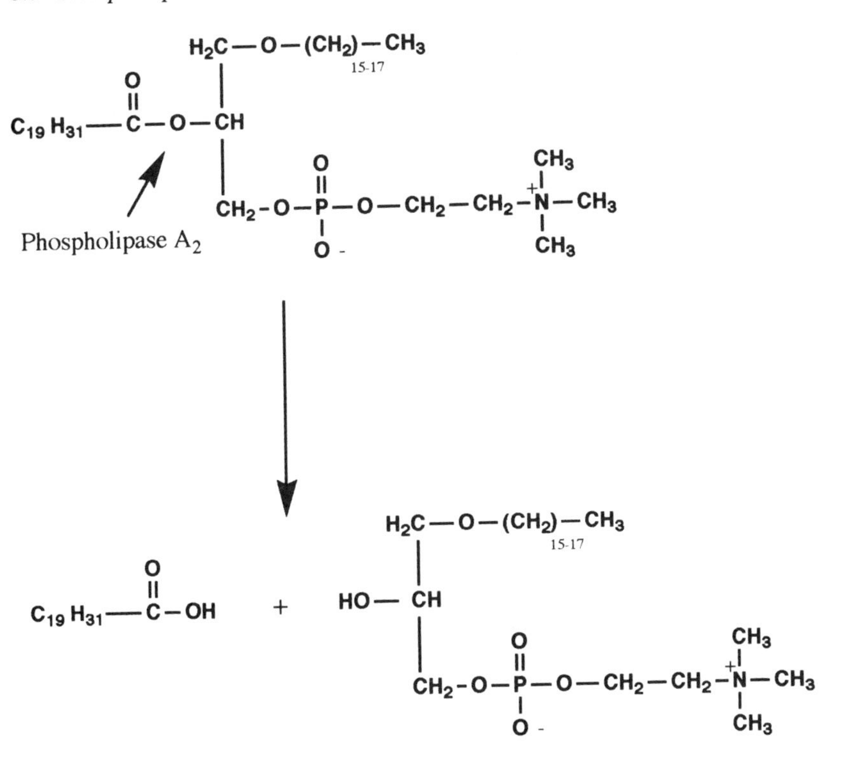

Figure 6.16. Hydrolysis of phosphatidylcholine by phospholipase A_2. This reaction yields two important products: arachidonic acid and lyso-PAF.

acid). Levels of LTA_4 increase maximally 1–2 min after stimulation, and then decrease because this molecule is unstable (being subject to nucleophilic substitution and non-enzymic conversions) or because it is enzymically converted into other products. The action of the enzyme LTA_4 hydrolase yields LTB_4 (5,12-dihydro-6,8,10,14-eicosatetraenoic acid), whereas glutathione transferase generates LTC_4 (5-hydroxy-6-*S*-glutathioyl-7,9,11, 14-eicosatetraenoic acid).

Stimulation of neutrophils with A23187 results in the release of large quantities of arachidonic acid, LTA_4 and non-enzymic derivatives of LTA_4. This is because, in neutrophils, the activity of LTA_4 hydrolase is limiting, and so the large quantities of LTA_4 that are generated are non-enzymically degraded rather than enzymically converted into LTB_4. In contrast, fMet-Leu-Phe stimulates lower levels of release of arachidonic acid, and so the

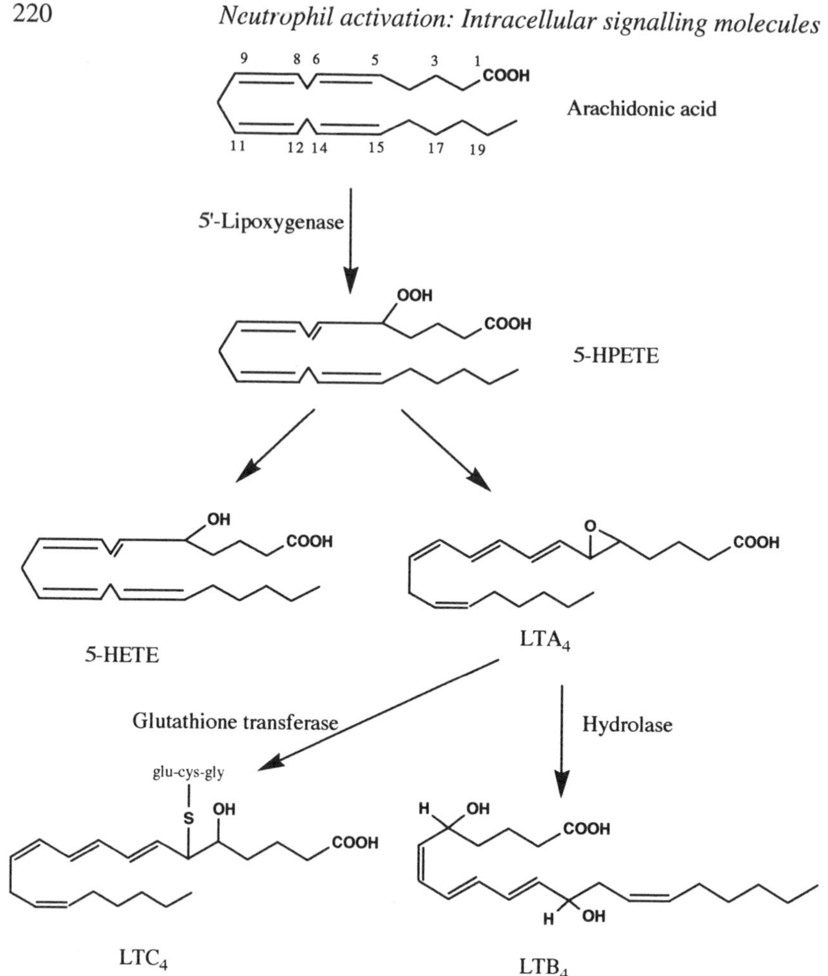

Figure 6.17. Leukotriene formation in neutrophils. Arachidonic acid, which is released from membrane phospholipids by the action of either phospholipase A_2 or diacylglycerol lipase (see Fig. 6.13), is oxygenated by 5-lipoxygenase to yield 5 hydroperoxy-6,8,11,14 eicosatetraenoic acid (5-HPETE). This is then converted into 5 hydroxy-6,8,11,14 eicosatetraenoic acid (5-HETE) and leukotriene (LT) A_4. LTA_4 may then be enzymically converted into LTC_4 and LTB_4. LTB_4 is the major product in activated neutrophils.

LTA_4 generated is almost all converted into LTB_4. Thus, LTB_4 is the major lipoxygenase product of fMet-Leu-Phe-stimulated neutrophils.

Neutrophil preparations stimulated with A23187 also generate products of 15´-lipoxygenase activity, such as LTC_4. However, most neutrophil preparations also contain eosinophils (usually about 2–5% contamination of the neutrophils). In contrast to neutrophils, which generate large quantities of

LTB_4 and very low quantities of LTC_4, eosinophils generate high concentrations of LTC_4 rather than LTB_4. Hence, products of 15′-lipoxygenase activity in neutrophil preparations are almost exclusively due to eosinophil contamination. (The effects of LTB_4 on neutrophils and the properties of the LTB_4 receptor are given in §3.1.)

Alternatively, arachidonic acid may be converted into prostaglandins via processes initially catalysed via the enzyme cyclooxygenase. This is not a major route of arachidonic acid metabolism in neutrophils, but the cells may produce substantial quantities of prostaglandin E_2 and thromboxane B_2 under some circumstances. Cyclooxygenase is present in almost all mammalian cells, and its action appears to require the presence of low amounts of hydroperoxide (such as its product, PGG_2). Thus, the activated enzyme abstracts a hydrogen atom from arachidonic acid and generates PGG_2, which is then converted into PGH_2 (Fig. 6.18). Both of these products are unstable (half-lives of only several minutes in physiological conditions) and serve as precursors of other prostaglandins, such as PGD_2, PGE_2 and $PGF_{2\alpha}$, and also of TXB_2, which is formed from an unstable precursor molecule, TXA_2. These conversions require the activities of specific enzymes, and in neutrophils the major products of the cyclooxygenase cascade are PGE_2 and TXB_2. PGE_2 can elevate cAMP levels via the activation of adenylate cyclase, and this second messenger can have an attenuating effect on functions such as chemotaxis. TXB_2 is generally regarded to be an inactive product of TXA_2, which has a very short biological half-life and can function as a vasoconstrictor and platelet activator.

Other roles of arachidonic acid include activation of the NADPH oxidase in the cell-free system (see §5.3.2.3). It may be required for membrane fusion during the process of recruitment of oxidase and receptor molecules from the membranes of specific granules. Finally, there is some evidence that arachidonic acid may:

activate protein kinase C (K_m = 15 μM arachidonic acid);
regulate intracellular Ca^{2+} levels by altering Ca^{2+} release from intracellular pools or prevent Ca^{2+} efflux;
activate the H^+ pump associated with NADPH oxidase activity.

These latter observations suggest a role for arachidonic acid as a genuine second messenger, and further work is necessary to assess if such a role exists during neutrophil activation. It has also been proposed that arachidonic acid may dissociate p21*rac* and GDI (§5.3.2.3), thus allowing the small GTP-binding protein to interact with p47-*phox* and p66-*phox* in the assembly of active NADPH oxidase complexes.

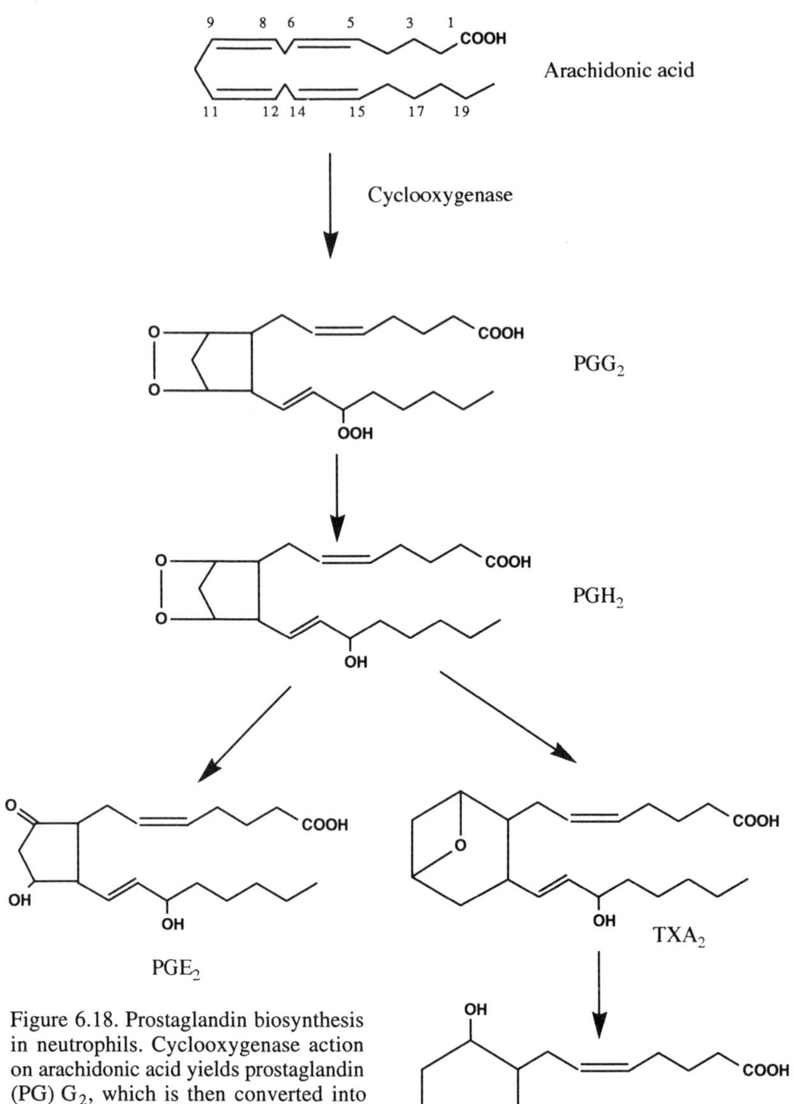

Figure 6.18. Prostaglandin biosynthesis in neutrophils. Cyclooxygenase action on arachidonic acid yields prostaglandin (PG) G_2, which is then converted into PGH_2. In neutrophils, PGH_2 is preferentially converted into PGE_2 and thromboxane (TX) A_2, which is unstable and converted into TXB_2.

6.3.3 Phospholipase D

6.3.3.1 Properties

The role of protein kinase C in many neutrophil functions is undisputed and has been recognised for some time. For many years it was believed that the source of DAG, the activator of protein kinase C, was derived from the activity of PLC on membrane phosphatidylinositol lipids. Whilst this enzyme undoubtedly does generate some DAG (which may then activate protein kinase C), there are many reasons to indicate that this enzyme activity is insufficient to account for all the DAG generated by activated neutrophils. More recently, experimental evidence has been provided to show that a third phospholipase (PLD) is involved in neutrophil activation, and that this enzyme is probably responsible for the majority of DAG that is formed during cell stimulation. The most important substrate for PLD is phosphatidylcholine, the major phospholipid found in neutrophil plasma membranes, which accounts for over 40% of the phospholipid pool. The *sn*-1 position of phosphatidylcholine is either acyl linked or alkyl linked, whereas the *sn*-2 position is invariably acyl linked. In neutrophils, alkyl-phosphatidylcholine (1-*O*-alkyl-PC) represents about 40% of the phosphatidylcholine pool (and is also the substrate utilised for PAF formation), whereas the remainder is diacyl-phosphatidylcholine. Both of these types of phosphatidylcholine are substrates for PLD and PLA_2.

The activity of PLD on phosphatidylcholine generates phosphatidic acid, and this may be further metabolised by the enzyme phosphatidate phosphohydrolase to form DAG (Fig. 6.19). Furthermore, the activity of DAG kinase can convert the DAG (generated either from phosphatidic acid or from the activity of PLC) back into phosphatidic acid. Both phosphatidic acid and DAG have functions as second messengers; thus the activities of PLD, phosphatidate phosphohydrolase and DAG kinase all play important roles in the generation of these intracellular signalling molecules.

6.3.3.2 PLD and neutrophil function

The first experiments implicating a role for PLD activity in neutrophil function were performed by Cockcroft and colleagues (Cockcroft & Stutchfield, 1989; Cockcroft, 1992) who measured phosphatidic acid accumulation in cells whose membrane phospholipids or ATP were radiolabelled. These experiments showed that phosphatidic acid accumulation during cell activation did not derive from DAG, but rather was directly generated from a phospholipid. Phosphatidic acid production from DAG (generated by PLC) requires ATP (see Fig. 6.13), but experiments with ^{32}P-labelled ATP indicated that the phosphate of phosphatidic acid was unlabelled, and so could not have been generated via this route.

Further evidence that the phosphatidic acid was not derived from a PLC-dependent process comes from the analysis of the fatty acid composition of phosphatidylinositol lipids and phosphatidic acid. The former lipids mainly have stearate and arachidonate side chains, whereas those of phosphatidic acid are generally more saturated. Other evidence for PLD involvement comes from the fact that this enzyme is able to catalyse a transphosphatidylation between phospholipids and aliphatic alcohols, such as ethanol and butanol, to produce phosphatidylalcohols (Fig. 6.19). Thus, alcohol treatment can decrease the levels of both phosphatidic acid and DAG that are generated by PLD, and the corresponding formation of phosphatidylethanol is a measure of PLD activity. Phosphatidic acid and phosphatidylethanol can be readily separated and identified by TLC. Furthermore, primary alcohols can be used to inhibit the formation of phosphatidic acid and DAG in intact neutrophils, and so the effects of inhibition of PLD activity on cell function can be assessed. Experiments by Garland and colleagues (Bonser et al., 1989; Thompson et al., 1990; Garland, 1992; Uings et al., 1992) have shown that treatment of neutrophils with concentrations of alcohols sufficient to inhibit PLD activity do not affect inositol phosphate formation via PLC. Thus, these alcohols do not act as non-specific phospholipase inhibitors and do not result in cellular toxicity.

These approaches have been used to show conclusively that the initial, low formation of DAG that occurs during activation with soluble agonists comes from PLC activity, and that the later, more sustained generation of DAG comes from PLD activity. Such experiments have also shown that primary alcohols can inhibit the activity of the NADPH oxidase under some conditions. When neutrophils are pretreated with cytochalasin B, primary alcohols are potent inhibitors of O_2^- secretion, and the kinetics of phosphatidic acid formation are rapid, peaking within about 20 s and coinciding with oxidase activation. However, in the absence of cytochalasin B, primary alcohols have little effect on the initiation of O_2^- secretion, but decrease the duration of oxidase activity; they also inhibit the later phase of luminol chemiluminescence, which is largely intracellular, and the kinetics of phosphatidic acid formation closely parallel the kinetics of this intracellular oxidase activity (Fig. 6.20). Thus, in cytochalasin-treated cells, PLD is activated rapidly, and this activation is required for O_2^- secretion: in the absence of cytochalasin, PLD is activated more slowly and its function is not for the activation of the oxidase, but rather for sustained (and intracellular) activity.

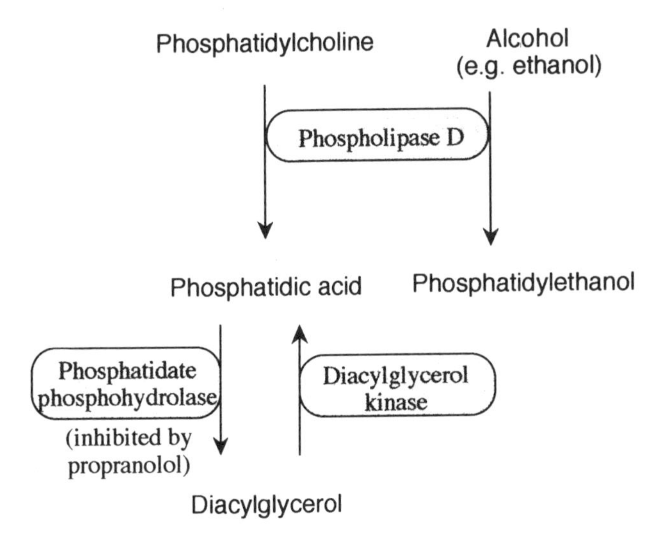

Figure 6.19. Products of phosphatidylcholine metabolism. Phosphatidylcholine is metabolised to phosphatidic acid via the activity of phospholipase D. The phosphatidic acid generated in this way may then be converted into diacylglycerol via phosphatidate phosphohydrolase (which is inhibited by propranolol), and the enzyme diacylglycerol kinase may regenerate the phosphatidic acid. Phospholipase D may also catalyse the transphosphatidylation of primary alcohols, such as ethanol and butanol, at the expense of the natural substrate, phosphatidylcholine. Thus, primary alcohols can prevent phosphatidic acid production via this route.

Cytochalasin B inhibits the exchange of actin monomers at the fast-growing ends of actin fibres and thus results in the shortening of the filaments (see §4.1.3.1). When added to neutrophil suspensions, it prevents phagocytosis (as predicted, because an active cytoskeletal network is required); however, exocytosis and oxidase activity in response to soluble agonists such as fMet-Leu-Phe are enhanced by cytochalasins. The precise mechanisms responsible for these enhancements are unknown, but when the cytoskeleton is disrupted, the cytoplasmic granules (and perhaps some other cytoplasmic components) can integrate more freely with the plasma membrane. This would explain enhanced degranulation and – because so many neutrophil functions require translocation of granule and cytoplasmic components to the plasma membrane during activation – may also explain why oxidase activation is enhanced and why some signalling processes are altered. Interestingly, cytochalasin B treatment does not affect PLC activity but does alter PLA_2 and PLD activity, indicating that translocation events may be necessary for the activation of these latter signalling systems.

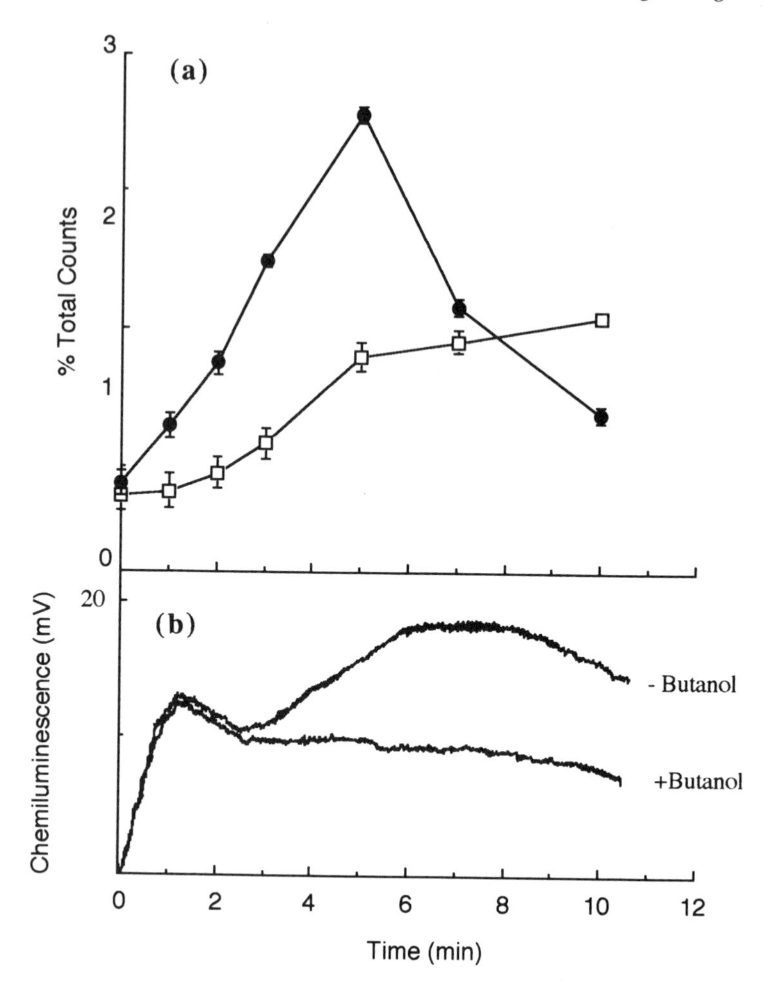

Figure 6.20. Role of phospholipase D in NADPH oxidase activation. In (a), neutrophils were preincubated with [^{3}H]-alkyl-lyso-PAF (5 μCi/ml) for 60 min at 37 °C. The cells were then washed twice with RPMI 1640 medium and finally resuspended at 2×10^6 cells/ml in the presence (❑) and absence (●) of 100 mM ethanol. The cells were then stimulated with 1 μM fMet-Leu-Phe and, at time intervals, aliquots were removed for analysis of phosphatidic acid (●) and phosphatidylethanol (❑) by thin layer chromatography (TLC). In (b), neutrophils were incubated in the presence and absence of 10 mM butanol, and luminol chemiluminescence (10 μM, final concentration of luminol) was measured after stimulation by 1 μM fMet-Leu-Phe. *Source:* Experiment of Gordon Lowe and Fiona Watson.

Primary alcohols inhibit the up-regulation of CR3 (complement receptor) molecules on the plasma membrane during neutrophil activation, indicating that PLD products are required for the translocation of specific granules,

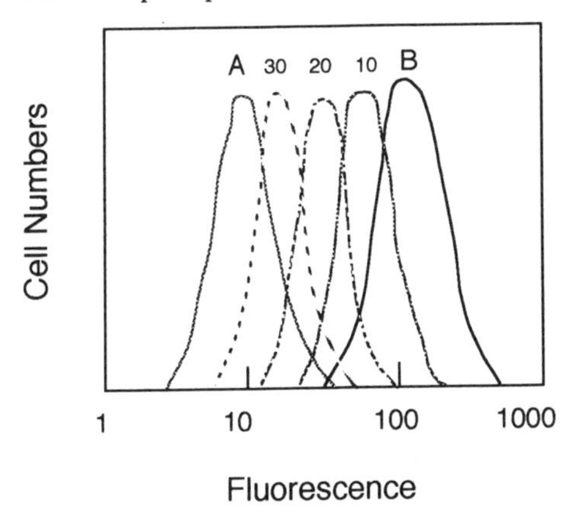

Figure 6.21. Role of phospholipase D in receptor up-regulation. Neutrophils were incubated in the presence or absence of fMet-Leu-Phe and butanol for 15 min prior to analysis of expression of CD11b by FACS analysis. In (a), neutrophils were not stimulated and suspensions did not contain butanol. In (b), suspensions did not contain butanol, but were stimulated with fMet-Leu-Phe. The hatched lines show the receptor expression of suspensions incubated with 10, 20 and 30 mM butanol for 5 min prior to stimulation by fMet-Leu-Phe. *Source:* Experiment of Fiona Watson.

gelatinase-containing granules or secretory vesicles (Fig. 6.21). The membranes of these granules and vesicles also contain cytochrome b molecules, and thus inhibition of PLD will prevent the up-regulation of NADPH oxidase molecules to the cell surface. This may explain why, in unprimed cells, alcohols do not affect the initial activation of the oxidase, which requires cytochrome b molecules already on the plasma membrane. Activation of these can occur in the absence of PLD-derived products; after their activation, more cytochrome molecules must be recruited via the translocation from subcellular stores to the plasma membrane. Primary alcohols inhibit this process, and so the respiratory burst is terminated more rapidly when PLD activity is blocked. PLD is present in the plasma membrane and in the cytosol; thus, translocation of either the enzyme itself or factors that regulate its activity (e.g. soluble tyrosine kinases; see §6.4.2) are probably involved in its activation. Thus, if these factors are translocated during cytochalasin B treatment, then subsequent cell stimulation will activate PLD more rapidly and to higher levels; hence, the oxidase will become more dependent on this system for its activation.

The enzyme phosphatidate phosphohydrolase can be inhibited by propranolol, although this inhibitor is not completely specific. Thus, propranolol treatment of neutrophils results in the increased formation of phosphatidic acid and the decreased formation of DAG. There is increasing evidence to support the idea that phosphatidic acid can serve as a signalling molecule. In unprimed cells, propranolol results in increased formation of phosphatidic acid in response to fMet-Leu-Phe, PAF and LTB_4, and this is acompanied by an increased formation of O_2^-. Also, addition of phosphatidic acid to neutrophil membranes can result in oxidase activation, and this PLD product is more effective than arachidonate in activating the oxidase in a cell-free system.

6.3.3.3 Activation of PLD

Experiments with cell-free systems and with permeabilised neutrophils show that (i) PLD activity requires Ca^{2+} and (ii) PMA can activate this enzyme. Thus, it has been proposed that initial DAG formation occurring during cell stimulation (via PLC activity) can function to activate low levels of protein kinase C activity, which can then lead to PLD activation. Once activated, PLD will generate greater and more sustained levels of DAG, and this will result in enhanced and prolonged activation of protein kinase C. There is much evidence to show that PLD activation occurs by the coupling of receptors via G-proteins. Again, this evidence comes from the use of non-hydrolysable analogues of GTP (e.g. GTPγS), the use of G-protein activators (e.g. fluoride) and sensitivity to pertussis toxin. Interestingly, whilst PLC and PLD activation are both inhibited by pertussis toxin, PLC activation is much more sensitive. It is known that neutrophils possess at least two pertussis-toxin-sensitive G-proteins ($G\alpha_{i\text{-}2}$ and $G\alpha_{i\text{-}3}$), both of which are coupled to fMet-Leu-Phe receptors. It may thus be that one of these G-proteins is coupled to PLC whilst the other is coupled to PLD.

There are many reports of neutrophil functions being regulated by protein tyrosine kinases, and in cells treated with cytochalasin B, agonists such as fMet-Leu-Phe, PAF and LTB_4 result in tyrosine phosphorylation and activation of PLD. Conversely, inhibitors of tyrosine kinase activity (e.g. erbstatin, ST271, ST638) also inhibit PLD activity. Because such inhibitors do not affect PLD activation by PMA, they must prevent a step involved in the activation of PLD, rather than inhibit the activity of PLD itself. These inhibitors do not affect fMet-Leu-Phe-stimulated inositol phosphate formation. Thus, it has been proposed that PLD activation in unprimed cells is G-protein coupled, but during priming with cytokines or after cytochalasin B treatment, tyrosine kinase activity becomes associated with the receptor–G-protein complex. Subsequent phosphorylation at tyrosine residues will then increase the efficiency of coupling of the receptor with the G-protein. Because the fMet-Leu-Phe receptor does not possess intrinsic tyrosine kinase activity, it is likely that a non-receptor or *src*-like tyrosine kinase is recruited. On the other hand, rather than activating a tyrosine kinase, cell stimulation may inhibit the activity of a tyrosine phosphatase that normally dephosphorylates at tyrosine residues. Inhibition of this enzyme will thus lead to enhanced phosphorylation at tyrosine residues, which will mimic the effects of increased tyrosine kinase activity. Interestingly, vanadate (an inhibitor of tyrosine phosphatases) can also activate PLD. The activity of *src*-like tyrosine kinases may be modulated by the phosphotyrosine phosphatase CD45, which is present in the cytoplasm but which becomes translocated to the plasma membrane during activation.

Agents that stimulate elevations in cAMP production also inhibit PLD activation. Thus, it may be that phosphorylation by protein kinase C can activate PLD, whereas phosphorylation at other sites on the enzyme by PKA can inhibit its activity.

6.4 Protein kinases

6.4.1 Protein kinase A

As discussed in section 6.3.1.5, the Ca^{2+}-activated kinase, protein kinase C, plays many important roles in neutrophil function and is activated by DAG, which may be generated from the activities of PLC and PLD. In addition, cAMP-dependent protein kinase (PKA) also regulates neutrophil function. This kinase, which is inactive in the absence of cAMP, possesses two catalytic sites and two regulatory sites. Elevations in the intracellular concentrations of cAMP occur after activation of the enzyme adenylate cyclase, which catalyses the formation of one molecule of cAMP from one molecule of ATP. The activity of adenylate cyclase is regulated by receptor-mediated G-protein coupling, some receptors stimulating adenylate cyclase and others inhibiting its activity, depending upon the type of the $G\alpha$-subunit (see Fig. 6.1). The cAMP produced after receptor occupancy binds to regulatory portions on the kinase (two molecules of cAMP bind per enzyme molecule), and the activated kinase can phosphorylate its target proteins. This phosphorylation can result in either the activation or the inhibition of the target. Levels of cAMP are controlled by the activity of cAMP phosphodiesterase, which degrades the molecule to 5′-AMP, which cannot activate the kinase.

In neutrophils, agents that activate adenylate cyclase or result in the elevation of cAMP levels (e.g. cAMP analogues or PGE_2) can inhibit agonist-stimulated degranulation, chemotaxis and respiratory-burst activity but do not affect PMA-stimulated activities. Thus, the inhibitory step must be prior to protein kinase C activation. Effects mediated by concanavalin A (which are reportedly not mediated via G-protein coupling) are also inhibited when cAMP levels are elevated. Cyclic adenosine monophosphate may play a role in preventing the slow increase in intracellular Ca^{2+} that is presumed to arise from enhanced Ca^{2+} influx. As described above (§6.3.3.3), elevated cAMP levels can inhibit PLD activity. Whether this is due to a PKA-dependent phosphorylation or to decreased intracellular Ca^{2+} levels is not known at present.

6.4.2 Protein tyrosine kinases

It has been shown that, upon stimulation, many neutrophil proteins become phosphorylated on tyrosine residues. Protein tyrosine kinases, which are responsible for this activity, are recognised to play important roles in the regulation of function of many types of cells. There are basically two types of protein tyrosine kinases: those whose activity is intrinsically linked to receptors (e.g. the insulin receptor) and those whose activity is not so linked. In neutrophils, tyrosine kinases can be found in membrane fractions and in the cytosol. Upon stimulation with agonists such as fMet-Leu-Phe, PAF, LTB_4, opsonised zymosan and GM-CSF, the phosphorylation on tyrosine residues of a number of cytosolic and membrane proteins increases. Furthermore, the activity of PLD (but not PLC) appears to be regulated by protein tyrosine kinase activity. The level of phosphorylation on tyrosine residues is also regulated by the activity of phosphotyrosine phosphatases. Thus, changes in the levels of phosphorylation can result from changes in activities of the tyrosine kinases and/or the tyrosine phosphatases. One such protein tyrosine phosphatase present in neutrophils is CD45 (the leukocyte common antigen). This is normally found in the cytosol, but during activation it translocates to the plasma membrane, where it can dephosphorylate membrane proteins. CD45 contains an intracellular tyrosine phosphatase region, a single transmembrane domain and an extracellular domain comprising a cysteine-rich motif, which may function to bind ligands.

6.4.3 Other kinases

In addition to these kinases, there are undoubtedly others in neutrophils, including Ca^{2+}/calmodulin-activated kinases, which regulate function. Furthermore, there are a variety of phosphatases that can reverse the effects of these kinases. As is observed in other cells, the activities of these kinases and phosphatases are regulated by their own phosphorylation status, which in turn is regulated by the activities of other enzymes that can catalyse phosphorylation/dephosphorylation reactions. Presently, very little is known of these systems in neutrophils; but, undoubtedly, these will shortly be characterised and their roles in the regulation of neutrophil functions defined. The identity of these enzymes will add further layers of complexity to what is already a multifunctional network of intracellular signalling systems that co-ordinately act to regulate the varied functions of neutrophils.

6.5 Ion pumping

Maintenance of the concentration of various ions, both inside and outside the neutrophil, is crucial for cell function. Concentrations of ions such as Na^+, K^+, Cl^-, Ca^{2+} and H^+ outside the neutrophil (i.e. in extracellular fluids) are 140, 5, 140, 2 and 100 mM, respectively. Inside the neutrophil, the concentrations of these ions are 20 mM, 120 mM, 80 mM, 100 nM and 100 mM, respectively. The transport of these ions into and out of the neutrophil is regulated by ion channels (which may exist in the 'open' or 'closed' states), ion pumps and ion transporters.

If the neutrophil plasma membrane is depolarised, then Ca^{2+} influx is increased (i.e. the influx is electrogenic). The channel is voltage regulated but not voltage operated. Changes in intracellular Ca^{2+} can also regulate the activity of a K^+ channel. In neutrophils, a Ca^{2+}-sensitive K^+ channel is activated upon stimulation with agonists such as fMet-Leu-Phe, ionophore and occupancy of integrin receptors, and a voltage-activated K^+ channel is also thought to exist. Both of these function to pump K^+ out of the neutrophil.

Experiments by Chappell and co-workers (Henderson & Chappell, 1992) have shown that a Zn^{2+}/Cd^{2+}-sensitive H^+ channel is stimulated during NADPH oxidase activation. The oxidase, by transferring an electron to O_2, must also pump a H^+ across the plasma membrane to compensate for this charge transfer – otherwise, activity would rapidly cease. Thus, the NADPH oxidase is electrogenic, and charge compensation via increased H^+ pumping sustains activity. Recently, this group has shown that whereas neither voltage gating nor pH variations are apparently involved in the regulation of the H^+ channel, arachidonic acid may play a role in opening this channel.

6.6 Bibliography

Agwu, D. E., McCall, C. E., & McPhail, L. C. (1991). Regulation of phospholipase D-induced hydrolysis of choline-containing phosphoglycerides by cyclic AMP in human neutrophils. *J. Immunol.* **146**, 3895–903.

Agwu, D. E., McPhail, L. C., Sozzani, S., Bass, D. A., & McCall, C. E. (1991). Phosphatidic acid as a second messenger in human polymorphonuclear leukocytes: Effects on activation of NADPH oxidase. *J. Clin. Invest.* **88**, 531–9.

Aharoni, I., & Pick, E. (1990). Activation of the superoxide-generating NADPH oxidase of macrophages by sodium dodecyl sulfate in a soluble cell-free system: Evidence for involvement of a G protein. *J. Leuk. Biol.* **48**, 107–15.

Berkow, R. L., & Dodson, R. W. (1991). Alterations in tyrosine protein kinase activities upon activation of human neutrophils. *J. Leuk. Biol.* **49**, 599–604.

Berridge, M. J., & Irvine, R. F. (1984). Inositol trisphosphate, a novel second messenger in cellular signal transduction. *Nature* **312**, 315–21.

Berridge, M. J., & Irvine, R. F. (1989). Inositol phosphates and cell signalling. *Nature* **341**, 197–205.

Billah, M. M. (1993). Phospholipase D and cell signaling. *Curr. Opin. Immunol.* **5**, 114–23.

Billah, M. M., Eckel, S., Mullmann, T. J., Egan, R. W., & Siegel, M. I. (1989). Phosphatidylcholine hydrolysis by phospholipase D determines phosphatidate and diglyceride levels in chemotactic peptide-stimulated human neutrophils. Involvement of phosphatidate phosphohydrolase in signal transduction. *J. Biol. Chem.* **264**, 17069–77.

Bonser, R. W., Thompson, N. T., Randall, R. W., & Garland, L. G. (1989). Phospholipase D activation is functionally linked to superoxide generation in the human neutrophil. *Biochem. J.* **264**, 617–20.

Caldwell, C. W., Patterson, W. P., & Yesus, Y. W. (1991). Translocation of CD45RA in Neutrophils. *J. Leuk. Biol.* **49**, 317–28.

Cockcroft, S. (1992). G-protein-regulated phospholipases C, D and A_2-mediated signalling in neutrophils. *Biochim. Biophys. Acta* **1113**, 135–60.

Cockcroft, S., & Stutchfield, J. (1989). The receptors for ATP and fMet-Leu-Phe are independently coupled to phospholipases C and A_2 via G-protein(s). Relationship between phospholipase C and A_2 activation and exocytosis in HL60 cells and human neutrophils. *Biochem. J.* **263**, 715–23.

Downard, J. (1990). The ras superfamily of small GTP-binding proteins. *TIBS* **15**, 467–72.

Gabig, T. G., Eklund, E. A., Potter, G. B., & Dykes, J. R., II (1990). A neutrophil GTP-binding protein that regulates cellfree NADPH oxidase activation is located in the cytosolic fraction. *J. Immunol.* **145**, 945–51.

Garland, L. G. (1992). New pathways of phagocyte activation: The coupling of receptor-linked phospholipase D and the role of tyrosine kinase in primed neutrophils. *FEMS Microbiol. Immunol.* **105**, 229–38.

Gomez-Cambronero, J., Huang, C.-K., Bonak, V. A., Wang, E., Casnellie, J. E., Shiraishi, T., & Sha'afi, R. I. (1989). Tyrosine phosphorylation in human neutrophil. *Biochem. Biophys. Res. Commun.* **162**, 1478–85.

Hallett, M. B., & Campbell, A. K. (1983). Two distinct mechanisms for stimulation of oxygen-radical production by polymorphonuclear leucocytes. *Biochem. J.* **216**, 459–65.

Henderson, L. M., & Chappell, J. B. (1992). The NADPH-oxidase-associated H^+ channel is opened by arachidonate. *Biochem. J.* **283**, 171–5.

Kessels, G. C. R., Roos, D., & Verhoeven, A. J. (1991). fMet-Leu-Phe-induced activation of phospholipase D in human neutrophils. Dependence on changes in cytosolic free

Ca^{2+} concentration and relation with respiratory burst activation. *J. Biol. Chem.* **266**, 23152–6.

Khachatrian, L., Rubins, J. B., Manning, E. C., Dexter, D., Tauber, A. I., & Dickey, B. F. (1990). Subcellular distribution and characterization of GTP-binding proteins in human neutrophils. *Biochim. Biophys. Acta* **1054**, 237–45.

Knaus, U. G., Heyworth, P. G., Kinsella, T., Curnutte, J. T., & Bokoch, G. M. (1992). Purification and characterisation of Rac2. A cytosolic GTP-binding protein that regulates human neutrophil NADPH oxidase. *J. Biol. Chem.* **267**, 23575–82.

Koenderman, L., Tool, A., Roos, D., & Verhoeven, A. J. (1989). 1,2-Diacylglycerol accumulation in human neutrophils does not correlate with respiratory burst activation. *FEBS Lett.* **243**, 399–403.

Krause, K. H., Campbell, K. P., Welsh, M. J., & Lew, D. P. (1990). The calcium signal and neutrophil activation. *Clin. Biochem.* **23**, 159-166.

Lackie, J. M. (1988). The behavioural repertoire of neutrophils requires multiple signal transduction pathways. *J. Cell Sci.* **89**, 449-452.

Lew, D. P., Andersson, T., Hed, J., Di Virgilio, F., Pozzan, T., & Stendahl, O. (1985). Ca^{2+}-dependent and Ca^{2+}-independent phagocytosis in human neutrophils. *Nature* **315**, 509–11.

Lew, D. P., Wollheim, C. B., Waldvogel, F. A., & Pozzan, T. (1984). Modulation of cytosolic-free calcium transients by changes in intracellular calcium-buffering capacity: correlation with exocytosis and O_2^- production in human neutrophils. *J. Cell Biol.* **99**, 1212–20.

McLeish, K. R., Gierschik, P., Schepers, T., Sidiropoulos, D., & Jakobs, K. H. (1989). Evidence that activation of a common G-protein by receptors for leukotriene B_4 and *N*-formylmethionine-leucyl-phenylalanine in HL-60 cells occurs by different mechanisms. *Biochem. J.* **260**, 427-434.

Maridonneau-Parini, I., & de Gunzburg, J. (1992). Association of rap1 and rap2 proteins with the specific granules of human neutrophils. Translocation to the plasma membrane during cell activation. *J. Biol. Chem.* **267**, 6396–402.

Maridonneau-Parini, I., Yang, C.-Z., Bornens, M., & Goud, B. (1991). Increase in the expression of a family of small guanosine triphosphate-binding proteins, rab proteins, during induced phagocyte differentiation. *J. Clin. Invest.* **87**, 901–7.

Merritt, J. E., McCarthy, S. A., Davies, M. P. A., & Moores, K. E. (1990). Use of fluo-3 to measure cytosolic Ca^{2+} in platelets and neutrophils. *Biochem. J.* **269**, 513–19.

Michell, R. H. (1975). Inositol phospholipids in cell surface receptor function. *Biochim. Biophys. Acta* **415**, 81–147.

Miller, R. J. (1992). Voltage-sensitive Ca^{2+} channels. *J. Biol. Chem.* **267**, 1403–6.

Neer, E. J., & Clapham, D. E. (1988). Roles of G protein subunits in transmembrane signalling. *Nature* **333**, 129–34.

Nishizuka, Y. (1984). Turnover of inositol phospolipids and signal transduction. *Science* **225**, 1365–70.

Osaki, M., Sumimoto, H., Takeshige, K., Cragoe, E. J., Jr., Hori, Y., & Minakami, S. (1989). Na^+/H^+ exchange modulates the production of leukotriene B_4 by human neutrophils. *Biochem. J.* **257**, 751–8.

Petersen, M., Steadman, R., Hallett, M. B., Matthew, N., & Williams, J. D. (1990). Zymosan-induced leukotriene B_4 generation by human neutrophils is augmented by rhTNF-α but not chemotactic peptide. *Immunol.* **70**, 75–81.

Pulido, R., Alvarez, V., Mollinedo, F., & Sanchez-Madrid, F. (1992). Biochemical and functional characterisation of the leukocyte tyrosine phosphatase CD45 (CD45RO, 180 kD) from human neutrophils. In vivo upregulation of CD45RO plasma membrane expression on patients undergoing haemodialysis. *Clin. Exp. Immunol.* **87**, 329–35.

Quinn, M. T., Parkos, C. A., Walker, L., Orkin, S. H., Dinauer, M. C., & Jesaitis, A. J. (1989). Association of a ras-related protein with cytochrome b of human neutrophils. *Nature* **342**, 198–200.

Quinn, M. T., Mullen, M. L., Jesaitis, A. J., & Linner, J. G. (1992). Subcellular distribution of the rap1A protein in human neutrophils: Colocalization and cotranslation with cytochrome b_{559}. *Blood* **79**, 1563–73.

Rosales, C., & Brown, E. J. (1992). Signal transduction by neutrophil immunoglobulin G Fc receptors. Dissociation on intracytoplasmic calcium concentration rise from inositol 1,4,5-trisphosphate. *J. Biol. Chem.* **267**, 5265–71.

Satoh, T., Nakafuku, M., Miyajima, A., & Kaziro, Y. (1991). Involvement of *ras*p21 protein in signal-transduction pathways from interleukin 2, interleukin 3, and granulocyte/macrophage colony-stimulating factor, but not from interleukin 4. *Proc. Natl. Acad. Sci. USA* **88**, 3314–18.

Sekar, M. C., & Hokin, L. E. (1986). The role of phosphoinositides in signal transduction. *J. Membr. Biol.* **89**, 193–210.

Smolen, J. E., Stoehr, S. J., & Kuczynski, B. (1991). Cyclic AMP inhibits secretion from electroporated human neutrophils. *J. Leuk. Biol.* **49**, 172–9.

Stabel, S., & Parker, P. J. (1991). Protein kinase C. *Pharmacol. Therapeut.* **51**, 71–95.

Stephens, L. R., Hughes, K. T., & Irvine, R. F. (1991). Pathway of phosphatidylinositol (3,4,5)-trisphosphate synthesis in activated neutrophils. *Nature* **351**, 33–9.

Tauber, A. I. (1987). Protein kinase C and the activation of the human neutrophil NADPH-oxidase. *Blood* **69**, 711–20.

Thompson, N. T., Tateson, J. E., Randall, R. W., Spacey, G. D., Bonser, R. W., & Garland, L. G. (1990). The temporal relationship between phospholipase activation, diradylglycerol formation and superoxide production in the human neutrophil. *Biochem. J.* **271**, 209–13.

Uings, I. J., Thompson, N. T., Randall, R. W., Spacey, G. D., Bonser, R. W., Hudson, A. T., & Garland, L. G. (1992). Tyrosine phosphorylation is involved in receptor coupling to phospholipase D but not phospholipase C in the human neutrophil. *Biochem. J.* **281**, 597–600.

Ulrich, A., & Schlessinger, J. (1990). Signal transduction by receptors with tyrosine kinase activity. *Cell* **61**, 203–12.

Watson, F., Robinson, J. J., & Edwards, S. W. (1991). Protein kinase C dependent and independent activation of the NADPH oxidase of human neutrophils. *J. Biol. Chem.* **266**, 7432–9.

Watson, F., Robinson, J. J., & Edwards, S. W. (1992). Sequential phospholipase activation in the stimulation of the neutrophil NADPH oxidase. *FEMS Microbiol. Immunol.* **105**, 239–48.

Weingarten, R., & Bokoch, G. M. (1990). GTP binding proteins and signal transduction in the human neutrophil. *Immunol. Lett.* **26**, 1–6.

Wright, J., Maridonneau-Parini, I., Cragoe, E. J., Jr., Schwartz, J. H., & Tauber, A. I. (1988). The role of the Na^+/H^+ antiporter in human neutrophil NADPH-oxidase activation. *J. Leuk. Biol.* **43**, 183–6.

7
Neutrophil priming: Regulation of neutrophil function during inflammatory activation

7.1 Background

By the late 1970s, the idea that mature, bloodstream neutrophils were terminally differentiated, end-of-line cells was being questioned. In a series of experiments by McCall and colleagues (1973, 1979), the biochemical properties of 'toxic' neutrophils – that is, neutrophils isolated from the bloodstream of patients with acute bacterial infections – were characterised. Compared to normal neutrophils isolated from healthy controls, these 'toxic' neutrophils exhibited increased oxidative metabolism, phagocytosis, chemotaxis and 2-deoxyglucose transport. They also had increased cellular levels of alkaline phosphatase and showed toxic granulation. Two possibilities could account for these enhanced functions: either the bacterial infection caused the release from the bone marrow of a more active neutrophil population, or the infection somehow changed the activity of the circulating cells. If the latter explanation was the case, then the concept that neutrophils were terminally differentiated was probably incorrect.

At around this time, the synthetic formylated oligopeptides, thought to be analogous to bacterial-derived products, were being characterised. Remarkably, it was shown that the pretreatment of normal neutrophils in vitro with 2×10^{-8} M fMet-Leu-Phe for 15 min could alter the function of these cells so that they resembled 'toxic' neutrophils. The conclusion, therefore, was that factors induced by the bacterial infection (either derived from the bacteria themselves or else generated by the host in response to the infection) could up-regulate the function of mature, circulating neutrophils.

It was later shown that substimulatory concentrations of many types of neutrophil agonists could also induce this up-regulation of function. Because the concentrations of these compounds that were required did not activate the neutrophil per se, the phenomenon was referred to as 'priming'.

Primed neutrophils generate enhanced levels of reactive oxidants and have higher levels of degranulation and greater phagocytic activity compared to untreated cells. Thus, they are much more potent in killing many types of microbial pathogens. It was therefore proposed that priming also occurred in vivo and that this phenomenon regulated the function of neutrophils during inflammatory activation.

In the 1980s, many purified or recombinant cytokines were becoming available for in vitro studies of immune cell function. Berton and colleagues (1986) showed that γ-interferon could up-regulate, or prime, the ability of human neutrophils to generate reactive oxidants. Two important features were observed. Firstly, incubation times of a few hours were required for maximal priming. Secondly, the priming effects of γ-interferon could be abolished by cycloheximide, an inhibitor of protein biosynthesis. These observations thus strongly indicated that active protein biosynthesis was somehow involved in this enhancement of oxidase function. It was later shown that priming with low concentrations of fMet-Leu-Phe, γ-interferon, GM-CSF and G-CSF was associated with a rapid and selective increase in de novo biosynthesis (Fig. 7.1). These series of experiments thus raised two questions:

i. What is the role of protein biosynthesis in the function of the mature neutrophil?
ii. What are the molecular processes responsible for the up-regulation of neutrophil function by cytokines and other pro-inflammatory mediators?

7.2 Mechanisms of priming

7.2.1 Short-term effects

Incubation of neutrophils with priming agents such as low concentrations of fMet-Leu-Phe (10^{-8} M for 20 min) or cytokines such as GM-CSF (50 U/ml for 30–60 min at 37 °C), enhances their ability to mount a respiratory burst in response to subsequent stimulation by agonists such as fMet-Leu-Phe, C5a and LTB_4 (Fig. 7.2). When receptor-mediated activation is measured, greater levels of oxidants are produced in primed cells. However, when PMA is used to stimulate the respiratory burst in GM-CSF-primed neutrophils, the lag phase before oxidant production begins is decreased, and activity declines more rapidly than in control cells (Fig. 7.3). This indicates that the oxidase may be partially activated or assembled during cytokine treatment. Changes in cell shape and cytoskeletal alterations have also been

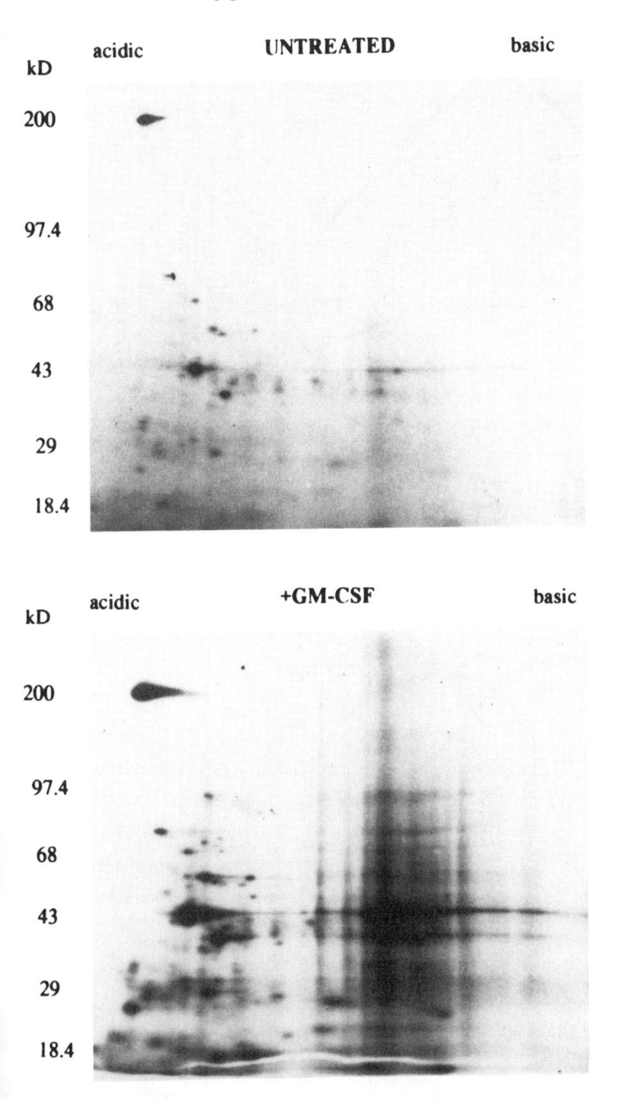

Figure 7.1. Effect of GM-CSF on neutrophil protein biosynthesis. Human neutrophils were incubated in RPMI 1640 medium supplemented with 2.5% foetal calf serum and 60 μCi/ml [^{35}S]-methionine, in the presence and absence of 50 U/ml GM-CSF. After 4 h incubation at 37 °C, the cells were pelleted by low-speed centrifugation. The proteins in the cell pellets were precipitated by 10% trichloracetic acid and then analysed by two-dimensional polyacrylamide gel electrophoresis, using iso-electrofocussing in the first dimension. Electrophoresis in the second dimension was performed in the presence of SDS and used a 12% acrylamide gel. *Source:* Experiment of Becky Stringer.

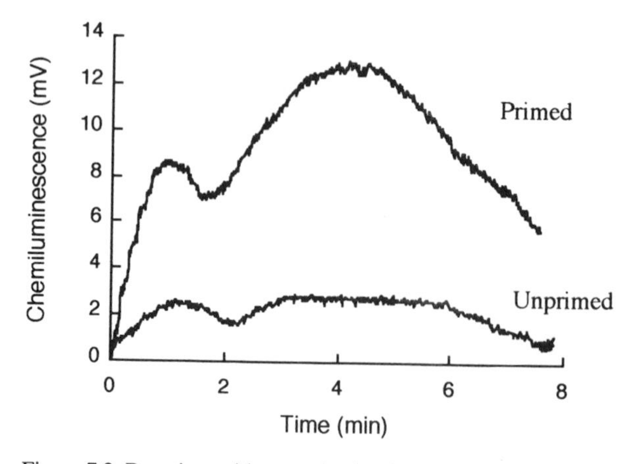

Figure 7.2. Reactive oxidant production by fMet-Leu-Phe-stimulated, primed and unprimed neutrophils. Neutrophils (5×10^5/ml) were incubated in the absence (unprimed) or presence (primed) of 50 U/ml GM-CSF for 60 min at 37 °C. After this incubation, luminol was added (to 10 μM, final concentration), and cells were stimulated by the addition of 1 μM fMet-Leu-Phe.

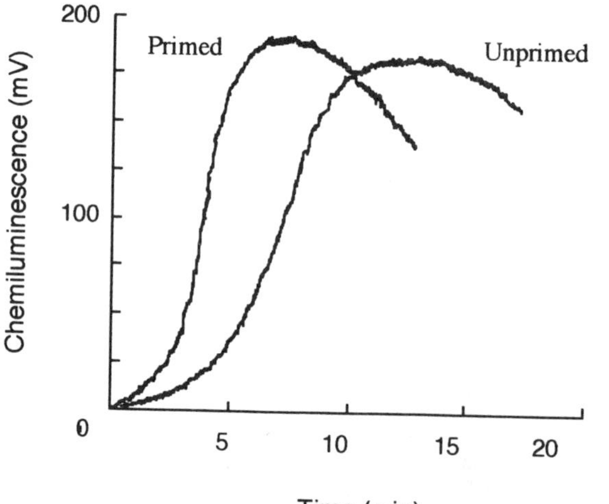

Figure 7.3. Reactive oxidant production by PMA-stimulated, primed and unprimed neutrophils. Experimental conditions were as described for Figure 7.2, except that after 60 min incubation in the presence or absence of GM-CSF, reactive oxidant production was stimulated by the addition of 0.1 μg/ml PMA.

observed during priming. In parallel with this up-regulation of NADPH oxidase function, phagocytosis and bacterial killing is enhanced in primed neutrophils, and the expression of several plasma membrane receptors (e.g. the fMet-Leu-Phe receptor, CR1 and CR3) is increased (Fig. 7.4).

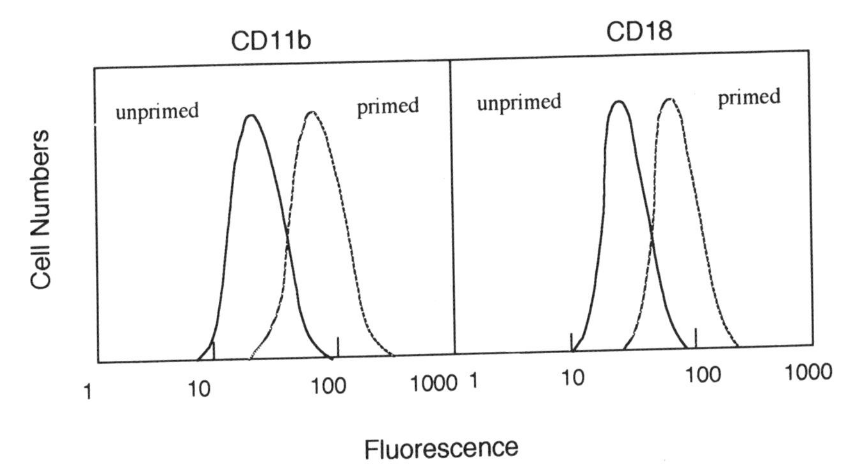

Figure 7.4. Up-regulation of CR3 expression during priming. Neutrophils were incubated in the absence (unprimed) or presence (primed) of 50 U/ml GM-CSF for 60 min at.37 °C. After this incubation, expression of CD11b and CD18 was measured by FACS analysis, using the method described in Edwards et al. (1990).

These changes in neutrophil function are rapid (typically observed within 15–60 min exposure to the priming agent) and are unaffected by the addition of inhibitors of macromolecular biosynthesis. Thus, they are completely independent of de novo biosynthesis. One mechanism that is undoubtedly involved in this process is the recruitment of receptors and cytochrome b molecules to the plasma membrane via the translocation of subcellular stores of preformed molecules (on the membranes of specific and gelatinase-containing granules or on secretory vesicles, Fig. 7.5). This process will account for the rapid increase in the number of certain receptors and cytochrome b molecules on the cell surface, which partly explains the ability of primed cells to generate increased levels of reactive oxidants. The mechanisms responsible for these molecular rearrangements are not completely defined, but they require (i) the generation of second messengers, (ii) the movement of granules and vesicles via changes in the cytoskeletal network and (iii) subsequent membrane fusion. It is also likely that such translocation events may alter either the intracellular signal transduction mechanisms responsible for cell activation (e.g. via the translocation of enzymes such as phospholipases or kinases) or else factors that affect the function of these systems. However, some experiments have shown that an increase in receptor number is not directly linked to certain functions, such as the enhanced ability of primed neutrophils to bind to endothelial cells or other surfaces. Thus, in addition to increases in the number of receptors on the plasma

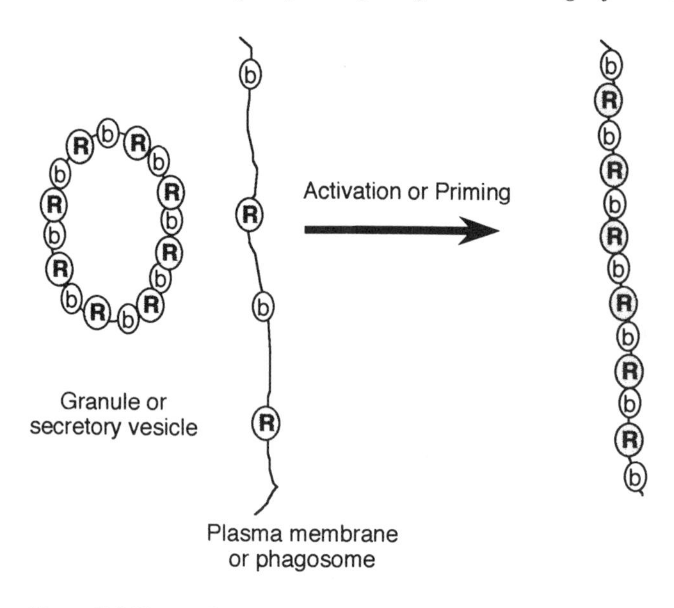

Figure 7.5. Up-regulation of cytochrome b and some receptors during activation or priming. In resting (unstimulated) neutrophils, few molecules of either cytochrome b or receptors such as CR3 are expressed on the plasma membrane. The major cellular pool of these molecules is present on the membranes of specific granules, gelatinase-containing granules or secretory vesicles. Upon stimulation (e.g. by agonists such as fMet-Leu-Phe or PMA) or priming (e.g. by cytokines such as GM-CSF) these subcellular pools translocate to the plasma membrane (or to the membrane of the phagocytic vesicle) so that the number of molecules expressed is increased.

membrane, priming may also alter the affinity of binding of these receptors to their ligands.

Recent technological developments for the assay of neutrophil functions have enabled many properties of these cells to be measured in unfractionated blood. Several important features have been observed:

i. Non-activated neutrophils are spherical and only adopt a typical amoeboid shape when they adhere to surfaces or are exposed to activating factors.
ii. The ability of neutrophils in whole blood to generate reactive oxidants is virtually undetectable, unless they have previously been primed (Fig. 7. 6).
iii. The expression of cell-surface molecules, such as CR3 and cytochrome b, is very low in blood neutrophils, but this expression may again be rapidly up-regulated during priming or activation (Fig. 7.7).

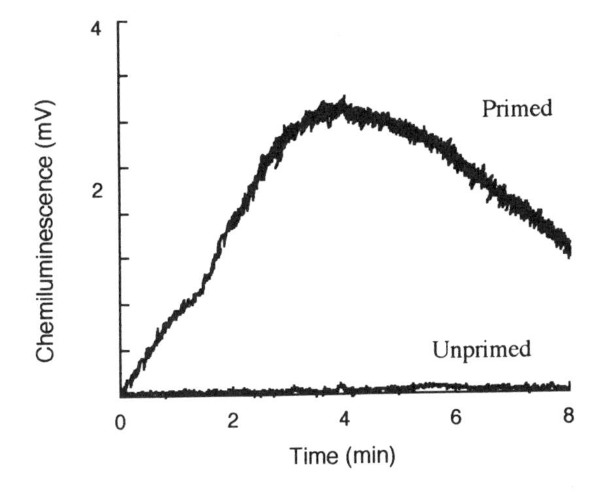

Figure 7.6. Activation of neutrophils in unfractionated blood. Unfractionated, whole blood (10 μl) was diluted to 1 ml in RPMI 1640 medium and incubated in the absence (unprimed) or presence (primed) of 50 U/ml GM-CSF for 60 min at 37 °C. After this incubation, luminol was added (to 100 μM, final concentration) and chemiluminescence measured after the addition of 1 μM fMet-Leu-Phe.

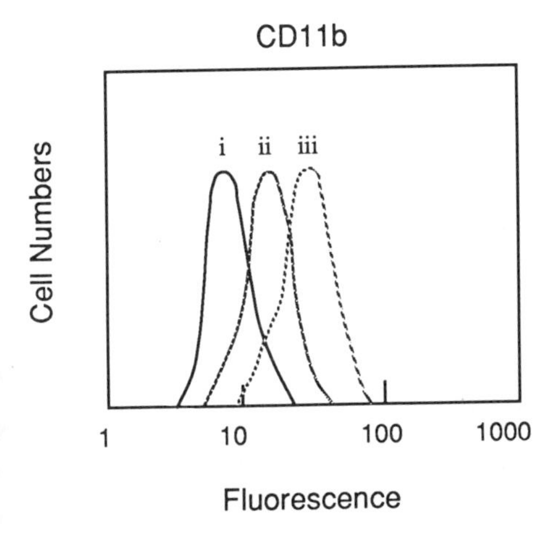

Figure 7.7. Expression of CR3 in neutrophils in unfractionated blood and after isolation. The expression of CD11b was measured in neutrophils in (i) unfractionated whole blood, (ii) after isolation using Mono-Poly resolving medium and (iii) after isolation using dextran sedimentation and centrifugation in Ficoll-Hypaque. Further experimental details may be found in Watson, Robinson & Edwards, 1992.

Observations of neutrophil function in whole blood and after purification have led to the realisation that many commonly-used purification procedures partially prime neutrophils, presumably by mobilising the more easily-activated subcellular stores of surface molecules (i.e. gelatinase-containing granules or secretory vesicles).

These findings imply a fundamental rethinking of the regulation of neutrophil function during inflammatory activation. Neutrophil function is inevitably partially primed after the cells have been purified from blood; hence, exogenously-added priming agents (such as cytokines) further up-regulate this partially-primed activity. The observed effect of a priming agent, under these circumstances, is thus to *increase* function. However, the observations of neutrophil function in unfractionated blood indicate that, within the circulation, neutrophils possess only very low (or perhaps negligible) ability to perform functions such as the respiratory burst. Thus, priming agents may not merely enhance function, but rather switch the neutrophil from a 'non-responder' state into a 'responder' state. If this is correct, then neutrophil function during inflammation is regulated in a two-step process. The circulating neutrophil has very limited ability to respond to certain agonists. The first step in the activation involves priming (e.g. by cytokines), which leads to receptor and oxidase up-regulation (and presumably some other cellular events). The second step is activation per se, in which the primed neutrophil becomes responsive to activating signals. Such a two-step activation process has obvious advantages: because two signals are required for full activation, the possibility of non-specific activation of, for example, reactive oxidant production by cells in the circulation is minimised.

7.2.2 Longer-term effects

In addition to these short-term priming effects, which do not involve de novo protein biosynthesis, priming agents such as cytokines induce some functional changes in neutrophils that are detectable many hours, or even days, after exposure. These effects certainly do require active gene expression. For example, cytokine-exposed neutrophils can survive for several days in culture, compared with cells not exposed to cytokines, which normally lose viability within 12–18 h. This enhanced viability is associated with increased functional activity. For example, whereas control neutrophil suspensions generate only barely-detectable levels of reactive oxidants after 24 h in culture, cells treated for the same time with GM-CSF can generate significant levels of oxidants. Furthermore, as is discussed in section 7.3.4, cytokine-exposed neutrophils can themselves secrete a range of secondary cytokines that may affect either their own function or the function of other cells and tissues at inflammatory sites. Both the enhanced functional viability and the secretion of secondary cytokines is blocked by inhibitors of macromolecular biosynthesis, confirming that these functions are due to activated gene expression. Thus, activated gene expression is required to enhance and extend the functional lifespan of neutrophils and also to generate molecules that will actively regulate the progress of an inflammatory reaction.

The following sections review the importance of macromolecular biosynthesis in neutrophil function, and describe how this gene activation is selectively up-regulated during priming with cytokines.

7.3. Protein biosynthesis in mature neutrophils

7.3.1 Background

For many years it was believed that mature, bloodstream neutrophils had little or no capacity for de novo macromolecular biosynthesis. The evidence for this conclusion was largely circumstantial and derived primarily from morphological studies. Electron micrographs of blood neutrophils showed that they had little endoplasmic reticulum and few ribosomes, and did not contain a nucleolus. Further, the condensed nature of the chromatin made it unlikely that the nucleus could be transcriptionally active. In view of the belief that these cells had a relatively-short lifespan (presumed to be as short as 12 h or less), it was further assumed that whilst protein biosynthesis might not be possible, it might not even be necessary. If these cells were functionally equipped for their rapid recruitment and activation, and provided that new cells could be released from the marrow very rapidly upon inflammatory challenge, then sustained neutrophil-mediated defence could occur by the rapid turnover of neutrophils followed by the recruitment of newly-released marrow cells. Thus, the picture of the short life of a neutrophil was that it either responded to an inflammatory challenge by the activation of phagocytic killing (which probably resulted in its self-destruction), or else it was cleared from the circulation within a few hours of its release from the marrow. Later, experiments measuring the rates of incorporation of radiolabelled precursors into macromolecules showed that biosynthesis was in fact detectable in these cells. However, this was thought not to be important for function but merely a residual activity carried over from earlier developmental stages in which de novo biosynthesis clearly was important for growth and differentiation.

It was noted, however, in experiments by Cline in the 1960s (e.g. Cline 1966) that RNA turnover (i.e. enhanced RNA biosynthesis and degradation) occurred during phagocytosis. Cline then set out to determine if this change in RNA metabolism was accompanied by changes in mRNA levels. If such changes did occur, then it was possible that the de novo biosynthesis of some protein(s) may be important for cell function. However, he found that inhibitors (i.e. actinomycin D to prevent transcription and puromycin to prevent translation) did not affect the phagocytosis of latex particles, even though these compounds inhibited RNA and protein biosynthesis by 90–95%. This observation thus appeared to substantiate the conclusions based on morphological studies – namely, that de novo biosynthesis was not necessary for neutrophil function.

However, Cline's experiments showed an important phenomenon that was not fully appreciated at the time, and even today is not widely quoted. Whilst the *initial* activation of neutrophil functions (e.g. phagocytosis, respiratory-burst activity and lactate production) was unaffected by inhibitors of macromolecular biosynthesis, *sustained* activity was markedly decreased in the presence of these inhibitors. For example, H_2O_2 production after 10 min incubation with actinomycin D was unaffected, but by 30 min incubation with the inhibitor, oxidant production was significantly decreased. Similarly, early lactate production was not affected, but by 60 min with actinomycin D, levels of production were inhibited. Thus, whilst it is true to say that the mature neutrophil may indeed be functionally equipped for its rapid and lethal activation, sustained activity on the other hand (i.e. for periods >30–60 min) requires new biosynthesis, presumably of components that are subject to rapid turnover. If such components are not replenished, then they will become rate limiting. As neutrophils are likely to be functional for many hours (or perhaps several days) during inflammatory challenge within the body, such biosynthesis in order to sustain their function is likely to be of profound pathological importance. Perhaps the reasons why this phenomenon has been largely overlooked by neutrophil researchers arise from the desire to perform short-time-scale experiments in the laboratory and to use saturating concentrations of stimulants, so that maximal changes in activity can be measured. During activation within the body, neutrophils are likely to be exposed to submaximal concentrations of agonists (or low ratios of bacteria to neutrophils) and will be functional for many hours. Perhaps more care and thought should thus go into the design of experiments in vitro so that they more closely approximate in vivo conditions, and hence provide a clearer picture of the pathophysiology of neutrophil function.

Today, it is widely appreciated that mature neutrophils are biosynthetically active, and that this biosynthesis can be rapidly and selectively up-regulated during inflammatory challenge. It is now clear that neutrophils can perform many of the functions previously thought to be restricted to monocytes and macrophages, which have long been known to be active in the secretion of an array of regulatory molecules. In view of the enormous interest in the pharmacological modulation of neutrophil activity (either to enhance or to decrease neutrophil function in human disease; see Chapter 8) it seems remarkable that, to date, this aspect of neutrophil physiology has been largely ignored.

7.3.2 *Identity of actively-expressed proteins*

There are essentially three groups of proteins actively expressed by neutrophils:

i. *'House-keeping' proteins,* such as actin and phosphoglycerokinase, are endogenously expressed, and their expression does not alter significantly during inflammatory activation.
ii. Proteins synthesised by non-activated bloodstream neutrophils include key cellular components, such as some plasma membrane receptors, and some oxidase components, which are subject to high turnover rates and must be replenished via de novo synthesis in order to sustain functional activity. Expression of these is up-regulated (usually two- to threefold) upon activation.
iii. Proteins that are not expressed (or else are expressed at only very low rates) in blood neutrophils include a series of cytokines that are rapidly and transiently expressed in neutrophils exposed to cytokines or chemoattractants. Their expression increases markedly and speedily during inflammatory activation.

Several experimental approaches can be utilised to measure biosynthesis or determine if a particular process requires active transcription or translation:

(i) *Transcription/translation inhibitors.* Add inhibitors of transcription (e.g. actinomycin D) or translation (e.g. cycloheximide or puromycin) to cell suspensions to determine their effects on function. It has been shown that some stimulatory effects of certain cytokines on neutrophil function can be prevented by such inhibitors.

(ii) *Radiolabelled precursor molecules.* Incorporate radiolabelled precursor molecules, such as labelled uridine or labelled amino acids (e.g. [^{35}S]-methionine), into RNA and proteins, respectively. Biosynthetic activity can then be monitored by autoradiography of individual cells, by measuring total radioactive counts incorporated into acid-insoluble macromolecules, or by analysis of newly synthesised macromolecules by gel electrophoresis. One common technique is to label proteins (perhaps in the presence and absence of pro- or anti-inflammatory agonists) and to examine polypeptide profiles by two-dimensional polyacrylamide gel electrophoresis (2D-PAGE, see Fig. 7.1). Whilst it is possible to gain some insight into the nature of the polypeptides synthesised (e.g. their pI, relative molecular mass, whether or not they are secreted, their location in particular subcellular fractions), it is almost impossible to identify specific proteins unambiguously in this way.

(iii) *Immunoprecipitation.* An adaptation of the above technique is to label all newly synthesised proteins but then to immunoprecipitate a particular protein using a specific antibody. The labelled protein present in the immunoprecipitate can then be analysed by scintillation counting or PAGE. Using this technique, it is possible to obtain quite accurate measurements of the changes in rates of labelling of specific proteins during treatment with inhibitors or activating agents. This technique naturally requires the availability of a suitable antibody raised against the protein under investigation. A drawback is that some antibodies (particularly monoclonal antibodies) may readily recognise or label the proteins to which they were raised, but may not always efficiently immunoprecipitate their targets.

(iv) *Analysis of mRNA levels.* Using either Northern blots, RNA dot blots or in situ hybridisation, the analysis of mRNA levels is a means of quantifying levels of transcripts for a particular component. Again, it is then possible to determine the effects of inhibitors or stimulants on the levels of such transcripts. This technique requires a suitable (and labelled) probe to hybridise to the target transcript, and this probe is usually cloned DNA (either cDNA, genomic DNA or single-stranded RNA) or a short, synthetic oligonucleotide. Such methods provide fairly accurate means of quantifying changes in levels of specific mRNA molecules, changes that may be due to either altered rates of transcription (i.e. increased or decreased rates of mRNA production) or altered stability (i.e. a change in the rate of degradation of the mRNA). It must be pointed out, however, that in some cases increases in mRNA levels are not always accompanied by increases in rates of protein expression. Usually there is a lag of 1–2 h between increases in mRNA levels and protein expression. Sometimes, rates of protein expression can be enhanced without parallel increases in mRNA levels, either because of enhanced translation of existing mRNA or else because increases in mRNA production are paralleled by increases in the rate of degradation.

Also, mRNA levels may increase, but a second stimulus may be required before this message is translated into active protein.

All of these techniques have been used in neutrophil studies to show that the expression of certain cellular components can be up-regulated by changes in rates of transcription, mRNA stability, enhanced translation (of newly-transcribed or pre-existing mRNA) and enhanced processing. Therefore, it is usually necessary to measure changes in expression of particular components using several techniques – for example, changes in mRNA levels *and* changes in amount of functional protein.

7.3.3 Plasma membrane receptors

Immunoprecipitation experiments have indicated that bloodstream neutrophils actively express several plasma membrane receptors, such as FcγRIII, CR1, CR3 (the α-chain or CD11b, but not the β-chain or CD18) and major histocompatibility complex (MHC) class I. The MHC class I molecules are surface glycoproteins and heterodimers consisting of a 43-kDa MHC-encoded H chain that is non-covalently linked to a 12-kDa L chain (β_2-microglobulin). The biosynthesis of all of these receptors can be prevented by incubation with actinomycin D and cycloheximide, indicating that active transcription and translation are both required for expression. The reason for this appears to be that these plasma membrane receptors are subject to high turnover rates – that is, they may be continuously removed from the cell surface via shedding or internalisation and need to be replenished. This replenishment of receptors can occur via receptor recycling, translocation of preformed subcellular pools (see §7.2.1) or de novo biosynthesis. Parallel experiments have indicated that granule proteins such as elastase, lactoferrin and myeloperoxidase are *not* actively synthesised by blood cells, indicating that only certain cellular components must be synthesised. De novo biosynthesis of granule proteins may not be necessary because these are present in the cell at very high concentrations and are utilised only when the cell is fully activated.

In contrast to the enhanced receptor expression that occurs during activation and priming, the expression of some receptors actually *decreases* during this process. This is particularly evident with receptors such as CD16 (the low-affinity receptor for IgG) and the ligand that binds monoclonal antibody 31D8 (a surface molecule of undefined nature whose expression is closely linked to fMet-Leu-Phe responsiveness). These molecules are GPI-linked and are shed from the cell surface during activation (see Fig. 3.10).

Thus, their continued expression on the plasma membrane requires translocation from preformed pools and/or de novo biosynthesis.

The fact that receptors need to be replaced via de novo biosynthesis can be demonstrated in experiments where neutrophils are cultured in the presence and absence of cycloheximide. When protein biosynthesis is blocked by this inhibitor, the expression of FcγRIII on the cell surface cannot be maintained as the cells age in culture (Fig. 7.8), indicating that continued expression of this receptor is a balance between the amount shed and the amount replaced via new biosynthesis. Newly synthesised receptors appear to be functional, because they can be detected within the biosynthetic machinery of the cell, and newly made (i.e. newly labelled) receptors are detected in the plasma membrane.

In addition to this active biosynthesis in non-stimulated cells, incubation of neutrophils with agents such as GM-CSF results in a two- to fivefold enhancement of biosynthesis of these receptors. Again, it has been observed that these newly synthesised receptors are functionally active and are also inserted into the plasma membrane. Perhaps the most dramatic enhancement in receptor expression that occurs during cytokine exposure is that of FcγRI expression. Because this receptor is not expressed in blood neutrophils, these neutrophils do not bind monomeric IgG (§3.10.2). However, during incubation of neutrophils with γ-interferon, expression of this receptor is observed, firstly (within 2 h) as an increase in levels of mRNA (detected by Northern analyses) and later (>8–12h) by detection of this receptor on the plasma membrane. Interestingly, expression of FcγRI has been observed on blood neutrophils of patients who have been receiving G-CSF therapy. This may be a direct effect of G-CSF on activation of expression of this receptor, or else due to the fact that administration of G-CSF has stimulated a cytokine cascade and that secondary production of γ-interferon has actually resulted in the activation of expression. Furthermore, neutrophils isolated from the synovial fluid of some patients with rheumatoid arthritis express FcγRI, indicating that protein biosynthesis has been activated in these cells within the inflamed joint (Fig. 7.9). Cells in which FcγRI expression has been stimulated by γ-interferon have greater phagocytic activity and higher rates of ADCC than control cells. Cotreatment of cells with γ-interferon and cycloheximide prevents the expression of this receptor, and these enhanced functions are not observed.

The regulation of expression of receptors on the plasma membrane is thus controlled by several factors:

i. Some receptors are shed from the plasma membrane during either aging of the cells or activation and priming. Such receptors include

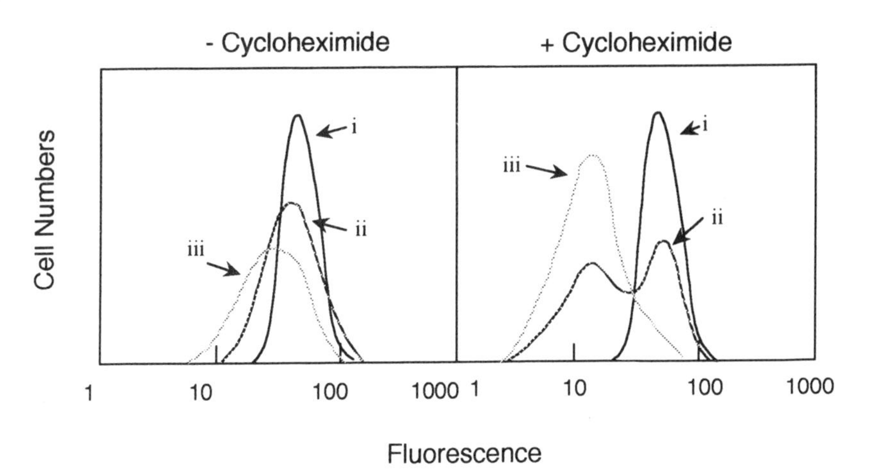

Figure 7.8. Role of protein biosynthesis in expression of FcγRIII. Neutrophils were incubated for (i) 15 min, (ii) 3 h and (iii) 4 h in the absence or presence of 30 μg/ml cycloheximide. Afterwards, CD16 (FcγRIII) expression was determined by FACS analysis.

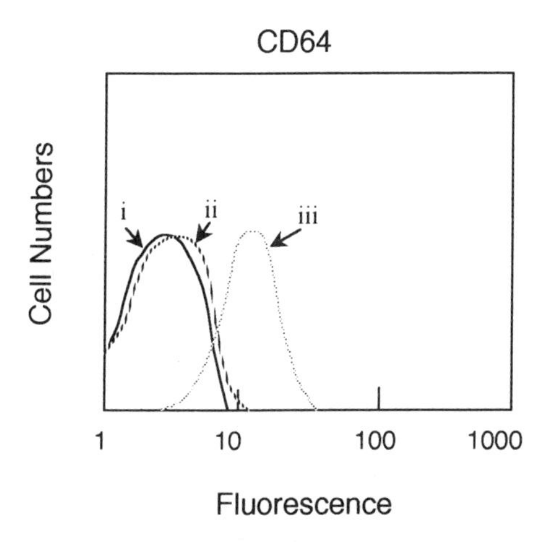

Figure 7.9. FcγRI expression in blood and synovial-fluid neutrophils. Neutrophils were isolated from the blood (trace ii) and synovial fluid (trace iii) of a patient with rheumatoid arthritis. Expression of CD64 (FcγRI) was then measured by FACS analysis. Trace (i) indicates the fluorescence distribution of a non-immune isotype control monoclonal antibody. Similar results were obtained in 6 out of 11 patients with rheumatoid arthritis.

those attached to the plasma membrane via GPI linkages, such as FcγRIII and the receptor recognised by the monoclonal antibody 31D8. Shedding occurs via the activities of phospholipases or proteases, such as pronase or elastase, that may be released by activated neutrophils (see Fig. 3.10).

ii. Receptors may be internalised, but the fate of the internalised receptor depends upon whether or not the receptor has bound its ligand. For example, if FcγRII binds its ligand, then it is internalised and transported to lysosomes, where both the ligand and the receptor are degraded. However, if this receptor has not bound its ligand, then it is internalised but recycled to the plasma membrane, where it is reinserted and remains functional.

7.3.4 Cytokines

7.3.4.1 Interleukin-1 (IL-1)

Tiku and colleagues (Tiku, Tiku & Skosey, 1986) demonstrated that the treatment of neutrophils with opsonised zymosan or PMA led to the secretion of a factor(s) with Interleukin-1 (IL-1)-like activity. This activity was characterised by its ability to:

i. induce the differentiation of mouse thymocytes,
ii. stimulate the proliferation of normal mouse synovial fibroblasts and
iii. stimulate the release of neutral proteases from synovial cells.

Incubation periods in excess of 2 h were required before this activity was detected in cell-free supernatants. More recently, the use of cDNA probing of Northern transfers (to detect specific mRNA levels), the use of ELISA techniques (to detect protein levels immunologically) and the development of more specific bioassays (culture techniques in which a biomolecule stimulates proliferation in a particular cell line) have resulted in a more thorough analysis of IL-1 production by neutrophils. IL-1 is only poorly expressed in blood neutrophils because mRNA for this cytokine is detectable only at very low levels (if at all), and protein production is usually below the level of detection of most assays. However, exposure of neutrophils to lipopolysaccharide (LPS), or to cytokines such as GM-CSF, TNF or IL-1 itself, results in a rapid but transient increase in IL-1 expression.

Maximal levels of IL-1β mRNA are detected within 1 h exposure to GM-CSF; levels then decline to about 50% of maximal by 2 h and then fall to near base-line levels by 8 h (Fig. 7.10). Similarly, levels of IL-1β protein are transiently expressed, broadly following the changes in mRNA levels. Thus, levels of antigenically-detectable protein are maximal by 2–4 h (in cell extracts) and are detected extracellularly in the culture medium by 4 h. Whilst IL-1 (IL-1α and IL-1β at 1 ng/ml) and TNF (at 5 ng/ml) are capable of stimulating an increase in IL-1β mRNA levels, they can synergise with

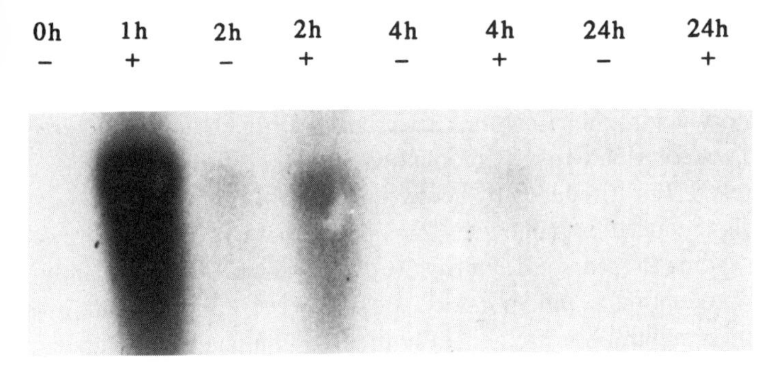

Figure 7.10. IL-1β expression by GM-CSF-stimulated neutrophils. Human neutrophils were incubated in the absence (–) and presence (+) of 50 U/ml GM-CSF, and at time intervals total RNA was extracted. After electrophoresis on formamide-containing gels, IL-1β mRNA levels were probed. *Source:* Experiment of Julie Quayle.

each other in the induction of expression. The cytokines IL-2–6 only stimulate low levels of IL-1β expression. This increased expression results from increased rates of transcription of the gene, and also from an enhanced stability of the transcribed mRNA. Cycloheximide can also enhance levels of IL-1β mRNA, indicating that levels of this mRNA species are controlled both by enhanced transcription and also by posttranscriptional modifications. Whilst both IL-1α and IL-1β mRNA levels are detected after stimulation (IL-1β » IL-1α), only IL-1β is translated; in some instances a second stimulus may be required to enhance the rate of translation of pre-existing mRNA molecules into active proteins.

It has been appreciated for many years that monocytes and macrophages are important producers of IL-1 at inflammatory sites. Thus, it is now recognised that neutrophils can also secrete this cytokine following appropriate stimulation, but there are two important differences in the mechanisms that regulate IL-1 production in these cell types:

i. Neutrophils rapidly synthesise IL-1 in response to stimulation, but this expression is only transient. Maximal increases in mRNA levels are attained within 1 h, falling to base-line levels within 8–12 h. This is in contrast to stimulation of monocytes, in which mRNA levels peak by 3–4 h and remain high for periods of up to 24 h. This fits the generalised picture of neutrophils being the first immune cells to be attracted into inflammatory sites and then becoming rapidly but transiently activated. Because of this transient activation in gene expression, if a population of neutrophils from an inflammatory site (e.g. a rheumatoid joint) is analysed, only a small percentage of the cells will have active biosynthesis of IL-1; the vast majority will have biosynthesis levels that have returned to base-line. The study of populations of inflammatory neutrophils may thus give the false impression that these cells do not contribute to IL-1 production in vivo.

ii. The levels of IL-1 produced by neutrophils are only about 10–20% of the levels produced by monocytes (on a per-cell basis). This finding is taken by some to mean that neutrophils do not contribute significantly to IL-1 production at inflammatory sites. However, during acute inflammation, neutrophils are generally the first immune cells to arrive and often do so in large numbers. Thus, neutrophil-derived IL-1 production can be significant, and probably sufficient to augment an inflammatory response and initiate a cytokine cascade.

IL-1 has a variety of effects during inflammation, such as:

increasing the production of collagenase and loss of proteoglycan into arthritic joints;

regulating the activation of fibroblasts, osteoclasts and synovial cells;

recruiting and regulating leukocyte function;

priming neutrophils for oxidative activity, phagocytosis, degranulation and stimulation of the production of thromboxanes, 6-keto$PGF_{1\alpha}$ and PGE_2.

7.3.4.2 *Tumour necrosis factor (TNF)*

Yamazaki, Ikenami and Sugiyama (1989) showed that neutrophils treated with plant or animal lectins secreted a factor that could kill tumour cells in vitro. This cytotoxic activity was distinct from reactive oxygen metabolites and was a heat-labile, trypsin-sensitive protein. The factor was tentatively identified as a *tumour necrosis factor (TNF)*-like molecule because it was antigenically indistinguishable from monocyte-derived TNF, although of a slightly different size to that secreted from monocytes. Analysis of mRNA levels (in Northern transfers using a specific TNF-α cDNA probe) has since shown that TNF-α is not expressed by bloodstream neutrophils, but its expression can be rapidly and transiently induced by LPS. Levels of mRNA are detected 60 min after stimulation and reach maximal amounts by 4 h. The supernatants of unstimulated neutrophil suspensions contain very low TNF activity (35–60 pg/ml), but stimulated cells secrete much higher levels (160–190 pg/ml). GM-CSF stimulates an increase in TNF-α mRNA, but these transcripts are not actively translated into protein. Again, monocytes produce four to five times more TNF than do activated neutrophils (on a per-cell basis), but the high numbers of neutrophils that may be found at inflammatory sites indicate that neutrophil-derived TNF in vivo may be significant.

TNF production by neutrophils can have profound paracrine or autocrine effects. TNF induces fever, augments collagenase and PGE_2 formation by fibroblasts, stimulates IL-1 and PGE_1 production by macrophages, induces osteoclast bone resorption, inhibits lipoprotein lipase activity and immunostimulates lymphocytes. Its effects on neutrophils include increased adherence, enhancement of phagocytosis and priming of both the respiratory burst and degranulation. High concentrations of this cytokine can directly activate the production of reactive oxidants by neutrophils (see §3.5.4).

7.3.4.3 Interleukin-6 (IL-6)

IL-6 expression can be activated in neutrophils upon exposure to GM-CSF and TNF, whilst IL-3, G-CSF, γ-interferon and lymphotoxin do not induce expression of this cytokine. Expression of IL-6 has been identified by analysis of mRNA levels and by protein analysis. Following GM-CSF exposure, expression is detectable by 2 h, maximal by 6 h and then returns to base-line levels. LPS, PMA (but not fMet-Leu-Phe) and cycloheximide also induce IL-6 transcripts. The finding that this protein-synthesis inhibitor increases mRNA levels suggests either that transcription is regulated by a short-lived repressor, or else that decay of its mRNA may be regulated by a short-lived RNase.

IL-6 induces terminal maturation of B cells, promotes the growth of hybridoma/myeloma cells and T cells, and acts upon haematopoietic progenitors in synergy with IL-1 and IL-3. It also induces the production of acute-phase proteins by liver cells. In addition to its production by neutrophils, IL-6 is also synthesised by monocytes, T and B cells, fibroblasts, cardiac myxoma cells, some bladder carcinomas, cervical cancer cells and glioblastomas. Stimulated neutrophils generate about ten times less IL-6 (on a per-cell basis) than do monocytes.

7.3.4.4 Interleukin-8 (IL-8)

IL-8, like the other cytokines described above, is not constitutively expressed by blood neutrophils, but its expression may be rapidly induced upon adherence to plastic or by exposure to factors such as LPS (10 ng/ml), TNF-α (20 ng/ml) and IL-1β (20 ng/ml). The stimulants C5a, fMet-Leu-Phe and LTB_4 do not by themselves induce expression, but they synergise with LPS in induction of expression of this cytokine. Expression is detectable both by increases in mRNA levels and by induction of the protein.

Incubation times of 0.5–1 h are required for detection of mRNA levels (which return to base-line levels by 24 h), and incubation times in excess of 4 h are needed before protein secretion is detected. This transient expression of IL-8 is in contrast to the pattern of production by monocytes, epithelial cells and fibroblasts, which show expression that persists for up to 24 h after stimulation. IL-8 (and related NAP-like peptides) is a powerful neutrophil chemoattractant and, in combination with cytochalasin B, can induce degranulation and activation of the respiratory burst (§3.5.5).

7.3.4.5 Other cytokines

Other cytokines synthesised by activated neutrophils (but not bloodstream neutrophils) include interferon-α, G-CSF and M-CSF. Interferon-α expression is stimulated by exposure of blood neutrophils to G-CSF (incubation times >3 h are required), but expression of this cytokine is not activated by exposure to either LPS or fMet-Leu-Phe. G-CSF induces 0.95- and 1.2-kb mRNA molecules, the latter transcript containing the message plus a 3′ non-coding region. Synthesis of interferon-α protein (as determined by a radio-immunoassay) requires incubation times in excess of 6 h, and the functions of this cytokine include inhibition of neutrophil-colony formation and platelet formation in vitro.

GM-CSF stimulates the expression of both G-CSF and M-CSF by blood neutrophils. These latter cytokines play important roles in haematopoiesis (Chapter 2), and G-CSF can also prime mature neutrophil function (see §7.2.1). Levels of production of these CSFs are about five to ten times lower than those produced by monocytes (on a per-cell basis).

Thus, the production by mature neutrophils of interferon-α (which inhibits granulocyte-colony formation) and G-CSF (which stimulates neutrophil-colony formation) provides mechanisms by which these cells can generate factors capable of both positive and negative regulation of the rate of granulocyte formation. Other neutrophil-derived cytokines can (i) function directly in inflammation, (ii) alter the function of other immune cells or other neutrophils or (iii) play a role in tissue remodelling. A summary of the cytokines synthesised by neutrophils is shown in Figure 7.11, which highlights some of the biological effects of these molecules.

7.3.5 NADPH oxidase components

The O_2^--generating NADPH oxidase comprises five fundamental components:

gp91-*phox* and p22-*phox* (the large and small subunits, respectively, of the cytochrome b);

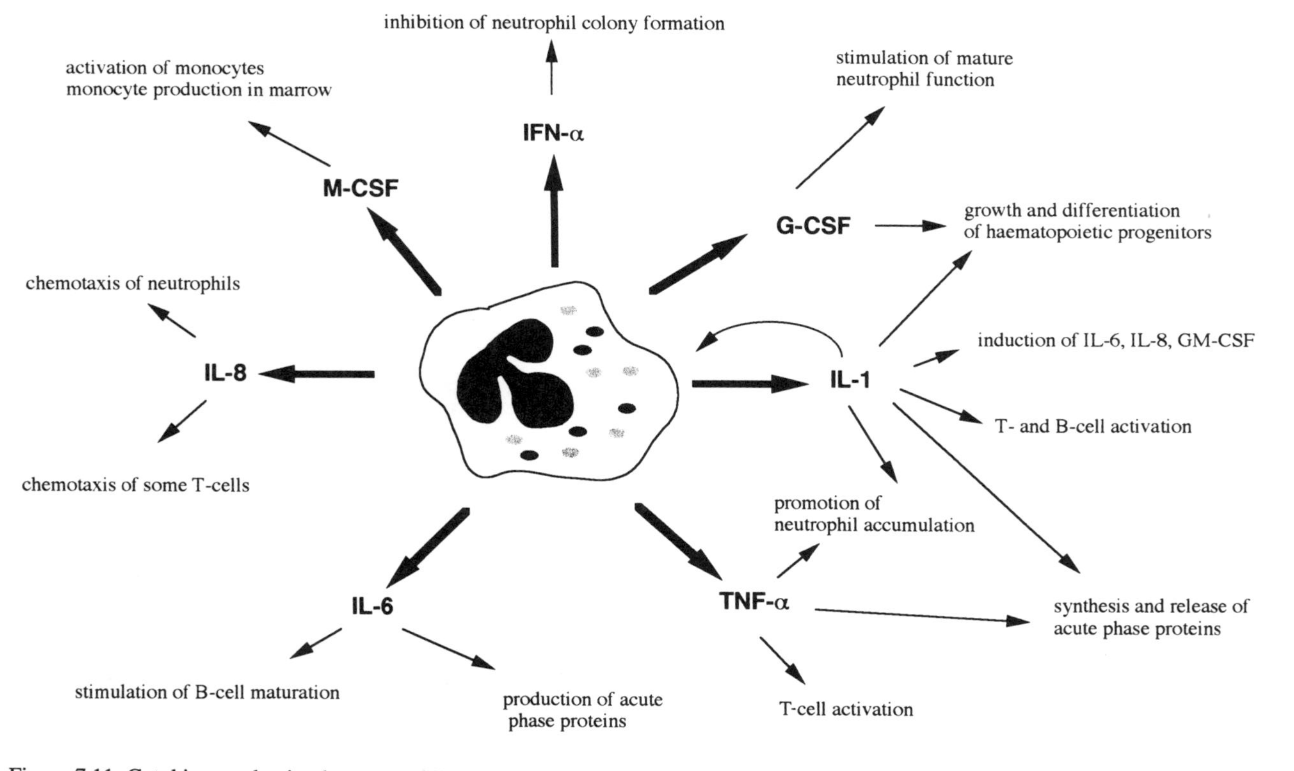

Figure 7.11. Cytokine production by neutrophils. This scheme shows the major cytokines produced by primed neutrophils, as well as some of the major biological effects of these neutrophil-derived cytokines.

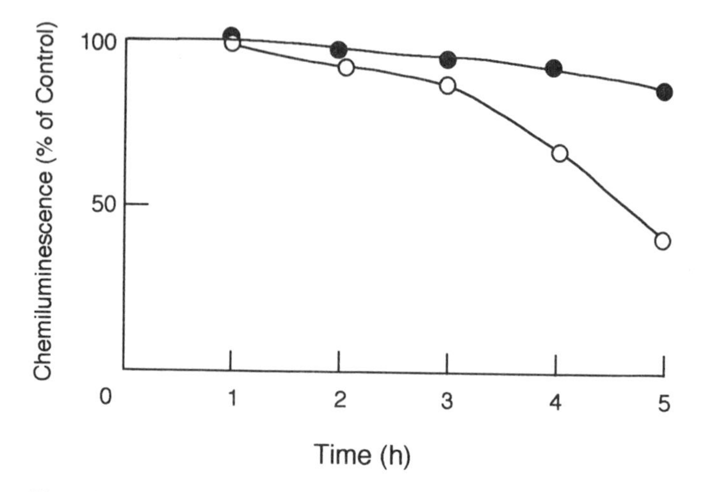

Figure 7.12. Role of protein biosynthesis in the ability of neutrophils to generate reactive oxidants. Neutrophils were incubated in the presence (○) and absence (●) of 30 μg/ml cycloheximide at 37 °C. At time intervals, portions were removed and luminol chemiluminescence measured after stimulation with 1 μM fMet-Leu-Phe. Values presented are a percentage of the control value measured at time zero.

p47-*phox* and p66-*phox* (the cytoplasmic components);
p21^{rac}, a small G-protein (see §5.3.1 for details).

Messenger RNA molecules for both subunits of the cytochrome and the two cytosolic components are detectable in unstimulated bloodstream cells. Experiments involving incubation of neutrophil suspensions with the protein synthesis inhibitor *cycloheximide* indicate that constitutive expression of one or more components of the oxidase is required for the neutrophil to maintain its ability to generate reactive oxidants. For example, when neutrophils are incubated in vitro with cycloheximide, their ability to generate reactive oxidants declines more rapidly than in control cells, as they age in culture (Fig. 7.12). This decline in oxidase activity when protein biosynthesis is blocked is not due to cell death, because cells treated with cycloheximide for this time still exclude trypan blue. Furthermore, when protein biosynthesis is stimulated in neutrophils by the addition of GM-CSF for 24 h in vitro, the ability to generate reactive oxidants is enhanced considerably above the levels observed in untreated cells.

Incubation with γ-interferon (2–50 U/ml) increases the ability of neutrophils to generate reactive oxidants in response to fMet-Leu-Phe, con A and PMA. Incubation times in excess of 2–4 h are required, and this enhanced activity is prevented if the cells are co-incubated with cycloheximide. This observation by Berton et al. (1986) was one of the first pieces of evidence to show that γ-interferon could stimulate the increased expression of one or more components of the NADPH oxidase. Direct analysis of expression of oxidase components has confirmed this idea. In blood neutrophils, expression of p21-*phox* (the small cytochrome b subunit) is constitutive but unaffected by incubation with γ-interferon and/or LPS. In contrast, mRNA levels for gp91-*phox* (the large cytochrome b subunit) are increased by such treatments, whilst expression of p47-*phox* is down-regulated.

7.3.6 Other components

In addition to the above components, several other factors, many of which play defined roles in neutrophil function during inflammation, are actively expressed by neutrophils. One of the first of these components to be identified was *plasminogen activator* (which activates the serum proenzyme plasminogen). The biosynthesis of this protein by neutrophils is stimulated by con A, and by low (8×10^{-10} M) but not high (1.6×10^{-8} M) concentrations of PMA. Expression is down-regulated by actinomycin D, cycloheximide, cAMP and glucocorticoids. Indeed, the biosynthesis of a number of neutrophil proteins is also regulated by glucocorticoids. For example, experiments by Blowers, Jayson and Jasani (1985, 1988) have shown that glucocorticoid treatment up-regulates the biosynthesis of at least seven endogenously expressed polypeptides and down-regulates the expression of two other components. Cellular effects of dexamethasone on neutrophils include the suppression of receptor-mediated phospholipase A_2-dependent release of arachidonic acid, and the concomitant decreased secretion of chemotactic factors from neutrophils. Such effects are in line with the known effects of dexamethasone on the stimulation of the phospholipase A_2 inhibitor *lipocortin* (or lipomodulin/macrocortin). Over the concentration range 10^{-9}–10^{-6} M, dexamethasone stimulates the expression of neutrophil proteins with relative molecular masses of 55, 49, 29, 27, 22 and 16 kDa.

Fibronectin is an extracellular matrix protein that mediates a variety of cellular effects. It is important in cell–cell and cell–substratum interactions (§3.9), mediates reticuloendothelial cell activity and binds both to C1q (the first component of complement) and to bacteria. It also increases the tumouricidal activity of macrophages and activates complement receptors, by regulating the binding of C3b-coated particles to neutrophils. It may mediate attachment of *Staphylococcus aureus* to neutrophils and may also play a role as an adhesion factor, promoting the adhesion of neutrophils to surfaces. Fibronectin mRNA (8.7–8.8 kb) is detected only at low levels in blood neutrophils, but at much higher concentrations (2–13-fold higher) in neutrophils isolated from the synovial fluid of patients with inflammatory joint diseases, such as rheumatoid arthritis and psoriatic arthritis. The active secretion of fibronectin by synovial-fluid neutrophils has been confirmed by affinity chromatography (binding to gelatin-sepharose 4B) and fluorography. These experiments have shown that a 230-kDa protein (fibronectin) and a 95-kDa gelatin-binding protein were actively synthesised and secreted. Fibronectin was only expressed by synovial neutrophils, whereas the 95-kDa protein was constitutively expressed in bloodstream cells. In fact, fibronectin is stored (but not synthesised) in blood neutrophils, and the 3′ untranslated region of its mRNA shares a common consensus sequence with IL-1, TNF and some interferons. Its production by neutrophils in inflamed joints may be important, because it can promote fibroblast adhesion to cartilage–collagen matrices and may thus promote cartilage destruction.

Exposure of cells to elevated temperatures, or *heat shock,* induces the transcription and translation of a set of proteins known as the *heat-shock proteins (HSPs)* or the stress proteins. This event usually occurs with the concomitant inhibition of biosynthesis of other cellular components. The response is believed to be an attempt by the cells to protect themselves from injury, and there is some evidence to indicate that it may also be linked to oxidative stress. Evidence in favour of this idea comes from observations such as the following:

i. Exposure of cells to H_2O_2 induces the formation of HSPs.
ii. Bacteria resistant to oxidative stress can overexpress HSP.
iii. Heat shock protects U-937 cells from oxidative injury.

Exposure of neutrophils to heat shock also results in the synthesis of HSPs and a decrease in their ability to generate reactive oxidants. This is not due to cell death, because the response is reversible and other cellular functions are unaffected. For example, after 20 min exposure at 45 °C, the NADPH oxidase is non-functional, but arachidonic acid release is 73% of control values and PGE_2 formation is unaffected. Oxidase activity returns to normal levels 150 min later. Several HSPs are synthesised (e.g. hsp70 and hsp85 at 41 °C, hsp48 and hsp60–65 at 43 °C). (The 'hsp' numbers refer to the relative molecular masses of the proteins.)

Involved in the control of cellular growth is c-*fos,* whose expression is transiently activated by certain growth factors and tumour promoters and may also be activated during cellular differentiation and by cAMP. In neutrophils, c-*fos* mRNA is expressed at low rates, but transcripts are greatly increased by fMet-Leu-Phe (2–50 nM). This expression is transient (peaking at 15 min and returning to base-line levels by 120 min) and inhibited by pertussis toxin, the protein kinase C inhibitor H7 and the calmodulin inhibitor W7. The protein kinase A inhibitor H9 does not affect mRNA levels. Expression of c-*fos* is also enhanced (two- to sixfold above constitutive levels) by TNF-α, G-CSF, GM-CSF and cycloheximide.

Other cellular components actively synthesised by neutrophils include a 3.1-kb mRNA encoding a serine-rich protein. This transcript encodes a 32-kDa protein that copurifies with the cytochrome b in some preparations of the oxidase from pig neutrophils. The protein becomes phosphorylated during cell activation; its identity, however, is unknown. The biosynthesis of cationic proteins with relative molecular masses of 37 and 57 kDa (identified by one-dimensional PAGE) has also been observed in GM-CSF-treated neutrophils.

7.4 Bibliography

Atkinson, Y. H., Lopez, A. F., Marasco, W. A., Lucas, C. M., Wong, G. G., Burns, G. F., & Vadas, M. A. (1988). Recombinant human granulocyte-macrophage colony-stimulating factor (rH GM-CSF) regulates fMet-Leu-Phe receptors on human neutrophils. *Immunol.* **64**, 519–25.

Beaulieu, A. D., Lang, F., Belles-Isles, M., & Poubelle, P. (1987). Protein biosynthetic activity of polymorphonuclear leukocytes in inflammatory arthropathies: Increased synthesis and release of fibronectin. *J. Rheumatol.* **14**, 656–61.

Bellavite, P., Bazzone, F., Cassatella, M. A., Hunter, K. J., & Bannister, J. V. (1990). Isolation and characterization of a cDNA clone for a novel serine-rich neutrophil protein. *Biochem. Biophys. Res. Commun.* **170**, 915–22.

Bender, J. G., McPhail, L., C., & van Epps, D. E. (1983). Exposure of human neutrophils to chemotactic factors potentiates activation of the respiratory burst enzyme. *J. Immunol.* **130**, 2316–23.

Berton, G., Zeni, L., Cassatella, M. A., & Rossi, F. (1986). Gamma interferon is able to enhance the oxidative metabolism of human neutrophils. *Biochem. Biophys. Res. Commun.* **138**, 1276–82.

Blowers, L. E., Jayson, M. I. V., & Jasani, M. K. (1985). Effect of dexamethasone on polypeptides synthesised in polymorphonuclear leucocytes. *FEBS Lett.* **181**, 362–6.

Blowers, L. E., Jayson, M. I. V., & Jasani, M. K. (1988). Dexamethasone modulated protein synthesis in polymorphonuclear leukocytes: Response in rheumatoid arthritis. *J. Rheumatol.* **15**, 785–90.

Cassatella, M. A., Bazzoni, F., Calzetti, F., Guasparri, I., Rossi, F., & Trinchieri, G. (1991). Interferon-γ transcriptionally modulates the expression of the genes for the high affinity IgG-Fc receptor and the 47-kDa cytosolic component of NADPH oxidase in human polymorphonuclear leukocytes. *J. Biol. Chem.* **266**, 22079–82.

Cassatella, M. A., Bazzoni, F., Flynn, R. M., Dusi, S., Trinchieri, G., & Rossi, F. (1990). Molecular basis of interferon-γ and lipopolysaccharide enhancement of phagocyte respiratory burst capability. *J. Biol. Chem.* **265**, 20241–6.

Cassatella, M. A., Flynn, R. M., Amezaga, M. A., Bazzoni, F., Vicentini, F., & Trinchieri, G. (1990). Interferon gamma induces in human neutrophils and macrophages expression of the mRNA for the high affinity receptor for monomeric IgG (FcγR-I or CD64). *Biochem. Biophys. Res. Commun.* **170**, 582–8.

Cassatella, M. A., Hartman, L., Perussia, B., & Trinchieri, G. (1989). Tumor necrosis factor and immune interferon synergistically induce cytochrome b_{-245} heavy-chain gene expression and nicotinamide-adenine dinucleotide phosphate hydrogenase oxidase in human leukemic myeloid cells. *J. Clin. Invest.* **83**, 1570–9.

Cicco, N. A., Lindemann, A., Content, J., Vandenbussche, P., Lubbert, M., Gauss, J., Mertelsmann, R., & Herrmann, F. (1990). Inducible production of Interleukin-6 by human polymorphonuclear neutrophils: Role of granulocyte-macrophage colony-stimulating factor and tumor necrosis factor-alpha. *Blood* **75**, 2049–52.

Cline, M. J. (1966). Phagocytosis and synthesis of ribonucleic acid in human granulocytes. *Nature* **212**, 1431–3.

Colotta, F., Wang, J. M., Polentarutti, N., & Mantovani, A. (1987). Expression of c-*fos* protooncogene in normal human peripheral blood granulocytes. *J. Exp. Med.* **165**, 1224–9.

Dewald, B., & Baggiolini, M. (1985). Activation of NADPH oxidase in human neutrophils. Synergism between fMLP and the neutrophil products PAF and LTB_4. *Biochem. Biophys. Res. Commun.* **128**, 297–304.

Djeu, J. Y., Serbousek, D., & Blanchard, D. K. (1990). Release of tumor necrosis factor by human polymorphonuclear leukocytes. *Blood* **76**, 1405–9.

Edwards, S. W., Holden, C. S., Humphreys, J. M., & Hart, C. A. (1989). Granulocyte–macrophage colony-stimulating factor (GM-CSF) primes the respiratory burst and stimulates protein biosynthesis in human neutrophils. *FEBS Lett.* **256**, 62–6.

Edwards, S. W., Say, J. E., & Hughes, V. (1988). Gamma interferon enhances the killing of *Staphylococcus aureus* by human neutrophils. *J. Gen. Microbiol.* **134**, 37–42.

Edwards, S. W., Watson, F., MacLeod, R., & Davies, J. M. (1990). Receptor expression and oxidase activity in human neutrophils: Regulation by granulocyte-macrophage colony-stimulating factor and dependence upon protein biosynthesis. *Biosci. Rep.* **10**, 393–401.

Eid, N. S., Kravath, R. E., & Lanks, K. W. (1987). Heat-shock protein synthesis by human polymorphonuclear cells. *J. Exp. Med.* **165**, 1448–52.

Granelli-Piperno, A., Vassalli, J.-D., & Reich, E. (1977). Secretion of plasminogen activator by human polymorphonuclear leukocytes. *J. Exp. Med.* **146**, 1693–1706.

Granelli-Piperno, A., Vassalli, J.-D., & Reich, E. (1979). RNA and protein synthesis in human peripheral blood polymorphonuclear leukocytes. *J. Exp. Med.* **149**, 284–9.

Hughes, V., Humphreys, J. M., & Edwards, S. W. (1987). Protein synthesis is activated in primed neutrophils: a possible role in inflammation. *Biosci. Rep.* **7**, 881–9.

Huizinga, T. W. J., Van Der Schoot, C. E., Roos, D., & Weening, R. S. (1991). Induction of neutrophil Fc-γ receptor 1 expression can be used as a marker for biologic activity of recombinant interferon-γ in vivo. *Blood* **77**, 2088–90.

Humphreys, J. M., Hughes, V., & Edwards, S. W. (1989). Stimulation of protein synthesis in human neutrophils by γ-interferon. *Biochem. Pharmacol.* **38**, 1241–6.

Jack, R. M., & Fearon, D. T. (1988). Selective synthesis of mRNA and proteins by human peripheral blood neutrophils. *J. Immunol.* **140**, 4286–93.

Jost, C. R., Huizinga, T. W. J., de Goede, R., Fransen, J. A. M., Tetteroo, P. A. T., Daha, M. R., & Ginsel, L. A. (1990). Intracellular localization and de novo synthesis of FcRIII in human neutrophil granulocytes. *Blood* **75**, 144–51.

Klebanoff, S. J., Vadas, M. A., Harlan, J. M., Sparks, L. H., Gamble, J. R., Agosti, J. M., & Waltersdorph, A. M. (1986). Stimulation of neutrophils by tumor necrosis factor. *J. Immunol.* **136**, 4220–5.

Kuijpers, T. W., Tool, A. T. J., van der Schoot, C. E., Ginsel, L. A., Onderwater, J. J. M., Roos, D., & Verhoeven, A. J. (1991). Membrane surface antigen expression on neutrophils: a reappraisal of the use of surface markers for neutrophil activation. *Blood* **78**, 1105–11.

La Fleur, M., Beaulieu, A. D., Kreis, C., & Poubelle, P. (1987). Fibronectin gene expression in polymorphonuclear leukocytes. *J. Biol. Chem.* **262**, 2111–15.

Lindemann, A., Riedel, D., Oster, W., Meuer, S. C., Blohm, D., Mertelsmann, R. H., & Herrmann, F. (1988). Granulocyte/macrophage colony-stimulating factor induces interleukin 1 production by human polymorphonuclear neutrophils. *J. Immunol.* **140**, 837–9.

Lindemann, A., Riedel, D., Oster, W., Ziegler-Heitbrock, H. W. L., Mertelsmann, R., & Herrmann, F. (1989). Granulocyte-macrophage colony-stimulating factor induces cytokine secretion by human polymorphonuclear leukocytes. *J. Clin. Invest.* **83**, 1308–12.

Lloyd, A. R., & Oppenheim, J. J. (1992). Poly's lament: The neglected role of the polymorphonuclear neutrophil in the afferent limb of the immune response. *Immunol. Today* **13**, 169–72.

McCall, C. E., Bass, D. A., DeChatelet, L. R., Link, A. S. J., & Mann, M. (1979). In vitro responses of human neutrophils to N-formyl-methionyl-leucyl-phenylalanine: correlation with effects of acute bacterial infection. *J. Infect. Dis.* **140**, 277–86.

McCall, C. E., DeChatelet, L. R., Cooper, M. R., & Shannon, C. (1973). Human toxic neutrophils: III. Metabolic characteristics. *J. Infect. Dis.* **127**, 26–33.

McColl, S. R., Paquin, R., Menard, C., & Beaulieu, A. D. (1992). Human neutrophils produce high levels of the interleukin 1 receptor antagonist in response to granulocyte/macrophage colony-stimulating factor and tumor necrosis factor α. *J. Exp. Med.* **176**, 593–8.

Maridonneau-Parini, I., Clerc, J., & Polla, B. (1988). Heat shock inhibits NADPH oxidase in human neutrophils. *Biochem. Biophys. Res. Commun.* **154**, 179–86.

Marino, J. A., Davis, A. H., & Spagnuolo, P. J. (1987). Fibronectin is stored but not synthesized in mature human peripheral blood granulocytes. *Biochem. Biophys. Res. Commun.* **146**, 1132–8.

Marucha, P. T., Zeff, R. A., & Kreutzer, D. (1990). Cytokine regulation of IL-1β gene expression in the human polymorphonuclear leukocyte. *J. Immunol.* **145**, 2932–7.

Mollinedo, F., Vaquerizo, M. J., & Naranjo, J. R. (1991). Expression of c-*jun*, *jun*-B and *jun*-D proto-oncogenes in human peripheral-blood granulocytes. *Biochem. J.* **273**, 477–9.

Neuman, E., Huleatt, J. W., & Jack, R. M. (1990). Granulocyte-macrophage colony-stimulating factor increases synthesis and expression of CR1 and CR3 by human peripheral blood neutrophils. *J. Immunol.* **145**, 3325–32.

Newburger, P. E., Dai, Q., & Whitney, C. (1991). In vitro regulation of human phagocyte cytochrome b heavy and light chain gene expression by bacterial lipopolysaccharide and recombinant human cytokines. *J. Biol. Chem.* **266**, 16171–7.

Perussia, B., Dayton, E. T., Lazarus, R., Fanning, V., & Trinchieri, G. (1983). Immune interferon induces the receptor for monomeric IgG1 on human monocytic and myeloid cells. *J. Exp. Med.* **158**, 1092–113.

Petroni, K. C., Shen, L., & Guyre, P. M. (1988). Modulation of human polymorphonuclear leukocyte IgG Fc receptors and Fc receptor-mediated functions by IFN-γ and glucocorticoids. *J. Immunol.* **140**, 3467–72.

Repp, R., Valerius, T., Sendler, A., Gramatzki, M., Iro, H., Kalden, J. R., & Platzer, E. (1991). Neutrophils express the high affinity receptor for IgG (FcγRI, CD64) after in vivo application of recombinant human granulocyte colony-stimulating factor. *Blood* **78**, 885–9.

Shirafuji, N., Matsuda, S., Ogura, H., Tani, K., Kodo, H., Ozawa, K., Nagata, S., Asano, S., & Takaku, F. (1990). Granulocyte colony-stimulating factor stimulates human mature neutrophilic granulocytes to produce interferon-α. *Blood* **75**, 17–19.

Steinbeck, M. J., Webb, D. S. A., & Roth, J. A. (1989). Role for arachidonic acid metabolism and protein synthesis in recombinant bovine interferon-γ-induced activation of bovine neutrophils. *J. Leuk. Biol.* **46**, 450–60.

Strieter, R. M., Kasahara, K., Allen, R. M., Standiford, T. J., Rolfe, M. W., Becker, F. S., Chensue, S. W., & Kunkel, S. L. (1992). Cytokine-induced neutrophil-derived interleukin-8. *Am. J. Pathol.* **141**, 397–407.

Tiku, K., Tiku, M. L., & Skosey, J. L. (1986). Interleukin 1 production by human polymorphonuclear neutrophils. *J. Immunol.* **136**, 3677–85.

Tosi, M. F., & Zakem, H. (1992). Surface expression of Fcγreceptor III (CD16) on chemoattractant-stimulated neutrophils is determined by both surface shedding and translocation from intracellular storage compartments. *J. Clin. Invest.* **90**, 462–70.

Watson, F., Robinson, J. J., & Edwards, S. W. (1992). Neutrophil function in whole blood and after purification: Changes in receptor expression, oxidase activity and responsiveness to cytokines. *Biosci. Rep.* **12**, 123–33.

Weisbart, R. H., Golde, D. W., Clark, S. C., Wong, G. G., & Gasson, J. C. (1985). Human granulocyte-macrophage colony-stimulating factor is a neutrophil activator. *Nature* **314**, 361–3.

Weisbart, R. H., Golde, D. W., & Gasson, J. C. (1986). Biosynthetic human GM-CSF modulates the number and affinity of neutrophil f-Met-Leu-Phe receptors. *J. Immunol.* **137**, 3584–7.

Yamazaki, M., Ikenami, M., & Sugiyama, T. (1989). Cytotoxin from polymorphonuclear leukocytes and inflammatory ascitic fluids. *Brit. J. Cancer* **59**, 353–5.

8
Disorders of neutrophil function

8.1 Introduction

The crucial role played by the neutrophil in protecting the host against bacterial and fungal infections is highlighted in patients with defects in neutrophil function. These individuals, either with low numbers of circulating neutrophils (i.e. *neutropenias*) or with specific defects in one or more key processes, are predisposed to life-threatening infections. Indeed, at neutrophil counts of <1000/μl blood, individuals are at risk from infections, with the risk of infection inversely proportional to the neutrophil count. Neutropenias may be due to disease, such as congenital neutropenia, aplastic anaemias, agranulocytosis, acute myeloid leukaemia and the myelodysplastic syndromes. In such cases, CSFs (e.g. GM-CSF, G-CSF) have been used clinically to improve the circulating neutrophil count. Whilst initial results have shown some promise, such CSF therapy is not without its drawbacks. Some malignant cells may be stimulated by CSF therapy, and accumulation of neutrophils in the lungs may occur. Circulating levels of CSFs may also increase the expression of adhesion molecules on the surface of circulating cells, which may result in increased adhesion of neutrophils to endothelial cells or in increased neutrophil–neutrophil or –platelet interactions.

Neutropenias may also arise as a side effect or deliberate consequence of therapy. For example, some drugs used in the treatment of inflammatory disorders are immunosuppressive, and if these decrease the number of circulating neutrophils to below the critical threshold level, then susceptibility to infection may result. During chemotherapy for the treatment of solid tumours, an inevitable consequence of cytotoxic therapy is that the bone marrow will be destroyed by the drugs; thus, patients will have a considerable risk of infection during this induction period. Similarly, during the treatment of haematological disorders (e.g. leukemias and myelodysplastic syndromes), the aim of therapy is to attack the bone marrow so as to destroy the malignant clone(s). The period of neutropenia that follows such cytotoxic therapy (even if it is followed by bone-marrow transplantation) is thus when patients are at greatest risk from infections. Again, the use of CSFs can decrease this period of posttherapeutic neutropenia, and this may lower the numbers of infective episodes and decrease hospitalisation times. However, more clinical trials are needed to establish the optimal therapeutic regimens and to identify which patients are most likely to receive maximal benefit from CSF therapy.

Other neutrophil disorders are related to specific defects in key neutrophil enzymes or processes; these are described in detail in the following sections. Many of these diseases are rare and have only recently been discovered. The survival of such patients is possible today only because these specific defects are now more readily identified, and because infections can be carefully managed using a wide range of potent antibiotics. In some patients (e.g. those with CGD), cytokine therapy may again prove useful.

On the other hand, there are a number of human diseases associated with an overactivity of neutrophil function. Many facets of the neutrophil antimicrobial arsenal, such as reactive oxygen metabolites and proteases, can attack host tissues as effectively as they can attack microbial targets. For this reason, activation of neutrophils under physiological conditions is carefully regulated and damage restricted for the following reasons:

i. Activation occurs via occupancy of specific receptors on the plasma membrane, and a two-step activation system (i.e. a priming agent and an activating agent) controls cytotoxic activation (§7.2.1).
ii. Many cells and tissues possess antioxidant protection systems and protease inhibitors.

However, it is now recognised that neutrophils can contribute to host-tissue damage if they are activated to secrete reactive oxidants and granule enzymes, and if the local concentrations of anti-oxidants and protease inhibitors within the tissue are low or defective. Thus, inappropriate neutrophil activation leading to host-tissue damage has been implicated in reperfusion injury, Crohn's disease, adult respiratory distress syndrome (ARDS) and rheumatoid arthritis. In these conditions, it is envisaged that neutrophils accumulate in tissues and become inappropriately activated to secrete their cytotoxic products, which then initiate or contribute to host-tissue damage.

8.2 Chronic granulomatous disease

Chronic granulomatous disease (CGD) is a recently-described disorder, first reported some 40 or so years ago as 'fatal granulomatosis of childhood', 'progressive septic granulomatosis' or 'chronic familial granulomatosis'. Descriptions of the disease emphasised a high incidence of severe and recurrent infections affecting lymph nodes, skin, lung and liver. The pathogens most frequently reported are *Staphylococcus aureus,* Gram-negative enteric bacteria and fungi. Histological studies reveal multiple abscesses, granulomata and characteristic lipid-filled histiocytes. Patients with CGD present early in life (sometimes within weeks of birth, but usually within the first year), although a less severe form of the disease may be diagnosed in young adults. Apart from increased susceptibility to bacterial and fungal infections, other complications (e.g. glomerulonephritis, polyarthritis and pulmonary fibrosis) are often reported. Until recent advances in the management of this disease, death usually occurred within the first ten years of life.

By the early 1960s it was appreciated that the oxidative metabolism of neutrophils during phagocytosis was unusual in that it was not associated with mitochondrial respiration. It was then discovered that H_2O_2 was generated during the respiratory burst, and this product of O_2 reduction was utilised in pathogen killing during phagocytosis. In 1966 Holmes and colleagues showed that neutrophils from patients with CGD could phagocytose bacteria normally but could not kill certain micro-organisms. A year later this group (Holmes, Page & Good, 1967) showed that neutrophils from CGD patients could not mount a respiratory burst during phagocytosis. This was a major discovery because it pin-pointed a biochemical defect underlying a life-threatening disease. Furthermore, this link between impaired bacterial killing and a respiratory burst defect was an invaluable guide to later studies aimed at the identification and characterisation of the respiratory burst enzyme: if a putative component is absent or defective in CGD neutrophils, then it is likely to be intimately associated with the respiratory-burst process. In this way, the CGD neutrophil has proved to be extremely important in the characterisation of the respiratory-burst enzyme and in evaluating its role in neutrophil function during infections.

8.2.1 Infections in CGD

CGD patients present with fungal and bacterial infections (e.g. *S. aureus*) and Gram-negative enteric bacteria (e.g. *Escherichia coli, Serratia marcescens, Proteus* and *Salmonella*). Only about 10% of fatal infectious episodes are attributed to *S. aureus;* the overwhelming majority (over 80%) are due to Gram-negative organisms. Interestingly, organisms that secrete H_2O_2 (e.g. catalase-negative organisms), such as *Streptococcus, Pneumococcus* and *Lactobacillus,* are not major pathogens in CGD patients. Presumably, this is because myeloperoxidase (which is present at normal levels in CGD patients) utilises this secreted H_2O_2 to produce toxic oxidants (see §5.4.1). Fungal infections by *Aspergillus* strains, *Candida albicans, Torulopsis glabrata* and *Hansenula anomala* are reported in about 20% of CGD patients. About half of these patients with fungal infections usually die from fungal pneumonia or disseminated fungal disease, whereas antifungal therapy or surgical removal of infected tissues protects the remainder.

Because O_2-independent killing mechanisms (§2.6) are present at normal levels in CGD neutrophils, some microbial pathogens are killed at normal rates. Furthermore, whilst reactive oxidants generated during the respiratory burst of normal neutrophils undoubtedly play an important role in microbial killing, the NADPH oxidase also plays a role in the regulation of intra-phagolysosomal pH. In normal neutrophils, the pH of the phagolysosome increases to 7.8–8.0 and then gradually becomes acidic (pH ≈ 6). In CGD neutrophils, this initial alkalisation does not occur. Because the efficacy of killing of some pathogens by O_2-independent proteins is pH dependent, the inability of CGD neutrophils to maintain adequate intraphagolysosomal pH control may also contribute to their inability to kill pathogens.

8.2.3 Incidence and genetics

The incidence of CGD is rare – about 1 : 300 000. As of the late 1980s, some 420 cases had been reported; almost all of these had suffered bacterial (and some fungal) pneumonias, with about three-quarters having dermatitis. The disease is about six times more common in males than in females, indicating an X-chromosome-linked transmission. It is often seen in brothers and sons of the same mother (even when there may be different fathers), and females of the family may be carriers. Autosomal recessive transmission of the disease also occurs, but this is often less severe than the X-linked form and is usually more diverse in terms of disease severity. Autosomal recessive CGD is often diagnosed later in life, and although the incidence of infection is frequent, episodes are usually less severe. Autosomal recessive CGD patients are susceptible to the same pathogens as those with X-linked CGD, and the same organs are targeted. Most female carriers are healthy and do not suffer recurrent infections, but a subpopulation of their phagocytes may be defective, as determined by the nitroblue tetrazolium (NBT) slide assay.

8.2.4 Biochemical and genetic defects in CGD

In 1978, Segal, Jones and colleagues published a landmark paper describing a novel cytochrome b within phagosomes of neutrophils that was not detected in neutrophils from patients with X-linked CGD. This cytochrome had previously been reported in horse and rabbit neutrophils by Hattori, Ohta and colleagues and Shinagawa and colleagues in Japan in the mid-1960s (see references in Segal 1988); however, the new finding was extremely important because this component was likely to be involved in the electron-transport processes leading to the reduction of O_2 to form O_2^-. Thus, Segal and Jones's observation opened the way to identify the molecular defects in CGD and also the processes by which normal neutrophils generate reactive oxidants. (The properties of this cytochrome are discussed in detail in §5.3.2.1.)

Extensive studies into the association of this cytochrome b with CGD neutrophils were performed by Segal and Jones, and by other workers, in the late 1970s and early 1980s. The cytochrome was completely absent (as determined by the absence of a distinctive absorption spectrum in spectroscopic studies) in almost all cases of X-linked CGD, but present at decreased levels in female relatives of these patients. In almost all cases of autosomal recessive CGD, the cytochrome was present but non-functional, in that it did not become reduced upon cellular activation. This indicated both the heterogeneous nature of the disease and also that some other biochemical defect was responsible for impaired function in these patients. Hence, the search was on for other components of the NADPH oxidase.

Further elucidation of the defects in CGD neutrophils awaited the purification of the cytochrome so that molecular (antibody and DNA) probes could be obtained and used to probe these defects. Yet the cytochrome was proving extremely difficult to purify, and many dubious descriptions of 'purified' preparations appeared in the literature. In the absence of antibody probes or amino-acid-sequence data to allow the synthesis of oligonucleotide probes, cDNA libraries could not be screened to isolate cDNA clones for the cytochrome; hence, information of its molecular structure could not be easily obtained. Instead, the gene for X-linked CGD was cloned by the elegant technique of *reverse genetics.* This technique was possible because Orkin and colleagues (Royer-Pokora et al. 1986) were studying patients with Duchenne muscular dystrophy (DMD) and identified patients with deletions within chromosome Xp21. One of these patients had an unusual deletion in this part of the chromosome that resulted in DMD, CGD, retinitis pigmentosis and MacLeod phenotype. Another patient had a deletion in this region that resulted in DMD and CGD. Genomic DNA from these patients failed to hybridise in Southern blots with a series of cDNA probes specific to regions within Xp21. Thus, the CGD gene was located to within 0.1% of the human genome (i.e. within a region of about 3000 kb of DNA).

Orkin's group then constructed a cDNA from mRNA isolated from HL-60 cells that had been induced to differentiate into mature cells by the addition of dimethyl formamide (DMF). During this differentiation, HL-60 cells acquire the ability to generate reactive oxidants, and the cytochrome b becomes expressed. Subtractive hybridisation was then performed to enrich the cDNA for sequences (which included the CGD gene) that were specific to phagocytic cells. This was done by hybridising the pooled cDNA to mRNA isolated from Epstein–Barr virus (EBV)-transformed B-cell lines that had been developed from the patients with DMD and CGD. Thus, the non-hybridised cDNA molecules from differentiated HL-60 cells (about 500) represented phagocyte-specific transcripts predicted to include transcripts for the CGD gene. The cDNA molecules were then hybridised to a collection of bacteriophage clones derived from Xp21 (originally used in an attempt to define the DMD locus), which represented about 250 kb of the Xp21 locus. Two of these bacteriophage cDNA clones hybridised with the subtracted cDNA molecules and were therefore predicted to represent phagocyte-specific genomic regions (probably the CGD gene); these clones were then used to screen a cDNA library of differentiated HL-60 cells.

After screening this library, a series of cDNA clones that spanned a 5-kb transcript was isolated. Northern analyses indicated that mRNA hybridising to these clones was absent from monocytes from five out of six X-linked patients. DNA sequencing of the putative CGD cDNA clones (X-CGD gene) predicted an amino acid sequence that showed no sequence homology to known b-cytochromes and contained no putative haem-binding site; the predicted size of the protein was 54 kDa. Originally, it was proposed that this protein was not the cytochrome b; however, at about this time Segal and colleagues (Teahan et al. 1987) had purified the cytochrome b, separated the two subunits and obtained sequence data on the 42 NH_2-terminal amino acids of the β chain. Analysis of this amino-acid-sequence data with the cDNA-sequence data of the X-CGD gene revealed a DNA sequencing error in the original data. Thus, the X-linked CGD gene was unequivocally identified as the result of a defect in the β-chain of the b cytochrome, as had been predicted by the biochemical studies.

Thus, it has been shown that, in the majority of X-linked CGD patients, the abnormality is due to the failure to transcribe the mRNA encoding the large (β) subunit of the b cytochrome. In these patients, both the heavy β-chain and the light α-chain are absent, even though the molecular defect appears to be restricted to the heavy chain; thus, expression of the heavy chain is somehow necessary for the expression/translation/stabilisation of the light chain. In some variant forms of X-linked CGD, the cytochrome b is present; such patients are termed X^+-CGD. The mRNA from one such patient has been converted to cDNA, amplified using the polymerase chain reaction and sequenced. It was found that a single nucleotide change, a C → A transversion, had occurred; this predicted a Pro → His substitution in amino acid residue 415 of the heavy chain protein. This mutation presumably renders the cytochrome inactive, so that while it is present in normal amounts, it cannot function in O_2^- generation. Several other mutations similar to this, but involving other base changes or deletions, have now been reported in other types of X^+-CGD.

The second major breakthrough in understanding the defect in CGD neutrophils came through the development of assays in which the NADPH oxidase can be activated in a cell-free system in vitro (§5.3.2.3). In these systems, activation of the oxidase can be achieved by the addition of cytoplasm to plasma membranes in the presence of NADPH and arachidonic acid (or SDS or related substances). Interestingly, the oxidase cannot be activated in these cell-free systems using extracts from CGD neutrophils; however, cytosol and plasma membranes from normal and CGD neutrophils may be mixed, and in most cases activity is restored if the correct mixing pattern is used. For example, as may be predicted, in X-linked CGD it is the membranes that are defective (because the cytochrome b is deficient), whereas in autosomal recessive CGD the cytosol is defective in the cell-free system.

Nunoi and co-workers (1988) fractionated neutrophil cytoplasm by Mono Q anion-exchange chromatography and obtained three fractions (NCF-1, -2 and -3) that were active in the assembly of the oxidase. Independently, Volpp and colleagues (Volpp, Nauseef & Clark, 1988) prepared antiserum from cytosolic factors that eluted from a GTP-affinity column, and this antiserum (B1) recognised cytoplasmic factors of relative molecular masses 47 kDa and 66 kDa. It was later shown by this group that these cytosolic factors translocated to the plasma membrane during activation. NCF-1 was shown to contain the 47-kDa protein and NCF-2 the 66-kDa protein. Analysis of the defect in the cytosol of autosomal recessive CGD patients revealed that most of these (88%) lacked the 47-kDa protein (p47-*phox*), whereas the remainder lacked the 66-kDa protein (p66-*phox*). Both of these components have now been cloned and recombinant proteins expressed. Interestingly, in the cell-free system, recombinant p47-*phox* and p66-*phox* can restore oxidase activity of the cytosol from autosomal recessive CGD patients who lack these components.

The importance of p47-*phox,* which is a phosphoprotein, was appreciated after the discovery that a protein of this relative molecular mass fails to become phosphorylated after activation of neutrophils from most patients with autosomal recessive CGD. In normal cells, this protein is phosphorylated by protein kinase C; it is initially located in the cytoplasm, but translocates to the plasma membrane during activation. In patients with X-linked CGD, this translocation fails to occur (even though the protein is phosphorylated), indicating that p47-*phox* may normally bind this to the cytochrome during oxidase assembly and activation.

Thus, at present, there are four major types of CGD:

i. X^--CGD (X-linked, cytochrome b absent);
ii. X^+-CGD (X-linked, cytochrome b present but abnormal);
iii. autosomal recessive CGD, defective in p47-*phox;*
iv. autosomal recessive CGD, defective in p66-*phox*.

8.2.5 Treatment of CGD

The conventional treatment for CGD is good management of infections. This includes the identification of the causative pathogen, determination of its antibiotic sensitivity and administration of that antibiotic, usually at high dosage and for long periods. This approach is usually successful, but sometimes surgery is also required. Furthermore, because phagocytosis is normal in CGD neutrophils, and because some antibiotics do not permeate neutrophils, internalised pathogens in CGD neutrophils may be protected within phagolysosomes. For this reason, rifampin, which can kill *S. aureus* within neutrophil phagosomes, is often administered. Trimethoprim/sulphamethazole (TMP/SMX) is useful because it can reach high intracellular concentrations, and this combination is often used prophylactically. Improved use of antibiotics has resulted in a marked improvement in survival of CGD patients. For example, the average age of death from CGD was about 4 years during 1955–65, whereas by 1975–82 this had risen to about 12 years. Prophylactic administration of TMP/SMX significantly decreases the incidence of infections in CGD patients with *S. aureus* but does not improve the incidence of infections with Gram-negative enteric organisms or fungi.

In the 1970s several attempts were made to improve the efficacy of pathogen killing by CGD neutrophils via the restoration of H_2O_2, which could then be available for oxidative killing. The enzyme glucose oxidase reduces O_2 to H_2O_2, and in vitro experiments showed that if the enzyme was linked to latex particles and then phagocytosed by CGD neutrophils, their ability to kill *S. aureus* and *S. marcescens* was enhanced. Because latex particles cannot be administered to patients, an alternative approach – introducing glucose oxidase into Ig-coated liposomes – was attempted; but this led to only modest improvements in killing.

Perhaps the most striking improvements in CGD function has come from the use of γ-interferon, one of the major macrophage-activating factors secreted by activated T cells (§3.5.2). This cytokine enhances many macrophage functions, including their ability to generate reactive oxidants. Berton and colleagues (1986) have shown that γ-interferon also enhances the ability of normal neutrophils to generate reactive oxidants, and that this ability is dependent upon activated de novo biosynthesis (§7.3.5). Analysis of mRNA levels for the heavy chain of cytochrome b has shown that γ-interferon treatment (up to 24 h) can increase the number of transcripts for this gene in normal blood monocytes and neutrophils, largely due to an increase in the rate of transcription.

These observations were taken further by examining whether γ-interferon treatment could up-regulate NADPH oxidase function in CGD neutrophils and monocytes. It was found that 12 out of 13 patients with autosomal recessive CGD had increased oxidase activity upon γ-interferon exposure; the only patient not responding was the one devoid of the b cytochrome. In X-linked CGD, 9 of 13 showed no improvement, whereas 3 showed some improvement and 1 had oxidase activity increased to near-normal levels. Patients with 'atypical' X-linked CGD (i.e. low oxidase activity and some cytochrome b) appear to respond best to γ-interferon treatment. Interferons-α and -β are without affect. This enhancement of oxidase function (detected by NBT slide tests and O_2^- production) is due, at least in part, to increased levels of mRNA for the heavy chain of cytochrome b. In the absence of γ-interferon treatment, monocyte-derived macrophages have extremely low or undetectable levels of mRNA for the cytochrome b heavy chain; however, this is increased about fivefold (to about 5% of normal) after γ-interferon treatment.

In vivo administration of γ-interferon to 'atypical' X-CGD patients also led to improvements in monocyte and neutrophil respiratory-burst activity and killing, and in some cases spectroscopically-detectable cytochrome b has been observed (occasionally present at up to 50% of normal levels). This improvement followed two subcutaneous doses (0.1 mg/m^2) on consecutive days and lasted for up to a month. Because TNF can act synergistically with γ-interferon in increasing the expression of the heavy chain of cytochrome b, perhaps the combined use of γ-interferon with TNF or some other cytokine(s) will prove even more beneficial.

Bone-marrow transplantation from siblings or compatible non-related donors has also proved useful in CGD treatment. The possibility also exists that gene transfer and somatic-cell therapy may be used in the near future. Furthermore, the introduction and expression of NADPH oxidase genes (defective in CGD patients) into haematopoietic stem cells and subsequent transplantation may provide a new way of treating CGD. B lymphocytes can be immortalised by infection with Epstein–Barr virus (EBV), and these cells produce O_2^- via an NADPH oxidase apparently similar to that of phagocytes. Rates of O_2^- production, however, are considerably lower than those observed in neutrophils. B-cell lines from CGD patients, established after EBV infection, fail to generate O_2^- because of a lack of the oxidase component that is also defective in the neutrophils from these patients. Recently, Segal's group (Thrasher et al., 1992) has developed B-cell lines from autosomal recessive CGD patients lacking p47-*phox*. These cell lines were then transfected with retrovirally-encoded p47-*phox*, and this genetic transfer was successful in restoring O_2^- generating ability to these cell lines. If such an approach can be used to transfect myeloid precursor cells, then gene-transfer therapy for CGD remains a distinct possibility.

8.3 Myeloperoxidase deficiency

Myeloperoxidase is an extremely potent, antimicrobial protein that is present in neutrophils at up to 5% of the total cell protein. Its role in the killing of a wide range of bacteria, fungi, viruses, protozoa and mammalian cells (e.g. tumour cells) is well established from in vitro studies. It also plays an important role in the inactivation of toxins and the activation of latent proteases, as well as in other functions described in section 5.4.1. In view of this apparent central role in neutrophil function during host defence, one would think that any deficiencies in this enzyme would have disastrous consequences on the ability of the host to combat infections. Until the early 1980s, this key role for myeloperoxidase in host protection seemed substantiated by the extremely low incidence of reports of patients with deficiencies of this enzyme. Indeed, up to this time, only 15 cases from 12 families had been reported worldwide. Sometimes these patients were asymptomatic but often suffered *Candida* infections, particularly if their myeloperoxidase deficiency was also associated with diabetes mellitus.

It was therefore somewhat surprising when, in the early 1980s, a number of independent reports worldwide indicated that myeloperoxidase deficiencies may be quite common in the population of apparently-healthy individuals. These observations came from the use of the Hemalog D or Technicon H6000, flow-cytochemical systems intended to automate differential white blood counts. Part of this system uses a peroxidase stain, and hence neutro-

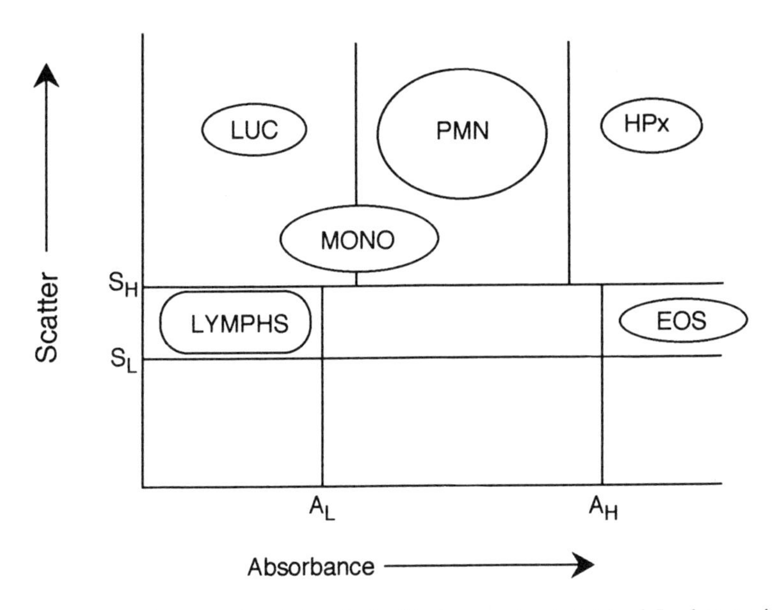

Figure 8.1. Flow-cytochemical analysis of leukocytes in whole blood. Leukocytes in whole blood may be stained either for peroxidase (using 4-chloro-1, naphthol plus H_2O_2), which stains monocytes, neutrophils and eosinophils, or for esterase (using α-naphthol butyrate), which stains monocytes. The diagram represents the scatter and peroxidase-staining channel. *Abbreviations:* LYMPHS, lymphocytes (peroxidase-negative); MONO, monocytes; PMN, neutrophils; EOS, eosinophils; LUC, large, unstained cells; HPx, high peroxidase-staining cells; S_H, S_L, high and low scatter; A_H, A_L, high and low absorbance.

phils are counted in a gate of large, peroxidase-positive cells (Fig. 8.1). The absence of this population may be indicative of a genuine absence of neutrophils (i.e. neutropenia), but conventional microscopic analysis of stained blood films indicated in some individuals the presence of normal numbers of neutrophils. Hence, the results of the flow-cytochemical analyses indicated normal numbers of myeloperoxidase-deficient neutrophils, a conclusion confirmed by conventional assays for the presence of myeloperoxidase. Incidences of partial myeloperoxidase deficiencies were thus reported to be as high as 1:2000 in some populations, with complete deficiency as high as 1:10000.

It was also discovered that myeloperoxidase is not present in the neutrophils of chickens, but this deficiency is not normally associated with an increased prevalence of infections. Thus, if many healthy individuals (or chicken neutrophils) possess low or zero levels of myeloperoxidase but have no increased risk of infection, what really is the importance of the myeloperoxidase system in neutrophil function during host protection?

The significance of partial myeloperoxidase deficiencies is difficult to evaluate because the enzyme is present in such vast quantities in normal neutrophils. Thus, even if there is a substantial decrease in the total amount of enzyme present per cell, there may still be sufficient enzyme to function efficiently during phagocytosis. Coupled with the fact that neutrophils possess a considerable degree of 'overkill' by being able to produce and deliver large quantities of toxic products to the targeted organism, partial myeloperoxidase deficiency may be without effect on the overall potency of the myeloperoxidase–H_2O_2 system.

Analysis of neutrophil function in patients with a *total* myeloperoxidase deficiency has revealed that whilst these cells kill some targets as efficiently as control neutrophils, other targets as not killed as effectively. For example, killing of organisms such as *Staphylococcus aureus, Escherichia coli* and *Serratia marscesens* is retarded in vitro, being slower than normal over the initial incubation, but becoming normal 2–3 h after phagocytosis (Fig. 8.2). Furthermore, candidacidal activity is absent in these deficient neutrophils, as judged by their inability to kill *Candida albicans, C. krusei* and *C. tropicana*. This suggests that the myeloperoxidase system may be important for the killing of *Candida* spp. or for the initial, rapid killing of certain organisms, whereas other myeloperoxidase-independent systems can kill such pathogens, albeit at a slower rate. Furthermore, myeloperoxidase-deficient or -inhibited neutrophils generate more O_2^- and H_2O_2 during the respiratory burst than do control cells, because a product of this enzyme self-regulates the duration of reactive oxidant production. Some experimental evidence favours the idea that myeloperoxidase activity may down-regulate ·OH production by decreasing the availability of H_2O_2 for the metal-catalysed Haber–Weiss reaction (§5.2.4). Conversely, during myeloperoxidase deficiency, because of the increased availability of H_2O_2 in the absence of myeloperoxidase activity, more O_2^- and H_2O_2 are available, which may then promote ·OH formation. However, some reports doubt the possibility that any ·OH is generated by activated neutrophils under any circumstances. In any case, an increased accumulation of H_2O_2 because of its lack of utilisation by myeloperoxidase may result in levels reaching a sufficient threshold such that H_2O_2-dependent killing processes are effective. Furthermore, phagocytosis of IgG- and C3b-coated bacteria by myeloperoxidase-deficient neutrophils may be more efficient, because the myeloperoxidase–H_2O_2 system can adversely affect the function of surface IgG and C3b receptors.

Thus, in vitro evidence suggests that if myeloperoxidase is completely absent from neutrophils (or else its activity is inhibited), then certain bacteria (e.g. *S. aureus*) are killed more slowly, whereas others (e.g. *Candida*)

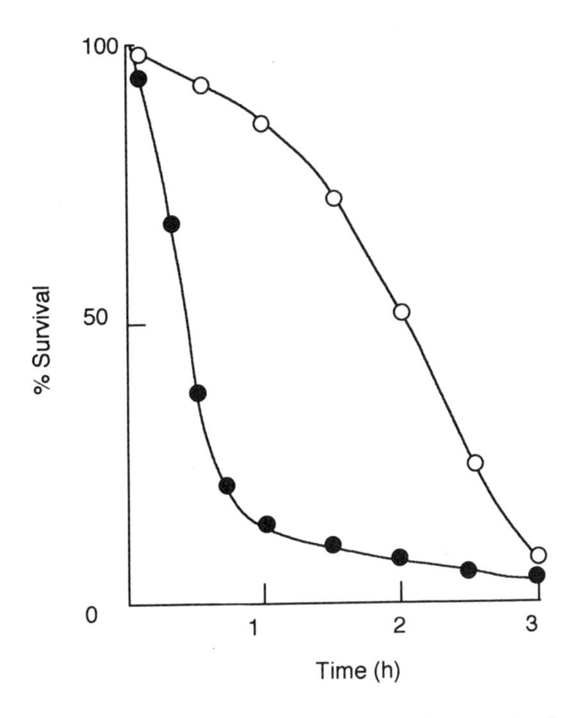

Figure 8.2. Killing of *Staphylococcus aureus* by normal and myeloperoxidase-deficient neutrophils. Kinetics of killing of opsonised *S. aureus* by normal (●) and myeloperoxidase-deficient neutrophils (○).

are not killed at all. This deficiency may be partially offset by an increase in the levels of O_2^- and H_2O_2 that are generated during the respiratory burst. Furthermore, the activities of myeloperoxidase-independent antimicrobial systems may be elaborated in patients with this deficiency. *S. aureus* infections are a major problem in patients with CGD (whose neutrophils do not generate any reactive oxidants; see §8.2), and this organism does not secrete sufficient H_2O_2 to reconstitute an effective myeloperoxidase–H_2O_2 system. Although cytoplasts (neutrophils devoid of cytoplasmic granules and hence deficient in myeloperoxidase) generate an amount of H_2O_2 equivalent to that produced by neutrophils, they do not effectively kill *S. aureus.* An elegant series of experiments by Odell and Segal (1988) showed that cytoplasts could kill these bacteria only if they were first coated with myeloperoxidase prior to phagocytosis. This implies a direct role for myeloperoxidase in the killing of this organism. Perhaps patients with complete myeloperoxidase deficiency do not suffer from infections with this organism because other, non-neutrophil defensive processes offer protection.

These myeloperoxidase deficiencies may be hereditary or acquired. The genetics of hereditary myeloperoxidase deficiency are unknown, but a survey of patients by Nauseef and co-workers (Nauseef 1988) revealed (by immunoblotting) that, in many patients, the mature enzyme was absent but promyeloperoxidase was present. Such observations suggest that in these patients the gene is normal and may be normally transcribed; the defect resides in how this protein is processed and perhaps packaged.

There are a number of conditions that have been reported to increase the incidence of acquired myeloperoxidase deficiencies. These include pregnancy, lead intoxication, Hodgkin's disease, prolonged bacterial infections and disseminated cancers. In the latter case, it may be that drug therapy contributes to the deficiency. Such deficiencies may be detected as a decreased level of enzyme per cell, or else a decrease in the percentage of peroxidase-positive cells visualised in a blood film. There is also a considerable incidence of myeloperoxidase deficiency in patients with either acute or chronic *myelocytic leukaemia* (48% and 20% of cases, respectively); there are no reports of such deficiencies in patients with acute or chronic lymphocytic leukaemia. Whilst it is difficult to assess whether this decreased level of myeloperoxidase is associated with the increased risk of infection that is observed in these patients, normalisation of enzyme levels after induction is usually a good sign of achieving remission. Of patients with myelodysplastic syndromes (§8.8), 25% have an increased number of myeloperoxidase-deficient neutrophils, and there is an association between the appearance of such cells and the development of cytogenetic abnormalities.

A study of 148000 subjects by Lanza and colleagues (1987) reported 36 individuals with myeloperoxidase deficiency, 10 of whom were completely deficient; a further 2 individuals with total deficiency were identified from familial studies. Of the 12 patients with total deficiency, 7 had either benign or malignant tumours, but none was undergoing radio- or chemotherapy at the time of analysis; hence, the myeloperoxidase deficiency could not be attributed, in these cases, to the therapy used for treatment of the malignant disease. These findings imply either that the tumour somehow affects the expression of myeloperoxidase in these patients, or that myeloperoxidase-deficient individuals perhaps have an increased incidence of tumours. Much follow-up work is needed for these studies on myeloperoxidase-deficient neutrophils in order to evaluate these proposals.

8.4 Specific-granule deficiency

Specific-granule deficiency is a very rare neutrophil defect that has only been described in five patients worldwide. Patients present with recurrent episodes of deep and superficial skin abscesses, otitis and mastoiditis, and infections of the skin and lungs. *Staphylococcus aureus, Candida albicans* and *Pseudomonas aeroginosa* are common pathogens. Examination of neutrophils from these patients indicates a bilobed rather than a polymorphic nucleus and the absence of specific granules, as determined by staining techniques. In some instances, however, it may be that specific granules are present but empty, and hence undetectable by such methods. The plasma membrane is defective in alkaline phosphatase, and while azurophilic granules are present, these stain abnormally. Carbohydrate distribution (as determined by staining techniques) is abnormal, and this may lead to more generalised defects in membrane synthesis or glycoprotein biosynthesis.

Isolated neutrophils of patients with this disease lack specific-granule contents such as lactoferrin and vitamin B_{12}-binding protein, and gelatinase is also absent. Myeloperoxidase is present at normal levels, and non-stimulated adherence, aggregation and respiratory-burst activity are normal or enhanced. However, up-regulation of receptors for C3bi, fMet-Leu-Phe and laminin does not occur during activation, because these are present on mobilisable pools on the specific-granule membranes of normal neutrophils. In vitro, chemotaxis is abnormal, and the efficiency of killing of pathogens such as *E. coli* and *S. aureus* is about half that of normal neutrophils.

In view of the fact that the myeloperoxidase–H_2O_2 system should be operational in neutrophils of patients with this disease, their inability to kill pathogens appears difficult to explain. However, it has recently been shown that neutrophils from patients with specific-granule deficiency also lack defensins, the antimicrobial peptides that are normally present at high concentrations in azurophilic granules (§2.6.3). This finding indicates that this disease is not restricted to the specific granule per se but, as just described, may be more closely related to a deficiency in membrane/glycoprotein biosynthesis.

Patients with thermal injury are also deficient in specific granules. In this case, however, it appears that this deficit is due to activation of the cells such that they discharge their specific granules. Neutrophils from patients with burns have an increased expression of plasma-membrane markers (as would be predicted if specific-granule membranes have fused with the plasma membrane), and serum levels of lysozyme and lactoferrin are elevated. These patients have impaired chemotaxis and defective oxygen metabolism.

8.5 Chediak–Higashi syndrome (CHS)

This is a rare, autosomal recessive disease that was first described when giant cytoplasmic granules were observed in neutrophils, monocytes and lymphocytes of a patient with a history of infections. Patients with CHS suffer recurrent infections and have a generalised cellular dysfunction because of the fusion of cytoplasmic granules. For example, melanocytes contain giant melanosomes, and these lead to pigmentation disorders such as partial albinism of the hair, skin and ocular fundi. Platelet function (but not number) is impaired due to a deficiency in storage pools of serotonin and ADP, and so aggregation is defective, leading to increased bleeding times. There is a tendency for lymphocyte–histiocyte proliferation in the liver, spleen and marrow, which leads to neutropenia and the so-called acceleration phase of the disease, during which there are serious bacterial and viral infections (e.g. by Epstein–Barr virus) that often result in death.

Circulating neutrophils of patients with CHS contain giant azurophilic and specific granules that form during haematopoiesis, and stem cells with these abnormalities are seen in the marrow. Chemotaxis, degranulation and bactericidal activity of circulating neutrophils are all impaired (often delayed), and these defects, coupled with the neutropenia that may occur, result in increased susceptibility to infections. Lymphocyte and NK cell functions are also defective.

Patients with CHS suffer recurrent infections of the skin, respiratory tract and mucus membranes, such as recurrent pyoderma, subcutaneous abscesses, otitis media, sinusitis, periodontal disease, bronchitis and pneumonia. Infections occur from wide variety of Gram-positive and -negative bacteria, with *Staphylococcus aureus* being a particularly-common pathogen, but infections with *Streptococcus pyogenes, Haemophilus influenzae,* and the fungi *Aspergillus* and *Candida* are also common. Several of the neutrophil dysfunctions observed may account for impaired killing. Generally, events associated with membrane fusion/activation are defective, and endogenous reactive oxidant production is elevated whilst adherence is decreased. The abnormally-sized granules contain lower levels of some antibacterial proteins, which may also contribute to impaired function. For example, levels of lysozyme, defensins and acid phosphatase are normal but levels of cathepsin G, elastase, myeloperoxidase and β-glucuronidase are decreased. Phagocytosis of organisms such as *S. aureus, Serratia marcescens, Escherichia coli* and *Streptococcus pneumoniae* may be normal, but killing is usually delayed. This may be because whilst activation of reactive oxidant production is normal, the discharge of granule contents is delayed. In some cases, the production of H_2O_2 is reported to be elevated. Other defects that have been reported include membrane fluidity, elevated cAMP levels, disordered assembly of microtubules and defective interaction of microtubules with granule membranes. This abnormal microtubule structure leads to abnormal orientation of CHS neutrophils within chemotactic gradients.

Disorders similar to CHS in humans have been reported in beige mice, Aleutian mink, Hereford cattle and killer whales, and neutrophil dysfunctions similar to those reported in humans are apparent in these animals. Therapy of CHS in humans usually consists of careful management of infections with appropriate antibiotics, although improvements have been reported following administration of high doses of ascorbic acid. Bone-marrow transplantation offers some hope to these patients, especially those entering the accelerated phase of the disease.

8.6 Hyperimmunoglobulin E-recurrent infections (Job's syndrome)

The phenomenon of hyperimmunoglobulin E-recurrent infections (HIE) was first reported in 1966 in two female patients with fair skin, red hair and severe eczematoid dermatitis. The abscesses contained *Staphylococcus aureus* that were 'cold', that is, devoid of the signs of normal inflammation such as surrounding erythema. The term 'Job's syndrome' was coined based on the Book of Job (2: 7). Subsequently, male patients were described with coarse facies and a history of recurrent skin, lung and joint abscesses, with *Staphylococcus aureus, Haemophilus influenzae* and *Candida albicans* being the predominant pathogens. *Aspergillus* infections secondary to pneumonia are common, and about 50% of patients have mucocutaneous *Candida* infections. These patients all have extremely high levels of serum IgE.

The disease is now known to affect both sexes, irrespective of colouring and race, but the underlying feature is that of recurrent infections, particularly with a history of chronic dermatitis and high serum IgE levels. The patients invariably have coarse facial features, such as broad nasal bridge, prominent nose and irregularly-shaped cheeks and jaw. Levels of IgG and IgM are normal, but serum and secreted (e.g. salivary) levels of IgA are low, and the IgE recognises *S. aureus* and *C. albicans*. In these patients, serum IgE levels are often >2500 IU/ml and can be as high as 150 000 IU/ml. However, serum IgE levels may also be elevated in non-HIE patients with atopic dermatitis, but in these patients the IgE is usually raised against inhaled substances such as pollens, mites and moulds; in HIE the IgE is raised against *S. aureus* and *C. albicans.*

The inflammatory response in HIE patients is strikingly abnormal, despite the large local infections that occur. Neutrophils are present only at low numbers in these sites. It has been suggested that too few neutrophils arrive too late to deal with the pathogens, which thus have the opportunity to multiply. Indeed, chemotaxis of neutrophils and monocytes may be defective in these patients; however, this defect is variable and probably does not account fully for impaired host protection. In vitro, neutrophils from these patients ingest and kill pathogens such as *S. aureus* normally, and their respiratory-burst activity (as assessed by NBT reduction and chemiluminescence) is normal. Patients with HIE also often have low-level eosinophilia, and sometimes eosinophils may account for 40% of the white blood cells (6000/mm^2 blood). Eosinophil infiltration of infected sites may occur, and IgE stimulation of basophils and mast cells may result in histamine release. It has thus been proposed that histamine released from IgE-activated mast cells may down-regulate neutrophil function at these infected sites. Antihistamines (H2-blockers) have been reported to improve chemotaxis in some (but not all) HIE patients: stimulation of H2 receptors results in elevation of cAMP levels. A chemotaxis-inhibitory factor (a heterogeneous mixture of 61-, 45- and 30-kDa proteins) has also been isolated from monocytes of HIE patients.

Several patients with HIE also have osteogenesis imperfecta, a genetic disease localised to chromosome 7. Some patients who do not have HIE but do have a chemotaxis defect may also have a deletion in chromosome 7. Thus, it has been proposed that neutrophil migration may be regulated by a factor encoded in region 7q22–7qter. Deletions in this region may thus account for impaired neutrophil chemotaxis, susceptibility to infections and coarse facies. Treatment of HIE is by careful management of infections and use of antibiotics, especially those active against *S. aureus;* thus, dicloxicillin and TMP/SMX (trimethoprim/sulfamethoxazole) have proved beneficial.

8.7 CD11/CD18 leukocyte glycoprotein deficiency

This is a rare, inherited disorder characterised by recurrent and, often, life-threatening infections. This dysfunction results from absent or very low expression of the glycoprotein family, LFA-1, CR3 and gp150,95, which function in leukocyte adhesion (§3.9). These adhesion molecules are normally responsible for various functions, such as the following:

leukocyte–endothelium interaction;

adhesion of leukocytes to surfaces or to each other;

phagocytosis of opsonised or coated particles;

certain forms of cytotoxicity and chemotaxis.

This disease was first observed in the mid- to late-1970s when several patients presented with recurrent bacterial infections, primarily of the skin and subcutaneous tissues, middle ear and oropharyngeal mucosa. When examined in vitro, the neutrophils from these patients had defects in chemotaxis, phagocytosis, particle-stimulated respiratory-burst activity and granulation. Some patients also had a leukocytosis, and many had a delayed umbilical cord separation. Treatment is by prophylactic antibiotic therapy and aggressive antibiotic therapy during infections, but mortality rates are very high.

8.8 Myelodysplastic syndromes (MDS)

The myelodysplastic syndromes are a group of heterogeneous, haematological disorders characterised by a defect in the production of one or more haematopoietic cell lineages. MDS is often associated with the elderly (age >60 yr), and many patients progress into acute leukaemia. The French–American–British (FAB) group characterise MDS into five subgroups:

i. *RA* (refractory anaemia with <15% ringed sideroblasts), 28% of all cases;
ii. *RARS* (refractory anaemia with >15% ringed sideroblasts), 24% of all cases;
iii. *RAEB* (refractory anaemia with excess of blasts), 23% of all cases;
iv. *CMML* (chronic myelomonocytic leukaemia), 16% of all cases;
v. *RAEB-T* (refractory anaemia with excess blasts in transformation), 9% of all cases.

Anaemia is common (<11 g/dl, normal range 12–18 g/dl), and neutropenias, thrombocytopenias and monocytosis may be observed.

Diagnosis is made following careful examination of blood films, marrow aspiration and biopsy. Distinguishing marrow abnormalities include the presence of ringed sideroblasts, multinuclear fragments, bizarre nuclear shapes and abnormally-dense or very fine chromatin. *Sideroblasts* are erythroid precursors in which iron-containing granules can be identified. Sideroblasts in normal marrow usually contain less than five iron granules/cell; those with more iron are abnormal and are termed *ringed* if the iron granules cover more than one-third of the nuclear rim. Cytogenetic studies indicate a variety of chromosomal abnormalities, with genetic lesions giving rise to the development of abnormal clones. These genetic abnormalities appear distinct from those commonly observed in myeloproliferative disorders and acute non-lymphocytic leukaemia. Mutations in *ras* (the small GTP-binding protein family; §6.2.2) and *fms* (the M-CSF receptor) are common in MDS patients.

Infection is the most common cause of morbidity and mortality in MDS patients, accounting for 40–60% of deaths in various studies. The common infections are those normally associated with neutropenias, such as Gram-negative septicaemia and bacterial bronchopneumonias. Indeed, most MDS patients are neutropenic at some stage in their disease. Even those who do not have a neutropenia may have a defect in their neutrophil function. Many patients have clearly-defined defects in T- and B-lymphocyte functions, and variable defects in monocyte numbers or function have been described. Disorders of neutrophil function are common. Many reports indicate that phagocytosis, chemotaxis, respiratory-burst activity and degranulation are defective in some MDS patients, and hypogranulation is often observed.

There has been a great deal of interest in the use of colony-stimulating factors to treat MDS. GM-CSF and G-CSF, which have been used in clinical trials, offer a potential dual benefit. Firstly, they can affect neutrophil development in the bone marrow, and so can improve the neutropenia that is associated with these disorders. Secondly, they have the potential to increase or repair the function of circulating neutrophils. Indeed, there are some reports to indicate that these CSFs can result in enhanced function of peripheral blood neutrophils in these patients. Most patients show improvements in neutrophil counts after GM-CSF or G-CSF administration. In some cases, this has been associated with a decrease in the number of infective episodes.

There has been some concern expressed regarding the use of CSFs to treat MDS patients. Because these cytokines have proliferative activity, they have the potential to induce a leukaemic transformation in the malignant clone. However, the combined use of CSFs with cytotoxic drugs such as cytosine arabinoside (ara-C) appears promising. If leukaemic clones are induced to proliferate by the cytokine, then they are killed by ara-C as they enter the cell cycle. Other forms of differentiation therapy, such as treatment with retinoids, 1,25-dihydroxyvitamin D_3 and interferons, have also been tested, but results have been variable.

8.9 Rheumatoid arthritis

Rheumatoid arthritis (RA) is the most common type of rheumatological disorder, with a prevalence of about 1% worldwide. The onset of the disease may be any time between puberty and old age, with peak times of onset be-

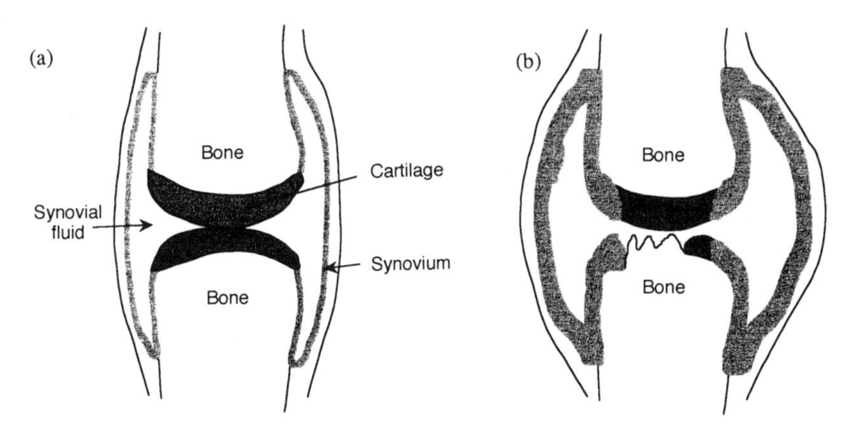

Figure 8.3. Morphological features of a normal and rheumatoid joint: (a) the essential features of a healthy synovial joint (e.g. a knee joint); (b) the pathological features of a joint of a patient with rheumatoid arthritis.

tween 30 and 50 years. In young adults, women are six times more likely to be affected than males, but in old age the ratio between afflicted men and women becomes closer to 1:1. RA principally affects peripheral synovial joints, but because other manifestations of the disease are apparent, it is classified as one of the diseases of connective tissue. Disease expression varies greatly, and definite diagnosis of RA is only made when several of the classical features of the disease (e.g. symmetry of affected joints, subcutaneous nodules, radiographic joint erosions, presence of rheumatoid factor in the serum) are observed.

The normal synovial joint comprises a number of important features (Fig. 8.3a). The *cartilage,* which caps the opposing ends of articulating bones, has a smooth surface, and the *chondrocytes* within this structure (which secrete proteoglycans) account for only about 0.1% of its volume. The function of the cartilage is to absorb energy by deformation; thus, healthy cartilage is highly compressible and highly elastic, features provided by the water-absorbing proteoglycans and the inelastic collagen fibrils. The cartilage merges at the margins with the *synovial membrane,* which functions to exchange nutrients and metabolites with the chondrocytes, to secrete hyaluronic acid and other lubricating glycoproteins and to phagocytose particulate matter. The synovial membrane is thus a meshwork of collagen fibres containing fibroblasts, fat cells and macrophage-like cells, and it secretes *synovial fluid,* which is a joint lubricant. Normal synovial fluid is virtually acellular and extremely viscous. In a normal knee joint, its volume is usually 1 ml or less.

A diseased, rheumatoid joint has several distinguishing features (Fig. 8.3b). The synovial membrane becomes inflamed and invaded by T and B lymphocytes. The lining cells proliferate and thicken to form a pannus, which grows from the joint margins first to cover and then to digest the cartilage. The cartilage thus loses its mechanical properties, becomes eroded and cannot protect the articular bones from physical damage. There is an excess of synovial fluid production by the synovium (an inflamed knee joint may contain 100 ml of fluid), and this fluid has low viscosity and may be heavily infiltrated by neutrophils (up to 5×10^7 cells/ml).

The underlying cause of RA is unknown, but genetic factors are undoubtedly involved. There is an increased prevalence of the disease in first-degree relatives of patients, and an association between HLA-D4 / HLA-DR4 and seropositive RA (i.e. the presence of rheumatoid factor). *HLA-DR4* (regions of the class II antigens of the major histocompatibility complex) confers on individuals a fourfold increased risk of developing RA. *Rheumatoid factor* is an auto-antibody that recognises epitopes on the Fc region of IgG, and IgM-rheumatoid factor is found in 80% of patients with RA. A high titre of this factor is a poor prognosis, and IgG-rheumatoid factors may also be present. Several mechanisms have been implicated in the development of antigenic IgG, including oxidative damage and levels of asparagine-linked galactosylation. The stimulus for the onset of the disease in susceptible individuals is not known, but there have been many suggestions as to a microbial source. Microbial infection may result in a weak but persistent immune response that becomes self-sustaining, or perhaps genetically-susceptible individuals possess abnormal lymphocyte clones that allow persistent inflammation to occur in response to foreign or host antigens. More recently, the theory of *molecular mimicry* has been proposed, which accommodates both of these possibilities. It may thus be envisaged that an infection results in the generation of antibodies, which recognise both microbial antigens and host-tissue antigens. Several microbial agents have been suggested as the aetiological trigger, such as Epstein–Barr virus, parvovirus and some bacteria.

8.9.1 Pathogenesis

In response to an unknown stimulus, lymphocytes migrate into the synovium and secrete cytokines. These then activate lymphocyte proliferation, stimulate antibody production by B lymphocytes and activate macrophages. The activated macrophages themselves secrete their own repertoire of cytokines and can also secrete proteases, which then contribute to cartilage de-

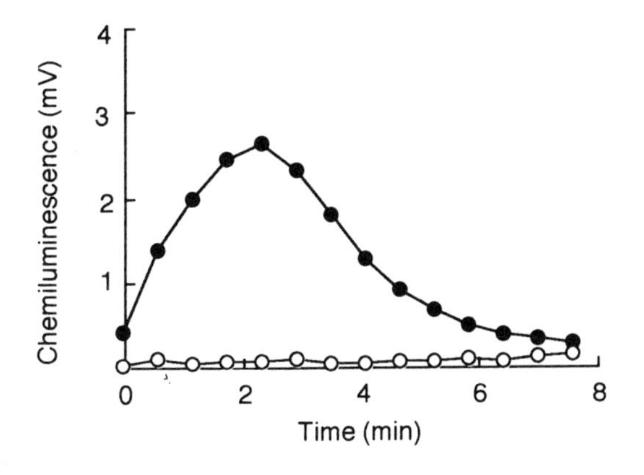

Figure 8.4. Activation of neutrophils by soluble immune complexes isolated from the synovial fluid of a patient with rheumatoid arthritis. Neutrophils were isolated from control blood and incubated in the absence (○) or presence (●) of 50 U/ml GM-CSF for 60 min at 37 °C. Thereafter, luminol chemiluminescence was measured following the addition of soluble immune complexes isolated from the synovial fluid of a patient with rheumatoid arthritis.

struction. Antibody deposition results in the activation of complement and the generation of complement fragments, and these (e.g. C5a) together with macrophage-derived cytokines (e.g. IL-8), can then stimulate the infiltration of neutrophils. The cytokines present in the inflamed joint (e.g. IL-1, IL-8, GM-CSF) then prime the infiltrating neutrophils, which have enhanced expression of some plasma membrane receptors and activated gene expression. This enhanced protein biosynthesis can both up-regulate and extend the ability of neutrophils to secrete cytotoxic products, and may also lead to the secretion of secondary cytokines, which can amplify the cytokine cascade associated with this disease (§7.3.4).

The major neutrophil-activating factors within rheumatoid joints are immune complexes. These predominantly IgG-containing complexes vary considerably in size. Curiously, blood neutrophils do not generate reactive oxidants in response to soluble immune complexes, needing to be primed before they can be activated in this way (Fig. 8.4). It is therefore of interest to note that neutrophils isolated from the synovial fluid of patients with rheumatoid arthritis have been primed in vivo and can secrete substantial quantities of reactive oxidants in response to these soluble complexes. Thus, it is extremely likely that these soluble immune complexes are responsible for activation of the secretion of reactive oxygen metabolites by infiltrating synovial-fluid neutrophils. This phenomenon may contribute to the events that lead to the destructive joint processes characteristic of this disease.

8.10 Bibliography

Alvaro-Gracia, J. M., Zvaifler, N. J., Brown, C. B., Kaushansky, K., & Firestein, G. S. (1991). Cytokines in chronic inflammatory arthritis: VI. Analysis of the synovial cells involved in granulocyte-macrophage colony-stimulating factor production and gene expression in rheumatoid arthritis and its regulation by IL-1 and tumor necrosis factor-α *J. Immunol.* **146**, 3365–71.

Anderson, D. C., Schmalsteig, F. C., Finegold, M. J., Hughes, B. J., Rothlein, R., Miller, L. J., Kohl, S., Tosi, M. F., Jacobs, R. L., Waldrop, T. C., Goldman, A. S., Shearer, W. T., & Springer, T. A. (1985). The severe and moderate phenotypes of heritable Mac-1, LFA-1 deficiency: Their quantitative definition and relation to leukocyte dysfunction and clinical features. *J. Infect. Dis.* **152**, 668–89.

Anderson, D. C., & Springer, T. A. (1987). Leukocyte adhesion deficiency: An inherited defect in the Mac-1, LFA-1, and p150,95 glycoproteins. *Ann. Rev. Med.* **38**, 175–94.

Arend, W. P., & Dayer, J.-M. (1990). Cytokines and cytokine inhibitors or antagonists in rheumatoid arthritis. *Arthrit. Rheum.* **33**, 305–15.

Arnaout, M. A., Spits, H., Terhorst, C., Pitt, J., & Todd, R. F., III (1984). Deficiency of a leukocyte surface glycoprotein (LFA-1) in two patients with Mo1 deficiency. *J. Clin. Invest.* **74**, 1291–1300.

Berton, G., Zeni, L., Cassatella, M. A., & Rossi, F. (1986). Gamma interferon is able to enhance the oxidative metabolism of human neutrophils. *Biochem. Biophys. Res. Commun.* **138**, 1276–82.

Blackburn, W. D. J., Koopman, W. J., Schrohenloher, R. E., & Heck, L. W. (1986). Induction of neutrophil enzyme release by rheumatoid factors: Evidence for differences based on molecular characteristics. *Clin. Immunol. Immunopathol.* **40**, 347–55.

Bohler, M.-C., Seger, R. A., Mouy, R., Vilmer, E., Fischer, A., & Griscelli, C. (1986). A study of 25 patients with chronic granulomatous disease: A new classification by correlating respiratory burst, cytochrome b, and flavoprotein. *J. Clin. Immunol.* **6**, 136–45.

Bolscher, B. G. J. M., van Zwieten, R., Kramer, I. M., Weening, R. S., Verhoeven, A. J., & Roos, D. (1989). A phosphoprotein of Mr 47,000, defective in autosomal chronic granulomatous disease, copurifies with one of two soluble components required for NADPH:O_2 oxidoreductase activity in human neutrophils. *J. Clin. Invest.* **83**, 757–63.

Borregaard, N., Cross, A. R., Herlin, T., Jones, O. T. G., Segal, A. W., & Valerius, N. H. (1983). A variant form of X-linked chronic granulomatous disease with normal nitroblue tetrazolium slide test and cytochrome b. *Eur. J. Clin. Invest.* **13**, 243–7.

Brennan, F. M., Zachariae, C. O. C., Chantry, D., Larsen, C. G., Turner, M., Maini, R. N., Matsushima, K., & Feldmann, M. (1990). Detection of Interleukin 8 biological activity in synovial fluids from patients with rheumatoid arthritis and production of Interleukin 8 mRNA by isolated synovial cells. *Eur. J. Immunol.* **20**, 2141–4.

Breton-Gorius, J., Houssay, D., & Dreyfus, B. (1975). Partial myeloperoxidase deficiency in a case of preleukaemia. *Brit. J. Haematol.* **30**, 273–8.

Brown, K. A. (1988). The polymorphonuclear cell in rheumatoid arthritis. *Brit. J. Rheumatol.* **27**, 150–5.

Clark, R. A., Malech, H. L., Gallin, J. I., Nunoi, H., Volpp, B. D., Pearson, D. W., Nauseef, W. M., & Curnutte, J. T. (1989). Genetic variants of chronic granulomatous disease: prevalence of deficiencies of two cytosolic components of the NADPH oxidase system. *New Eng. J. Med.* **321**, 647–52.

Curnutte, J. T., Berkow, R. L., Roberts, R. L., Shurin, S. B., & Scott, P. J. (1988). Chronic granulomatous disease due to a defect in the cytosolic factor required for nicotinamide adenine dinucleotide phosphate oxidase activation. *J. Clin. Invest.* **81**, 606–10.

Curnutte, J. T., Scott, P. J., & Babior, B. M. (1989). Functional defect in neutrophil cytosols from two patients with autosomal recessive cytochrome-positive chronic granulomatous disease. *J. Clin. Invest.* **83**, 1236–40.

Dinauer, M. C., & Ezekowitz, R. A. B. (1991). Interferon-γ and chronic granulomatous disease. *Curr. Opin. Immunol.* **3**, 61–4.

Dinauer, M. C., Curnutte, J. T., Rosen, H., & Orkin, S. H. (1989). A missense mutation in the neutrophil cytochrome b heavy chain in cytochrome-positive X-linked chronic granulomatous disease. *J. Clin. Invest.* **84**, 2012–16.

Dogan, P., Soyuer, U., & Tanrikulu, G. (1989). Superoxide dismutase and myeloperoxidase activity in polymorphonuclear leukocytes, and serum ceruloplasmin and copper levels, in psoriasis. *Brit. J. Dermatol.* **120**, 239–44.

Dularay, B., Elson, C. J., & Dieppe, P. A. (1988). Enhanced oxidative response of polymorphonuclear leukocytes from synovial fluids of patients with rheumatoid arthritis. *Autoimmunity* **1**, 159–69.

Edwards, S. W., Hughes, V., Barlow, J., & Bucknall, R. (1988). Immunological detection of myeloperoxidase in synovial fluids from patients with rheumatoid arthritis. *Biochem. J.* **250**, 81–5.

Ezekowitz, R. A. B., Orkin, S. H., & Newburger, P. E. (1987). Recombinant interferon gamma augments phagocyte superoxide production and X-chronic granulomatous disease gene expression in X-linked variant chronic granulomatous disease. *J. Clin. Invest.* **80**, 1009–16.

Ezekowitz, R. A. B., Sieff, C. A., Dinauer, M. C., Nathan, D. G., Orkin, S. H., & Newburger, P. E. (1990). Restoration of phagocyte function by interferon-γ in X-linked chronic granulomatous disease occurs at the level of a progenitor cell. *Blood* **76**, 2443–8.

Feldman, M., Brennan, F. M., Chantry, D., Haworth, C., Turner, M., Abney, E., Buchan, G., Barrett, K., Barkley, D., Chu, A., Field, M., & Maini, R. N. (1990). Cytokine production in the rheumatoid joint: Implications for treatment. *Ann. Rheum. Dis.* **49**, 480–6.

Ford, D. K. (1991). The microbial causes of rheumatoid arthritis. *J. Rheumatol.* **18**, 1441–2.

Goulding, N. J., Knight, S. M., Godolphin, J. L., & Guyre, P. M. (1992). Increase in neutrophil Fcγ receptor I expression following interferon gamma treatment in rheumatoid arthritis. *Ann. Rheumat. Dis.* **51**, 465–8.

Halliwell, B., & Gutteridge, J. M. C. (1990). The antioxidants of human extracellular fluids. *Arch. Biochem. Biophys.* **280**, 1–8.

Holmes, B., Page, A. R., & Good, R. A. (1967). Studies of the metabolic activity of leukocytes from patients with a genetic abnormality of phagocytic function. *J. Clin. Invest.* **46**, 1422–32.

Holmes, B., Quie, P. G., Windhorst, D. B., & Good, R. A. (1966). Fatal granulomatous disease of childhood: An inborn abnormality of phagocytic function. *Lancet* **i**, 1225–8.

Jacobs, A. (1990). The myelodysplastic syndrome. *Eur. J. Cancer* **26**, 1115–18.

Kahle, P., Saal, J., Pawelec, G., Westacott, C., Swan, A., Dieppe, P., Whicher, J., Barnes, I., & Thompson, D. (1991). Cytokines in rheumatoid arthritis. *Ann. Rheumat. Dis.* **50**, 405–7.

Kitahara, M., Eyre, H. J., Simonian, Y., Atkin, C. L., & Hasstedt, S. J. (1981). Hereditary myeloperoxidase deficiency. *Blood* **57**, 888–93.

Lanza, F., Fietta, A., Spisani, S., Castoldi, G. L., & Traniello, S. (1987). Does a relationship exist between neutrophil myeloperoxidase deficiency and the occurrence of neoplasms? *J. Clin. Lab. Immunol.* **22**, 175–80.

Lehrer, R. I., & Cline, M. J. (1969). Leukocyte myeloperoxidase deficiency and disseminated candidiasis: The role of myeloperoxidase in resistance to *Candida* infection. *J. Clin. Invest.* **48**, 1478–87.

McKenna, R. M., Wilkins, J. A., & Warrington, R. J. (1988). Lymphokine production in rheumatoid arthritis and systemic lupus ethythematosus. *J. Rheumatol.* **15**, 1639–42.

Merry, P., Winyard, P. G., Morris, C. J., Grootveld, M., & Blake, D. R. (1989). Oxygen free radicals, inflammation and synovitis: The current status. *Ann. Rheum. Dis.* **48**, 864–70.

Nauseef, W. M. (1988). Myeloperoxidase deficiency. In *Phagocytic Defects,* vol. I (Hematology/Oncology Clinics of North America), pp. 135–58, W. B. Saunders, Philadelphia.

Newburger, P. E., Ezekowitz, R. A. B., Whitney, C., Wright, J., & Orkin, S. (1988). Induction of phagocyte cytochrome b heavy chain gene expression by interferon-γ. *Proc. Natl. Acad. Sci. USA* **85**, 5215–19.

Nunoi, H., Rotrosen, D., Gallin, J. I., & Malech, H. L. (1988). Two forms of autosomal chronic granulomatous disease lack distinct neutrophil cytosol factors. *Science* **242**, 1298–1301.

Nurcombe, H. L., Bucknall, R. C., & Edwards, S. W. (1991a). Activation of the myeloperoxidase–H_2O_2 system by synovial fluid isolated from patients with rheumatoid arthritis. *Ann. Rheumat. Dis.* **50**, 237–42.

Nurcombe, H. L., Bucknall, R. C., & Edwards, S. W. (1991b). Neutrophils isolated from the synovial fluid of patients with rheumatoid arthritis: Priming and activation in vivo. *Ann. Rheumat. Dis.* **51**, 147–53.

Odell, E. W., & Segal, A. W. (1988). The bactericidal effects of the respiratory burst and the myeloperoxidase system isolated in neutrophil cytoplasts. *Biochim. Biophys. Acta* **971**, 266–74.

Parkos, C. A., Dinauer, M. C., Jesaitis, A. J., Orkin, S. H., & Curnutte, J. T. (1989). Absence of both the 91kD and 22kD subunits of human neutrophil cytochrome b in two genetic forms of chronic granulomatous disease. *Blood* **73**, 1416–20.

Parry, M. F., Root, R. K., Metcalf, J. A., Delaney, K. K., Kaplow, L. S., & Richar, W. J. (1981). Myeloperoxidase deficiency: Prevalence and clinical significance. *Ann. Intern. Med.* **95**, 293–301.

Pomeroy, C., Oken, M. M., Rydell, R. E., & Filice, G. A. (1991). Infection in the myelodysplastic syndromes. *Am. J. Med.* **90**, 338–44.

Ridderstad, A., Abedi-Valugerdi, M., & Moller, E. (1991). Cytokines in rheumatoid arthritis. *Ann. Med.* **23**, 219–23.

Robinson, J. J., Watson, F., Bucknall, R. C., & Edwards, S. W. (1992). Activation of neutrophil reactive-oxidant production by synovial fluid from patients with inflammatory joint disease: Soluble and insoluble immunoglobulin aggregates activate different pathways in primed and unprimed cells. *Biochem. J.* **286**, 345–51.

Royer-Pokora, B., Kunkel, L. M., Monaco, A. P., Goff, S. C., Newburger, P. E., Baehner, R. L., Cole, F. S., Curnutte, J. T., & Orkin, S. H. (1986). Cloning the gene for an inherited disorder – chronic granulomatous disease – on the basis of its chromosomal location. *Nature* **322**, 32–8.

Salmon, S. E., Cline, M. J., Schultz, J., & Lehrer, R. I. (1970). Myeloperoxidase deficiency: Immunologic study of a genetic leukocyte defect. *New Eng. J. Med.* **282**, 250–3.

Segal, A. W. (1988). Cytochrome b_{-245} and its involvement in the molecular pathology of chronic granulomatous disease. In *Phagocytic Defects,* vol. II (Hematology/Oncology Clinics of North America), pp. 213–23, W. B. Saunders, Philadelphia.

Segal, A. W., & Jones, O. T. G. (1978). Novel cytochrome b system in phagocytic vacuoles of human neutrophils. *Nature* **276**, 515–17.

Segal, A. W., Jones, O. T. G., Webster, D., & Allison, A. C. (1978). Absence of a newly described cytochrome b from neutrophils of patients with chronic granulomatous disease. *Lancet* **ii**, 446–9.

Solberg, C. O. (1980). Infections in the immunocompromised host. *Scand. J. Infect. Dis.* **S24**, 36–43.

Tauber, A. I., Borregaard, N., Simons, E., & Wright, J. (1983). Chronic granulomatous disease: A syndrome of phagocyte oxidase deficiencies. *Medicine* **62**, 286–309.

Teahan, C., Rowe, P., Parker, P., Totty, N., & Segal, A. W. (1987). The X-linked chronic granulomatous disease gene codes for the β-chain of cytochrome b_{-245}. *Nature* **327**, 720–1.

Thrasher, A., Chetty, M., Casimir, C., & Segal, A. W. (1992). Restoration of superoxide generation to a chronic granulomatous disease-derived B-cell line by retrovirus mediated gene transfer. *Blood* **80**, 1125–9.

Verhoef, G., & Boogaerts, M. (1991). *In vivo* administration of granulocyte-macrophage colony stimulating factor enhances neutrophil function in patients with myelodysplastic syndromes. *Brit. J. Haematol.* **79**, 177–84.

Volpp, B. D., Nauseef, W. M., & Clark, R. A. (1988). Two cytosolic neutrophil oxidase components absent in autosomal chronic granulomatous disease. *Science* **242**, 1295–7.

Westacott, C. I., Whicher, J. T., Barnes, I. C., Thompson, D., Swan, A. J., & Dieppe, P. A. (1990). Synovial fluid concentration of five different cytokines in rheumatic diseases. *Ann. Rheumat. Dis.* **49**, 676–81.

Woodman, R. C., Erickson, R. W., Rae, J., Jaffe, H. S., & Curnutte, J. T. (1992). Prolonged recombinant interferon-γ therapy in chronic granulomatous disease: Evidence against enhanced neutrophil oxidase activity. *Blood* **79**, 1558–62.

Yuo, A., Kitagawa, S., Okabe, T., Urabe, A., Komatsu, Y., Itoh, S., & Takaku, F. (1987). Recombinant human granulocyte colony-stimulating factor repairs the abnormalities of neutrophils in patients with myelodysplastic syndromes and chronic myelogenous leukemia. *Blood* **70**, 404–11.

Index